Data Fusion: Concepts and Ideas

H.B. Mitchell

Data Fusion: Concepts and Ideas

Springer

Author
Dr. H.B. Mitchell
Mazkaret Batya
Israel

ISBN 978-3-642-43730-4 ISBN 978-3-642-27222-6 (eBook)
DOI 10.1007/978-3-642-27222-6
Springer Heidelberg New York Dordrecht London

Softcover reprint of the hardcover 1st edition 2012

Printed on acid-free paper

Springer is part of Springer Science+Business Media (www.springer.com)

This book is dedicated to our children

DANIEL ASHER and NOA SOPHIA

For ye shall go out with joy, and be led forth with peace:
the mountains and the hills shall break forth before you in singing,
and all the trees of the field shall clap their hands.

Isaiah 55:12

Preface

The purpose of this book is to provide an introduction to the concepts and ideas of multi-sensor data fusion. It is an extensively revised second edition of the author's book "Multi-Sensor Data Fusion: An Introduction", which was originally published by Springer-Verlag in 2007. As in the original book, the present work has been designed as a textbook for a one-semester graduate course in multi-sensor data fusion. It should, however, also be useful to advanced undergraduates in electrical engineering or computer science who are studying data fusion for the first time and to practicing engineers who wish to use the concepts of data fusion in practical applications.

The main changes in the book are:

New Material. We have added a new chapter (Semantic Alignment), and approximately 30 new sections, 50 new examples, and 100 new references to the book. At the same time, material which is out-of-date, or has been superseded by new developments, has been eliminated and the remaining text has been rewritten for added clarity. Altogether the new edition is nearly 70 pages longer than the original book.

Matlab Code. Where appropriate we have given details of Matlab code which may be downloaded from the worldwide web. In a few places, where such code is not readily available, we have included Matlab code in the body of the text.

Problems. We have added problems at the end of each chapter. Most of these problems originally appeared on the author's website.

Figures. New figures have been added to the text. More than half of the remaining figures have been redrawn for added clarity.

Typography. The layout and typography of the book has been extensively revised. Examples and Matlab code now appear on a gray background for easy identification and advanced material is marked with an asterisk *.

Index. The index contains new headings and new entries. A considerable effort has been made to make the index comprehensive and useful for the reader.

The book is intended to be self-contained in so far as the subject of data fusion is concerned, although some prior exposure to the field of signal processing may be helpful to the reader. A clear understanding of data fusion can only be achieved with the use of a certain minimum level of mathematics. It is therefore assumed that the reader has a reasonable working knowledge of the basic tools of linear algebra, calculus and simple probability theory. More specific results and techniques which are required are explained in the body of the book.

Data fusion relies on a diverse collection of theories and techniques drawn from a wide of subjects, including neural networks, fuzzy logic, signal processing, statistical estimation as well as computer vision and signal processing. We have attempted to integrate this diverse collection of theories and techniques by adopting a Bayesian statistical framework. This has also enabled us to keep the book to a reasonable size while treating many new and important topics in great depth.

As with any other branch of engineering, data fusion is a pragmatic activity which is driven by practicalities. It is therefore important that the student is able to experiment with the different techniques presented in the book. For this purpose software code, written in Matlab, is particularly convenient. In this edition we have included Matlab code in the body of the text as well as including details of relevant Matlab code which may be downloaded from the worldwide web. For the professional engineer we have both illustrated the theory with many real-life applications and have provided him with an extensive list of up-to-date references.

The book is based on seminars, lectures and courses on data fusion which have been taught over several years. The structure and content of the book is based on material gathered and ideas exchanged with my colleagues. Particular thanks are extended to Dr. Ruth Rotman, who carefully reviewed the entire manuscript, and to my son, Daniel Mitchell, who helped draw many of the figures in the book. I am also indebted to my wife and daughter for the support and patience they have shown me while the book was being written.

Finally, to the reader. We hope you will enjoy reading this book and that it will prove to be an useful addition to the increasingly important and expanding field of data fusion.

Mazkeret Batya,
August 2011,
H.B. Mitchell

Contents

Chapter 1
Introduction

1.1 Definition

The subject of this book is *multi-sensor data fusion* which we define as "the theory, techniques and tools which are used for combining sensor data, or data derived from sensory data, into a common representational format". In performing data fusion, our aim is to improve the quality of the information, so that it is, in some sense, *better* than would be possible if the data sources were used individually.

The above definition implies that the sensor data, or the data derived from the sensory data, consists of multiple measurements which have to be combined. The multiple measurements may, of course, be produced by multiple sensors. However, the definition also includes multiple measurements, produced at different time instants, by a single sensor.

The general concept of multi-sensor data fusion is analogous to the manner in which humans and animals use a combination of multiple senses, experience and the ability to reason to improve their chances of survival.

The basic problem of multi-sensor data fusion is one of determining the best procedure for combining the multi-sensor data inputs. The view adopted in this book is that combining multiple sources of information with *a priori* information is best handled within a statistical framework. The main advantage of a statistical approach is that explicit probabilistic models are employed to describe the various relationships between sensors and sources of information taking into account the underlying uncertainties. In particular we restrict ourselves to the Bayesian methodology which provides us with a useful way to formulate the multi-sensor data fusion problem in mathematical terms and which yields an assessment of the uncertainty in all unknowns in the problem.

1.2 Synergy

The principal motivation for multi-sensor data fusion is to improve the quality of the information output in a process known as *synergy*. Strictly speaking, synergy does not require the use of multiple sensors. The reason being that the synergistic effect may be obtained on a temporal sequence of data generated by a single sensor. However, employing more than one sensor may enhance the synergistic effect in several ways, including: increased spatial and temporal coverage, increased robustness to sensor and algorithmic failures, better noise suppression and increased estimation accuracy.

Example 1.1. Multi-Modal Biometric Systems [17]. Biometric systems that rely on a *single* biometric trait for recognition are often characterized by high error rates. This is due to the lack of completeness or *universality* in most biometric traits. For example, fingerprints are not truly universal since it is not possible to obtain a good quality fingerprint from people with hand-related disabilities, manual workers with many cuts and bruises on their fingertips or people with very oily or very dry fingers. Multi-modal biometric sensor systems solve the problem of non-universality by fusing the evidence obtained from multiple traits.

Example 1.2. Multiple Camera Surveillance Systems [13]. The increasing demand for security by society has led to a growing need for surveillance activities in many environments. For example, the surveillance of a wide-area urban site may be provided by periodically scanning the area with a single narrow field-of-view camera. The temporal coverage is, however, limited by the time required for the camera to execute one scan. By using multiple cameras we reduce the mean time between scans and thereby increase the temporal coverage.

Broadly speaking, multi-sensor data fusion may improve the performance of the system in four different ways [3]:

Representation. The information obtained during, or at the end, of the fusion process has an abstract level, or a granuality, higher than each input data set. The new abstract level or the new granuality provides a richer semantic on the data than each initial source of information

Certainty. If V is the sensor data before fusion and $p(V)$ is the *a priori* probability of the data before fusion, then the gain in certainty is the growth in $p(V)$ after fusion. If $\widetilde{V}$ denotes data after fusion, then we expect $p(\widetilde{V}) > p(V)$.

Accuracy. The standard deviation on the data after the fusion process is smaller than the standard deviation provided directly by the sources. If data is noisy or erroneous, the fusion process tries to reduce or eliminate noise and errors. In general, the gain in accuracy and the gain in certainty are correlated.

Completeness. Bringing new information to the current knowledge on an environment allows a more complete the view on this environment. In general, if the information is redundant and concordant, we could also have a gain in accuracy

Example 1.3. Multi-Modal Medical Imaging: Gain in Completeness. We consider images obtained through Magnetic Resonance Imaging (MRI), Computed Tomography (CT) and Positron Emission Tomography (PET). The multi-sensor data fusion of all three images allows a surgeon to view "soft tissue" information (MRI) in the context of "skeleton" or "bone information" (CT) and in the context of "functional" or "physiological information" (PET).

1.3 Multi-sensor Data Fusion Strategies

As the above examples show, multi-sensor data fusion is a wide-ranging subject with many different facets. In order to understand it better, and to become familiar with its terminology, we shall consider it from three different points of view as suggested by Boudjemaa and Forbes [5], Durrant-Whyte [7] and Dasarathy [6].

1.3.1 Fusion Type

Boudjemaa and Forbes [5] classify a multi-sensor data fusion system according to what aspect of the system is fused:

Fusion across sensors. In this situation, a number of sensors nominally measure the same property, as, for example, a number of temperature sensors measuring the temperature of an object.

Fusion across attributes. In this situation, a number of sensors measure different quantities associated with the same experimental situation, as, for example, in the measurement of air temperature, pressure and humidity to determine air refractive index.

Fusion across domains. In this situation, a number of sensors measure the same attribute over a number of different ranges or domains. This arises, for example, in the definition of a temperature scale.

Fusion across time. In this situation, current measurements are fused with historical information, for example, from an earlier calibration. Often the current information is not sufficient to determine the system accurately and historical information has to be incorporated to determine the system accurately.

Example 1.4. Flood Forecasting [23]. Water companies are under constant pressure to reduce the frequency of combined sewer overflows to natural water courses from urban drainage systems (UDS). The management of storm-water through the UDS and similar applications require accurate real-time estimates of the rainfall. One way water companies have done this is to fuse measurements of the rainfall made by ground-based rain gauges and a weather radar system. This is "fusion across sensors" since, nominally, the two sensors measure the same property.

Example 1.5. Fire Detection [26] Urban-Rural Interface (URI) are zones where forest and rural lands interface with homes, other buildings and infrastructures. In the URI, fires are frequently due to the special nature of these zones: a fire can be the result of the action of a human on one side of the zone or could arrive from the other side (wild land). In such cases, the process of early detection and monitoring of a fire event is important for efficient control of the fire and if required, the evacuation of the area. In [26] a fire is detected using temperature, humidity and vision sensors. The temperature and humidity sensors are deployed "in-field" in the URI zone while the vision sensor is deployed "out-field" a distance from the URI zone. This is "fusion-across-attributes" since the sensors measure different quantities associated with the same experimental situation.

1.3.2 Sensor Configuration

Durrant-Whyte [7] classifies a multi-sensor data fusion system according to its sensor configuration. There are three basic configurations:

Complementary. A sensor configuration is called complementary if the sensors do not directly depend on each other, but can be combined in order to give a more complete image of the phenomenom under observation. Complementary sensors help resolve the problem of *incompleteness*.

Competitive. A sensor configuration is competitive if each sensor delivers an independent measurement of the same property. The aim of competitive fusion is to reduce the effects of *uncertain* and *erroneous* measurements.

Cooperative. A cooperative sensor configuration network uses the information provided by two, or more, independent sensors to derive information that would not be available from the single sensors.

The following example illustrates the complementary fusion of an infra-red and a visible-light image.

Example 1.6. Multispectral Bilateral Video Fusion [4]. A significant problem in night vision imagery is that while an infra-red (IR) image provides a bright and relatively low-noise view of a dark environment, it can be difficult to interpret due to inconsistencies with the corresponding visible-spectrum image. In bilateral fusion we enhance a visible video input using information from a spatially and temporally registered IR video input. Our goal is to create a video that appears as if it was imaged only in the visible spectrum and under more ideal exposure conditions than actually existed.

The following example illustrates noise suppression by competitive fusion of several noisy inputs.

Example 1.7. Noise Suppression. Given a scene S we model an image of S as an $M \times N$ matrix $\mathbf{G}$ whose elements are:

$$G(m,n) = S(m,n) + N(m,n) ,$$

where $S(m,n)$ denotes the (m,n)th element of a deterministic signal which represents the scene properties and $N(m,n)$ denotes the image noise at the (m,n)th element which we model as independent and identically distributed zero-mean Gaussian noise:

$$N(m,n) \sim \mathscr{N}(0,\sigma^2) .$$

By averaging K such images we obtain a fused image whose (m,n)th element is

$$\widetilde{G}(m,n) = S(m,n) + \widetilde{N}(m,n) ,$$

whose noise variance is reduced by a factor of K:

$$\widetilde{N}(m,n) \sim \mathscr{N}(0,\sigma^2/K) .$$

The following are two examples of cooperative fusion taken from the field of image fusion.

Example 1.8. Stereo Vision [22]. In stereo vision we combine two input images A and B which are two views of the same scene taken from slightly different viewing angles. The result is a new image which displays the height of the objects in the scene.

Example 1.9. Super-Resolution [22]. In super-resolution we cooperatively fuse together two or more images of the same scene and taken by the same sensor but from slightly different viewing angles. The result is a new image which has higher spatial resolution than any of the input images.

1.3.3 Input/Output Characteristics

Dasarathy [6] classifies a multi-sensor data fusion system according to its joint input/output characteristics. Table 1.1 illustrates the types of inputs/outputs considered in this scheme. *Note*: Not all of the possible input-output combinations are represented in the Dasarathy scheme. Instead only combinations in which the output is at the same semantic level, or one level higher than the input, are used.

Table 1.1 Dasarathy's Input/Output Data Fusion Model.

Symbol	Name	Description/Example
DaI-DaO	Data Input-Data Output	Input data is smoothed/filtered.
DaI-FeO	Data Input-Feature Output	Features are generated from the input data, e.g. edge detection in an image.
FeI-FeO	Feature Input-Feature Output	Input features are reduced in number, or new features are generated by fusing input features.
FeI-DeO	Feature Input-Decision Output	Input features are fused together to give output decision.
DeI-DeO	Decision Input-Decision Output	Multiple input decisions are fused together to give a final output decision.

It is sometimes useful to divide the DeI-DeO fusion model into two sub-models: a "soft" decision-input model (denoted as DsI-DeO) in which each input decision is accompanied by a degree-of-support, or reliability, value and a "hard" decision-input model (denoted as DhI-DeO) in which the input decisions are not accompanied by any degree-of-support values.

Example 1.10. Multi-Modal Face and Palmprint Fusion [20]. Biometric recognition systems often use face images or palmprint images. In general, the recognition rates obtained using only face or palmprint images is relatively low. However, significantly better recognition rates are obtained by separately extracting features from the face and palmprint images which are then fused together. The fusion thus takes place in the FeI-FeO domain.

1.4 Formal Framework

Multi-sensor data fusion systems are often of a size and a complexity that requires the use of a formal framework [18] around which we may organize our knowledge. One class of formal frameworks is the functional model. The functional model gives an overview of the main processes, or functions, in the system. They are very important in designing, building and understanding large scale multi-sensor data fusion systems. Table 1.2 lists some of the functional models which are used today [21]. For systems which are smaller in size an architectural model which shows the individual modules in the system and the flow of data between modules is often more useful. In this book we have opted for the architectual model of Luo-Kay [15] which consists of a *distributed* network of autonomous modules, in which each module represents a separate function in the data fusion system.

Table 1.2 Multi-Sensor Data Fusion Frameworks.

Name	Description
Joint Defence Laboratories (JDL)	The JDL model consists of several levels at which fusion can take place. The levels are: Pre-Processing (Level 0); Object Refinement (Level 1); Situation Assessment (level 2); Threat Assessment (Level 3) and Process Refinement (Level 4).
Thompoulos	The Thompoulos model consists of three modules, each module integrates input data at a different level. The modules are: Signal Level fusion, Evidence Level Fusion and Dynamics Level Fusion.
Pau	The Pau model consists of a sequence of modules that must be completed before the final fused output is established. The modules are: Feature Extraction, Association Fusion, Sensor Attribute Fusion, Analysis, and Aggregation and Representation.
Waterfall	The Waterfall model is similar to the Pau model except the system is continuously updated with feedback information from the decision-making module.
Omnibus	The Omnibus model consists of four modules: Signal Processing module, Pattern Recognition and Feature Extraction, Decision makng and Context Processing and Control Resource Tasking. The modules operate in a closed loop.

Example 1.11. Physiological Measurements [3]. We consider two physiological measurements made on a given patient: temperature and blood pressure. The measurements are provided by two sensors: a thermometer and a tensiometer. The data sources are *distributed* and *complementary*. In forming a diagnosis, the physician fuses together the two measurements. In the physiological space of the patient the physician is said to perform "fusion across attributes".

Apart from providing us with a structured framework, the modular design decreases the complexity involved in designing a data fusion system by compartmentalizing the design problem. Modularity also helps in the construction and maintenance of an operational multi-sensor data fusion system.

In analyzing a multi-sensor data fusion system it is useful to divide the system into three parts: the physical, informative and cognitive domains and to determine the flow of data between these parts [9, 18] (Fig. 1.1).

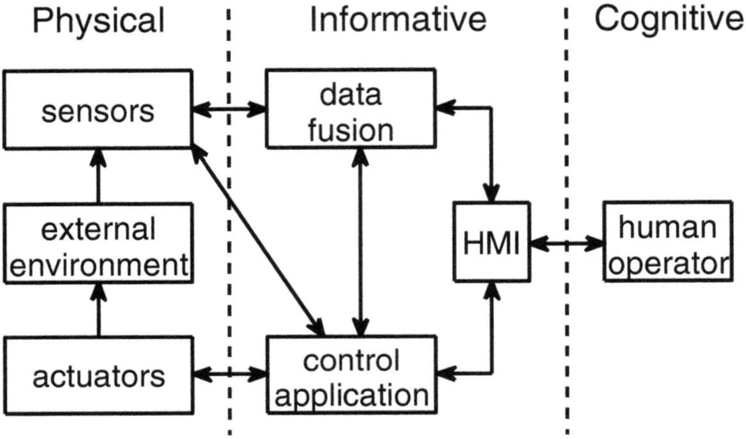

Fig. 1.1 Shows the division of a generic multi-sensor data fusion system into three parts or domains: physical, informative and cognitive and the flow between the parts. The figure is divided into three panels which correspond to the physical domain (leftmost panel), informative domain (middle panel) and cognitive domain (rightmost panel).

Physical Domain: Hardware. The physical domain contains the sensor modules each of which represents a sensor which physically interacts with the external environment. Each module contains a *sensor model* which provides us with a description of the measurements made by the sensor and of the local environment within which the sensor operates. In some applications we may wish to physically change the external environment. In this case the physical domain will also contain actuators which are able to modify the external environment.

Informative Domain: Software. The informative domain constitutes the heart of a multi-sensor data fusion system. It contains three blocks which are responsible for data fusion, control application/resource management and human-machine interface (HMI). The data fusion block is constructed as an autonomous network of "fusion" modules. This network is responsible for combining all the sensor data into a unified view of the environment in the form of an "environmental image". The control application/resource management block is constructed as autonomous networks of "control" modules. This network is responsible for all decisions which are made on the basis of the environmental image. In many applications the decisions are fed back to the sensor block. In this case the process is known as "sensor management".

Cognitive Domain: Human User. In many multi-sensor data fusion applications the human user is the final arbiter or decision maker. In this case it is important to design the system in such a way that all the information which is transmitted to the human user is transformed into a form which is intuitively usable by the user for his decision-making process.

Example 1.12. Control Loop [9]. Fig. 1.2 shows a control loop containing four blocks: sensor, actuator, data fusion and control application. The environment is observed with one or more sensors. The corresponding sensor observations are then passed to the data fusion block where they are combined to form a unified view of the environment ("environmental image"). The environmental image is, in turn, passed to the control application block. The loop is closed by feeding the decisions made by the control application block back to the environment.

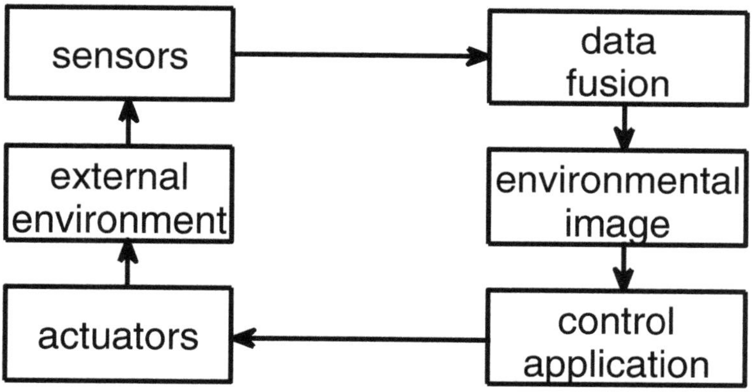

Fig. 1.2 Shows a simple control loop built from four blocks: sensors, actuators, data fusion and control application. A feedback mechanism is included by allowing the control application to act on the external environment via an actuator block.

1.4.1 Multi-sensor Integration

In the Luo and Kay model we make a clear distinction between multi-sensor data fusion and multi-sensor integration [15]. The former refers to the process of combining input data into a common representational format (Sect. 1.1) while the later refers to the use of multiple-sensor information to assist in a particular task. In the data fusion block only a few of the autonomous modules perform "multi-sensor data fusion" the remaining modules perform auxiliary functions. The modules which perform the data fusion will receive input data from the physical sensors $S_1, S_2, \ldots, S_M$ and from other modules. Figure 1.3 illustrates a system in which we simultaneously have multi-sensor integration and multi-sensor data fusion.

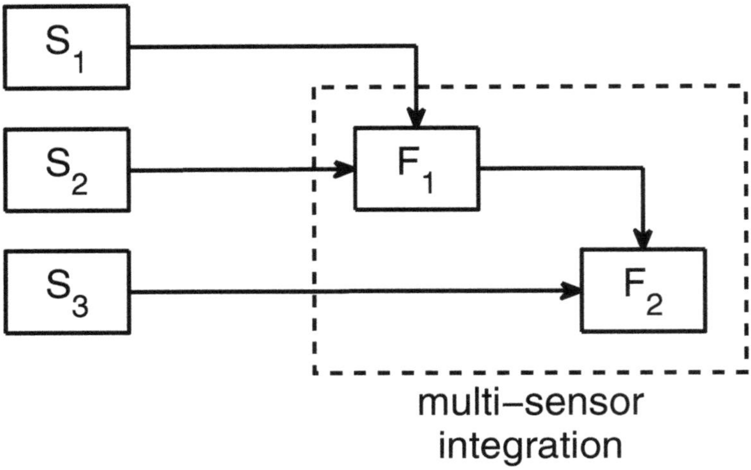

Fig. 1.3 Shows two multi-sensor data fusion blocks F_1 and F_2. F_1 performs data fusion on the output of sensors S_1 and S_2 and F_2 performs mult-sensor data fusion on the output of F_1 and the sensor S_3. Together F_1 and F_2 perform "multi-sensor integration".

1.5 Catastrophic Fusion*

The unsuspecting reader may conclude, on the basis of what has been presented so far, that a multi-sensor fusion system is *always* better than a single sensor system. This conclusion is, however, mistaken: Often the performance of a data fusion system F is below that of the individual sensors. This phenomenum is known as *catastrophic fusion* and clearly should be avoided at all times.

Let p_F denote the performance of the system F, then formally catastrophic fusion [16] is said to occur when

$$p_F << \min_m(p_m) ,$$

where p_m is the performance of the mth sensor S_m. In general, each sensor $S_m, m \in \{1,2,\ldots,M\}$, is designed to operate correctly only under specific conditions, or environment, E_m. Let E_F denote an environment in which all the sensors $S_m, m \in \{1,2,\ldots,M\}$, are operating correctly, then E_F is the "intersection" of the E_m which we write symbolically as

$$E_F = E_1 \wedge E_2 \wedge \ldots E_M . \tag{1.1}$$

Sometimes, however, F is used in an environment, E, which is inconsistent with one of the E_m, say E^*. When this happens, the signal from the corresponding sensor, S^*, may dominant the fused output with catastrophic results. To prevent this happening, multi-sensor fusion systems often employ *secondary classifiers* which monitor the performance of each sensor S_m.

Example 1.13. Automatic Speech Recognition: Preventing Catastrophic Fusion [16]. In ideal, or clean, conditions automatic speech recognition systems perform very well using a single audio sensor S_A. However, we often observe a substantial reduction in performance when background noise, channel distortion or reverberation are present. This has led to the development of automatic speech recognition systems which employ an audio sensor S_A *and* a visual sensor S_V (see Ex. 3.12). The two sensors are *complementary*: speech characteristics that are visually confusable are often acoustically distinct, and characteristics that are acoustically confusable are often visually distinct. The audio-visual system work as follows. Given a finite set of utterances $U_i, i \in \{1,2,\ldots,N\}$, we identify an unknown utterance, U, as

$$U^* = \arg\max_i \left(P(U_i|y_A, y_V)\right) , \tag{1.2}$$

where $P(U_i|y_A, y_V)$ is the conditional probability of U_i given an audio-visual observation (y_A, y_V). To a first approximation, the audio and visual features are conditionally independent. In this case, we may write $P(U_i|y_A, y_V)$ as

$$P(U_i|y_A, y_V) \propto P(U_i|y_A) \times P(U_i|y_V) , \tag{1.3}$$

where the conditional probabilities $P(U_i|y_A)$ and $P(U_i|y_V)$ are calculated off-line using a set of training samples D.

If the set of training samples and test samples are well matched, the solution represented by (1.2) and (1.3) is theoretically optimal: we automatically compensate for any noise or distortion and assign more importance to the classification which is more "certain". The underlying assumption is, however, that the conditional probabilities $P(U_i|y_A)$ and $P(U_i|y_V)$ generated during training, match the speech data that is being tested. When the speech data is contaminated with noise this assumption is no longer valid and the probability estimates are incorrect. In such cases, it is common to use a weighted integration scheme, where the influence of the noisy channel is attenuated. This weighting can be a function of the signal-to-noise ratio (SNR) in each channel, and can be implemented as follows:

$$P(U_i|y_A, y_V)) = (P(U_i|y_A))^{\alpha_A} \times (P(U_i|y_V))^{\alpha_V} ,$$

where

$$\alpha_A + \alpha_V = 1 .$$

The purpose of the weights $\alpha_m, m \in \{A, V\}$, is to "neutralize" a sensor S_m, whenever the secondary classifier C_m determines that S_m is operating outside its specified operating conditions, or environment, E_m. In the worst case, when environmental conditions make the sensor completely unreliable we set the corresponding weight to zero (see Fig. 3.13).

1.6 Organization

Although we shall discuss the physical, information and cognitive domains in a multi-sensor data fusion system, the emphasis will be on the information domain, and specifically on the data fusion block. The book is organized into five parts as follows.

Part I: Basic Concepts. This consists of three chapters. In Chapt. 1 we provide a general introduction to the subject of multi-sensor data fusion. This is followed by Chapts. 2 and 3 which provide overviews on sensors and data fusion architectures.

Part II: Common Representational Format. This consists of Chapts. 4-8. In Chapt. 4 we provide an overview of the basic concept of a common representational format and the different techniques used to create such a representation. The techniques may be broadly divided into spatial, temporal, semantic and radiometric calibration which are considered, respectively, in Chapts. 5, 6, 7 and 8.

Part III: Data Fusion. This consists of Chapts. 9-14. In Chapt. 9 we give an overview of the Bayesian approach to multi-sensor data fusion and the different techniques involved. This is followed by Chapts. 10-12 in which we deal with parameter estimation theory including sequential estimation theory. This is followed by Bayesian decision theory in Chapt. 13 and Ensemble learning in Chapt. 14.

Part IV. Sensor Management. This consists of a single chapter (Chapt. 15) in which we consider sensor management. Specifically we consider how the decisions made by the data fusion block may, if required, be fed back to the sensors.

Part V. Appendix. This consists of one appendix which is a summary of elementary results in probability theory, linear algebra and matrix theory with which the reader should be familiar.

1.7 Further Reading

General overviews on multi-sensor data fusion are [11, 12, 15]. For an extended discussion regarding the issues involved in defining multi-sensor data fusion and related terms, see [19, 24].

Problems

1.1. What is the motivation for data fusion?

1.2. What are the merits of using a Bayesian approach for data fusion?

1.3. When sensor measurements are transformed into a common representational format, their statistical distribution changes. Given some examples of these changes.

1.4. Multi-sensor fusion improves the quality of information in four ways (representation, certainty, accuracy and completeness). Use this framework to analyze the multi-sensor data fusion system for the multi-modal biometric system in Ex 1.10.

1.5. Multi-sensor fusion improves the quality of information in four ways (representation, certainty, accuracy and completeness). Use this framework to analyze the multi-sensor data fusion system for the multiple camera surveillance system in Ex. 4.4.

1.6. Three multi-sensor data fusion classification schemes are Boudjemaa-Forbes, Durrant-Whyte and Dasarathy). Use these schemes to analyze the multi-sensor data fusion flood forecasting system in Ex. 1.4.

1.7. Three multi-sensor data fusion classification schemes are Boudjemaa-Forbes, Durrant-Whyte and Dasarathy). Use these schemes to analyze to the triangulation algorithm in Ex. 4.2.

1.8. Only some input-output combinations are allowed in the Dasarathy classification scheme. Suggest reasons why this should be the case .

1.9. The fusion cell introduced in Sect. 3.3 has three inputs: input data, auxiliary data and external knowledge. (a) Explain the difference between these inputs. (b) Describe the triangulation algorithm in Ex. 4.2 using the fusion cell and identify the corresponding input data, auxiliary data and external knowledge.

1.10. Explain why catastrophic fusion occurs.

1.11. Explain how the auxiliary sensors C_A and C_V in the multi-sensor data fusion audio-visual automatic speech recognition system in Ex. 1.13 prevent catastrophic fusion.

1.12. In Ex. 1.13 the sensors C_A and C_V are regarded as sources of auxiliary data and not input data. (a) Could they have been classified as sources of input data? (b) Is there any difference between classifying C_A and C_V as sources of auxiliary data or input data?

References

1. D'Agostini, G.: Bayesian Reasoning in Data analysis. World Scientific, Singapore (2003)
2. Anthony, R.: Principles of Data Fusion Automation. Published by Artech House, Inc. (1995)
3. Bellot, D., Boyer, A., Charpillet, F.: A new definition of qualified gain in a data fusion process: application to telemedicine. In: Proc. 5th Int. Conf. on Information Fusion, Annapolis, Maryland (2002)
4. Bennett, E.P., Mason, J.L., McMillan, L.: Multispectral bilateral video fusion. IEEE Trans. Imag. Process. 16, 1185–1194 (2007)
5. Boudjemaa, R., Forbes, A.B.: Parameter estimation methods for data fusion, National Physical laboratory Report No. CMSC 38–04 (2004)
6. Dasarathy, B.V.: Decision Fusion. IEEE Computer Society Press (1994)
7. Durrant-Whyte, H.F.: Sensor models and multisensor integration. Int. J. Robotics Res. 7, 97–113 (1988)
8. Elmenreich, W.: An introduction to sensor fusion. Institut fur Technische Informatik, Technischen Universitat Wien, Research Report 47/2001 (2001)
9. Elmenreich, W.: Sensor fusion in time-triggered systems PhD thesis, Institut fur Technische Informatik, Technischen Universitat Wien (2002)
10. Elmenreich, W., Pitzek, S.: The time-triggered sensor fusion model. In: Proc. 5th IEEE Int. Conf. Intell. Enging. Syst., pp. 297–300 (2001)
11. Hall, D.L., Llinas, J. (eds.): Handbook of Multisensor Data Fusion. CRC Press (2001)
12. Hyder, A.K., Shahbazian, E., Waltz, E. (eds.): Multisensor Fusion. Kluwer Academic Publishers (2002)
13. Jones, G.D., Allsop, R.E., Gilby, J.H.: Bayesian analysis for fusion of data from disparate imaging systems for surveillance. Imag. Vis. Comp. 21, 843–849 (2003)
14. Klein, L.A.: Sensor and Data Fusion: Concepts and Applications. SPIE Optical Engineering Press (1993)
15. Luo, R.C., Yih, C.-C., Su, K.L.: Multisensor Fusion and Integration: Approaches, Applications and Future Research Directions. IEEE Sensors J. 2, 107–119 (2002)
16. Movellan, J.R., Mineiro, P.: Robust sensor fusion: analysis and applications to audio-visual speech recognition. Mach. Learn. 32, 85–100 (1998)
17. Nandakumar, K.: Integration of multiple cues in biometric systems. MSc thesis, Department of Computer Science and Engineering, Michigan State University (2005)
18. Opitz, F., Henrich, W., Kausch, T.: Data fusion development concepts within complex surveillance systems. In: Proc. 7th Int. Conf. Inform. Fusion, Stockholm, pp. 308–315 (2004)
19. Oxley, M.E., Thorsen, S.N.: Fusion or integration: what's the difference? In: Proc. 7th Int. Conf. Inform. Fusion Stockholm, pp. 429–434 (2004)
20. Raghavendra, R., Dorizzi, B., Rao, A., Kumar, G.H.: Designing efficient fusion schemes for multimodal biometric systems using face and palmprint. Patt. Recogn. 44, 1076–1088 (2011)
21. Shahbazian, E.: Introduction to DF: Models and processes, architectures, techniques and applications. In: Hyder, A.K., Shahbazian, E., Waltz, E. (eds.) Multisensor Data Fusion. Kluwer Academic Publications (2002)
22. Szeliski, R.: Computer Vision: Algorithms and Applications. Springer, Heidelberg (2011)
23. Todini, E.: A Bayesian technique for conditioning radar precipitation estimates to rain-gauge measurements. Hydrology Earth Sys. Sci. 5, 187–199 (2001)
24. Wald, L.: Some terms of reference in data fusion. IEEE Trans. Geo. Remote Sens. 37, 1190–1193 (1999)
25. Waltz, E., Llinas, J.: Multisensor Data Fusion. Artech House (1990)
26. Zervas, E., Mpimpoudis, A., Anagnostopoulos, C., Sekkas, O., Hadjiefthymiades, S.: Multisensor Data Fusion for Fire Detection. Inf. Fusion 12, 150–159 (2011)

Chapter 2
Sensors

2.1 Introduction

In this chapter we consider the sensors. These are special devices which interact *directly* with the environment and which are ultimately the source of all the input data in a multi-sensor data fusion system [12]. The physical element which interacts with the environment is known as the *sensor element* and may be any device which is capable of perceiving a physical property, or environmental attribute, such as heat, light, sound, pressure, magnetism or motion. To be useful, the sensor must map the value of the property or attribute to a quantitative measurement in a *consistent* and *predictable* manner.

In Chapt. 1 we introduced a formal framework in which we represented a sensor fusion system as a distributed system of autonomous modules. To support such a scheme, the sensors must not only measure a physical property, but must also perform additional functions. These functions can be described in terms of compensation, information processing, communication and integration:

Compensation. This refers to the ability of a sensor to detect and respond to changes in the environment through self-diagnostic tests, self-calibration and adaption.

Information Processing. This refers to processes such as signal conditioning, data reduction, event detection and decision-making, which enhance the information content of the raw sensor measurements.

Communication. This refers to the use of a standardized interface and a standardized communication protocol for the transmission of information between the sensor and the outside world.

Integration. This refers to the coupling of the sensing and computation processes on the same silicon chip. Often this is implemented using microelectromechanical systems (MEMS) technology.

A practical implementation of such a sensor is known as a smart, or intelligent, sensor [11].

2.2 Smart Sensor

A *smart sensor* is a hardware/software device that comprises in a compact small unit a sensor element, a micro-controller, a communication controller and the associated software for signal conditioning, calibration, diagnostic tests and communication. The smart sensor transforms the raw sensor signal to a standardized digital representation, checks and calibrates the signal, and transmits this digital signal to the outside world via a standardized interface using a standardized communication protocol.

Fig. 2.1 shows the measurement of a physical property by a smart sensor.

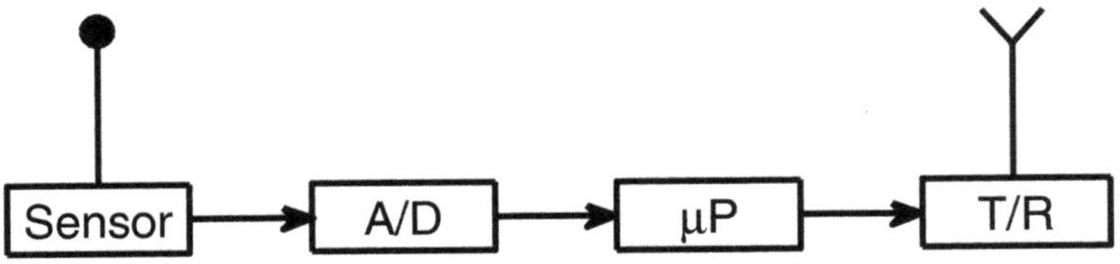

Fig. 2.1 The sensor element measures the physical property and outputs an analog signal which is amplified, filtered and then converted to a digital signal by the analog-to-digital, or A/D, unit. The digital signal is processed by the microprocessor, μP, where it is temporally stored before being transmitted by the transmitter/receiver. *Note*: The smart sensor may also receive command instructions via the transmitter/receiver unit.

The transfer of information between a smart sensor and the outside world is achieved by reading (writing) the information from (to) an interface file system (IFS) which is encapsulated in the smart sensor as shown in Fig. 2.2 [9].

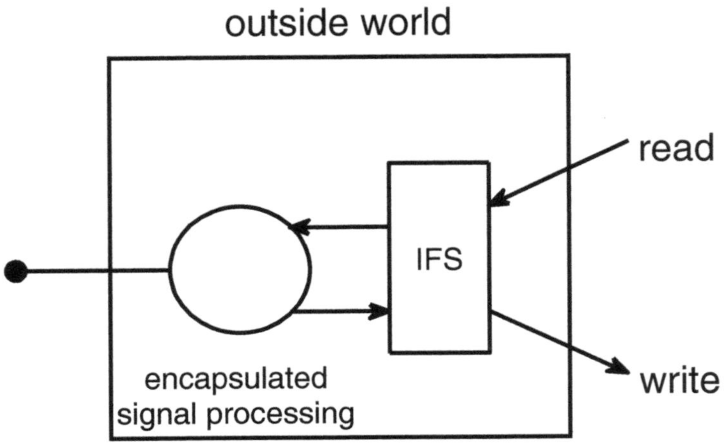

Fig. 2.2 Shows a smart sensor with a sensor element and the encapsulated signal processing functions and the encapsulated interface file system (IFS).

Example 2.1. Smart Image Sensor [2]. When a sensor is packaged together with the processing resources shown in Fig. 2.1 it is called a “smart sensor”. However, although more tightly packaged, the smart sensor is rarely smarter than its conventional counterpart. Brajovic and Kanade [2] show that a “smart sensor” only becomes really “smart” when the tight integration of sensing and processing results in an adaptive sensing system that can react to environmental conditions and consistently deliver useful measurements even under the harshest conditions.

2.3 Logical Sensors

A *logical sensor* is defined as any device which functions as a source of information for a multi-sensor data fusion node. Thus a logical sensor encompass both physical sensors and any fusion node whose output is subsequently fed into another fusion node. Unless stated otherwise, from now on the term “sensor” will refer to a logical sensor or a source-of-information. The following example illustrates a multi-sensor data fusion application which has one “physical” sensor but $N > 1$ “logical” sensors.

Example 2.2. Lesion Localization in Dermoscopy Images [3]. Dermoscopy is a major tool used in the diagnosis of melanoma and other pigmented skin lesions. Due to the difficulty of human interpretation, automated analysis of dermoscopy images has become an important research area. In [3] an accurate automatic lesion localization method is presented which relies on thresholding the dermoscopy image I_0 using several global thresholding algorithms $\{A_1, A_2, \ldots, A_N\}$. The thresholded images $\{I_1, I_2, \ldots, I_N\}$ are then fused together to a localization of the lesions which is more accurate than each thresholded image.

The camera which records the dermoscopy image is the “physical” sensor S_0. The threshold algorithms $\{A_1, A_2, \ldots, A_N\}$ constitute the “logical” sensors $\{S_1, S_2, \ldots, S_N\}$. The thresholded images $\{I_1, I_2, \ldots, I_N\}$ are the inputs to the data fusion block F where a more accurate localization of the lesions is calculated (Fig. 2.3).

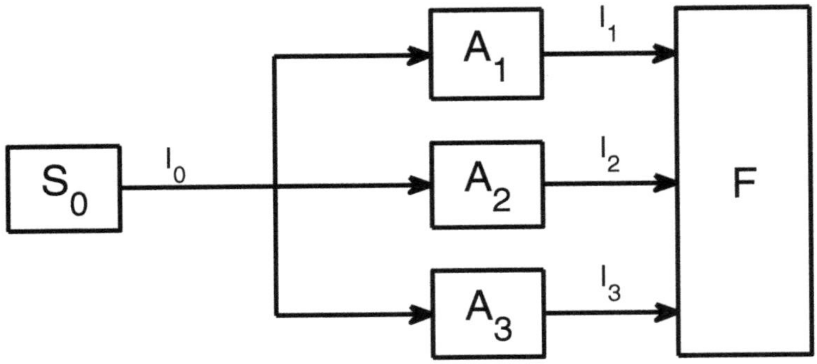

Fig. 2.3 Shows the accurate localization of a skin lesion. I_0 is an image of the lesion captured by a camera (sensor) S_0. The logical sensors $S_k, k \in \{1,2,\ldots,N\}$ are the different algorithms A_k which generate thresholded images I_k. The images $I_k, k \in \{1,2,\ldots,N\}$, are fused together in F to give an accurate lesion localization.

2.4 Interface File System (IFS)

The IFS provides a structured space which is used for communicating information between a smart sensor and the outside world [8]. The following example neatly illustrates the concept of an IFS.

Example 2.3. Brake Pedal in a Car: An Interface File System [15]. We consider a driver in a car. For the purposes of braking, the *brake pedal* acts as an IFS between the driver and the brakes. At this interface there are two relevant variables. The first variable is P: the pressure applied to the brake pedal by the driver and the second variable is R: the resistance provided by the brakes back to the driver. The relative position of the brake pedal uniquely identifies the values of P and R to both the driver and the brakes. The temporal association between sensing the information (e. g. by the driver) and receiving the information (e. g. by the brakes) is implicit, because of the mechanical connection between the driver and the brakes.

A record of the IFS can be accessed, both from the sensor and from the outside world (Fig. 2.2). Whenever one of the internal processes of a smart sensor requires information from the outside world or produces information for the outside world, it accesses the appropriate records of the IFS and reads (writes) the information to (from) this record. The internal processes of a smart sensor are thus not visible to the outside world.

Often we implement the IFS so that it acts as a temporal firewall between the sensors and the outside world. In this case, the IFS uses *local* interface file systems

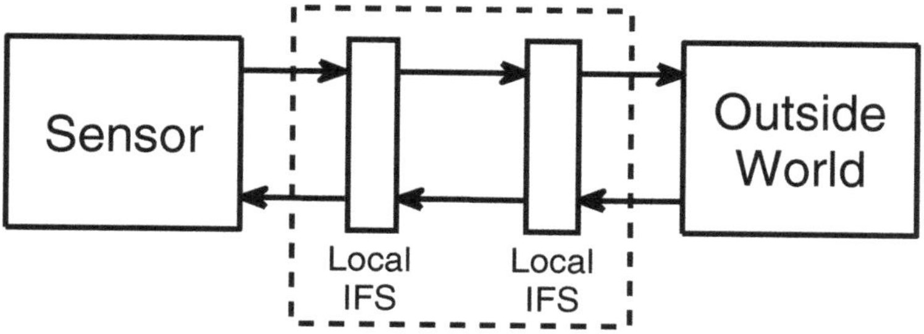

Fig. 2.4 Shows the Interface file system (IFS) built as a temporal firewall. The IFS contains two separate local interface file systems: one is accessed by the sensor and the other is accessed by the outside world.

that can be written by the sensor and read by the outside world, without having direct communication between the sensor and the outside world (Fig. 2.4).

2.4.1 Interface Types

In the smart sensor model we distinguish between three interface types: the real-time service interface, the diagnostic and maintenance interface and the configuration and planning interface. All information that is exchanged across these interfaces is stored in files of the IFS.

Real-time Service (RS) Interface. This interface provides time sensitive information to the outside world. The communicated data is sensor observations made by a sensor on real-time entities in the environment. The RS accesses the real-time IFS files within the smart sensor which is usually time-critical.

Diagnostic and Management (DM) Interface. This interface establishes a connection to each smart sensor. Most sensors need both parameterization and calibration at start-up and the periodic collection of diagnostic information to support maintenance activities. The DM interface accesses the diagnostic IFS files within the smart sensor which is not usually time-critical.

Configuration and Planning (CP) Interface. This interface is used to configure a smart sensor for a given application. This includes the integration and set-up of newly connected sensor modules. The CP interface accesses the configuration IFS files within the smart sensor which is not usually time-critical.

2.4.2 Timing

For reading (writing) information from (to) the different interfaces we could use any possible communication and computational schedule [5]. In distributed systems two different types of scheduling are used:

Event Triggered. In this schedule, all communication and processing activities are initiated whenever a significant change of state occurs.

Time Triggered. In this schedule, all communication and processing activities are initiated at pre-determined times.

For many safety-critical real-time applications the *time-triggered architecture* (TTA) [14] is preferred because its schedule is deterministic. The TTA relies on the synchronization [18] of all local clocks [1] to a global clock: in which all measurements, communication and processing activities are initiated at *pre-determined* time instants which are *a priori* known to all nodes.

The use of a global clock does not, in and of itself, ensure that in a distributed system we are able to *consistently* order events according to their global time. This is illustrated in the following example.

Example 2.4. Ordering Events in a Distributed Network [14]. Consider two local clocks c and C. Suppose the following sequence of events occurs: local clock c ticks, event e occurs, local clock C ticks. In such a situation the event e is time-stamped by the two clocks with a difference of one tick. The finite precision of the global time-base makes it impossible, in a distributed system, to order events consistently on the basis of their global time stamps.

The TTA solves this problem by using a *sparse time base* in which we partition the continuum of time into an infinite sequence of alternating durations of *activity* (duration π) and *silence* (duration Δ) (Fig. 2.5). The architecture must ensure that significant events, such as the sending of a message or the observation of an event, occur only during an interval of activity. Events occuring during the same segment of activity are considered to have happened at the same time. Events that are separated by at least one segment of silence can be consistently assigned to different time-stamps for all clocks in the system. This ensures that temporal ordering of events is always maintained.

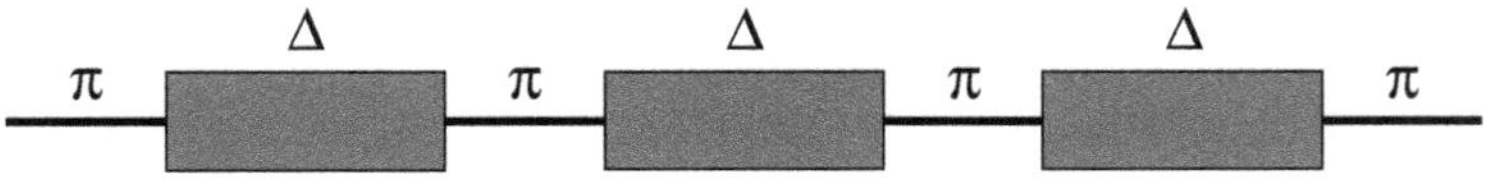

Fig. 2.5 Shows the sparse time based used in the time-triggered architecture. π denotes the intervals of activity and Δ denotes the intervals of silence. If δ is the precision of the clock synchronization, then we require $\Delta > \pi > \delta$.

---

[1] A clock is defined as a physical device for measuring time. It contains a counter and a physical oscillation mechanism that periodically generates an event which increases the counter by one. This event is the "tick" of the clock. Whenever the sensor perceives the occurrence of an event e it will instantaneously record the current tick number as the time of the occurrence of the event e.

2.5 Sensor Observation

We may regard a sensor as a small "window" through which we are able to view a physical property which is characteristic of the outside world or environment. In general the physical property is evolving continuously in time and value. However, the sensor only provides us with a "snapshot" of the process: Often the output of a sensor is reduced to a single scalar value.

The output of a sensor is known as a *sensor observation* and includes the following terms:

Entity-Name E. This includes the name of the physical property which was measured by the sensor and the units in which it is measured. Often the units are defined implicitly in the way the system processes the measurement value.

Spatial Location $\mathbf{x}$. This is the position in space to which the measured physical property refers. In many cases the spatial location is not given explicitly and is instead defined implicitly as the position of the sensor element [2].

Time Instant t. This is the time when the physical property was measured. In real-time systems the time of a measurement is often as important as the value itself.

Measurement $\mathbf{y}$. This is the value of the physical property as measured by the sensor element. The physical property may have more than one dimension and this is the reason we represent it as a vector $\mathbf{y}$. We often refer to it as the *sensor measurement* which may be discrete or continuous. We shall assume that all values are given in digital representation.

Uncertainty $\Delta\mathbf{y}$. This is a generic term and includes many different types of errors in $\mathbf{y}$, including measurement errors, calibration errors, loading errors and environmental errors. Some of these errors are defined *a priori* in the sensors data sheet and others may be calculated internally if the sensor is capable of validating its own measurements.

Symbolically we represent a sensor observation using the following five-tuple

$$O = \langle E, \mathbf{x}, t, \mathbf{y}, \Delta\mathbf{y} \rangle \ . \tag{2.1}$$

Sometimes not all the fields in (2.1) are present. In this case we say the observation is *censored* and we represent the missing fields by an asterix ($*$). For example, if the spatial location $\mathbf{x}$ is missing from the sensor observation, then we write the corresponding censored observation as

$$O = \langle E, *, t, \mathbf{y}, \Delta\mathbf{y} \rangle \ . \tag{2.2}$$

---

[2] We shall often use $\mathbf{u}$ to represent a spatial location in place of $\mathbf{x}$.

Example 2.5. Time-of-Flight Ultrasonic Sensor [13]. In a time-of-flight ultrasonic sensor *ToF*, the sonar transducer emits a short pulse of sound and then listens for the echo. The measured quantity is the time between the transmission of the pulse and reception of the echo. This is converted to a *range reading* r which is one-half of the distance traveled by the sound pulse. In this case, $\mathbf{y}$ is the range reading and the corresponding sensor observation is

$$O = \langle ToF, *, t, r, \Delta r \rangle ,$$

where Δr is the uncertainty in r.

Example 2.6. A Digital RGB (Color) Image. A digital RGB (color) image I is characterized by an array of pixels. At each pixel location, $\mathbf{x}$, there are three pixel values $R(\mathbf{x})$, $G(\mathbf{x})$ and $B(\mathbf{x})$. Let $\mathbf{X}$ denote the relative positions of the array of pixels, then the corresponding sensor observation is

$$O = \langle I, \mathbf{X}, t, \boldsymbol{\Theta}, * \rangle ,$$

where $\boldsymbol{\Theta}$ denotes the array of (R,G,B) values.

2.5.1 Sensor Uncertainty

Sensor measurements are uncertain, which means that they can only give an estimate of the measured physical property. The uncertainty can manifest itself in several different ways:

Random Errors. These errors are characterized by a lack of repeatability in the output of the sensor. They are caused by measurement noise. For example, they may be due to fluctuations in the capacity of the resistance of the electrical circuits in the sensor, or due to the limited resolution of the sensor.

Systematic Errors. These errors are characterized by being consistent and repeatable. There are many different types of systematic errors:

Calibration errors. These are a result of error in the calibration process, and are often due to linearization of the calibration process for devices exhibiting nonlinear characteristics.

Loading errors. These arise if the sensor is *intrusive* which, through its operation, alters the measurand.

Environmental Errors. These arise from the sensor being affected by environmental factors which have not been taken into account.

Common Representation Format Errors. These occur when we transform from the original sensor space to a common representational format (see Chapt. 4).

Spurious Readings. These errors are non-systematic measurement errors. An example of such an error occurs when a sensor detects an obstacle at a given location **x** when, in fact, there is no obstacle at **x**.

Example 2.7. Systematic Errors in a Tipping-Bucket Rain Gauge [19]. Rain-gauges, and in particular the most commonly used tipping-bucket rain-gauge (Fig. 2.6), are relatively accurate instruments. However, at high rain intensities, they suffer from "undercatchment": part of the water is not "weighed" by the tipping-bucket mechanism. This is a *systematic error* which, to a good approximation, may be corrected using the following calibration curve:

$$y = \alpha y_G^{\beta} ,$$

where y is the actual rain intensity, y_G is the rain-gauge measured intensity and α and β are two calibration parameters [19].

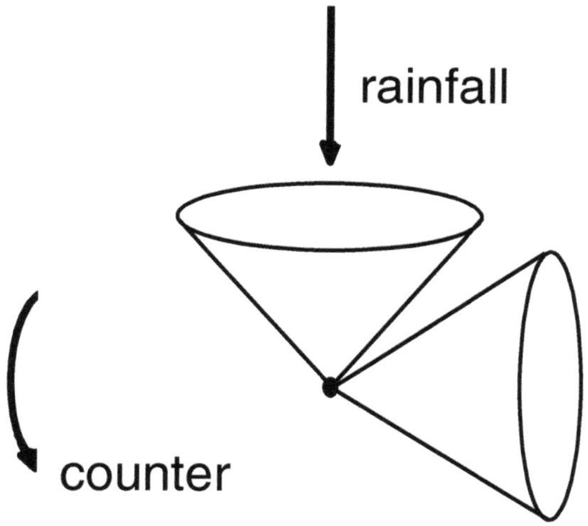

Fig. 2.6 Shows a tipping-bucket rain-gauge. This consists of two specially designed buckets which tip when the weight of 0.01 inches of rain falls into them. When one bucket tips, the other bucket quickly moves into place to catch the rain. Each time a bucket tips, an electronic signal is sent to a recorder. The tipping-bucket rain-gauge is especially good at measuring drizzle and very light rainfall. However it tends to underestimate the rainfall during very heavy rain events, such as thunderstorms.

Example 2.8. Spurious Readings in Time-of-Flight Ultrasonic Sensor. In a time-of-flight ultrasonic sensor (Ex. 2.5), the sensor calculates the distance r to the nearest obstacle from the measured time between the transmission of a pulse and the reception of its echo. Sometimes the sensor will detect an obstacle at a given range r when, in fact, there is no object at r. This can happen if the sensor receives a pulse emitted by second sensor and interprets the pulse as if it were its own pulse. This is a *spurious reading*.

2.6 Sensor Characteristics

In selecting an appropriate sensor, or set of sensors, for a multi-sensor data fusion application, we need to consider the characteristics of the individual sensors. For this purpose it is convenient to group the sensor characteristics into the following categories..

State. The sensors are classified as being internal or external. Internal sensors are devices used to measure "internal" system parameters such as position, velocity and acceleration. Examples of such sensors include potentiometers, tachometers, accelerometers and optical encoders. External sensors are devices which are used to monitor the system's geometric and/or dynamic relation to its tasks and environment. Examples of such sensors include proximity devices, strain gauges and ultrasonic, pressure and electromagnetic sensors.

Function. The sensors are classified in terms of their functions, i. e. in terms of the parameters, or *measurands*, which they measure. In the mechanical domain the measurands include displacement, velocity, acceleration, dimensional, mass and force.

Performance. The sensors are classified according to their performance measures. These measures include accuracy, repeatability, linearity, sensitivity, resolution, reliability and range.

Output. The sensors are classified according to the nature of their output signal: analog (a continuous output signal), digital (digital representation of measurand), frequency (use of output signal's frequency) and coded (modulation of output signal's frequency, amplitude or pulse).

Energy Type. The sensors are classified according to the type of energy transfered to the sensor. For example, thermal energy involves temperature effects in materials including thermal capacity, latent heat and phase change properties, or electrical energy involves electrical parameters such as current, voltage, resistance and capacitance.

For multi-sensor data fusion applications which involve multiple sensors we must also take into account the fusion type and sensor configuration as described in Sects. 1.3.1 and 1.3.2.

2.7 Sensor Model *

As we explained in Sect. 2.2, the smart sensor smoothes, checks and calibrates the sensor signal before it is transmitted to the outside world. In order to perform these functions, we require a sufficiently rich *sensor model* which will provide us with a coherent description of the sensors ability to extract information from its surroundings, i. e. to make meaningful sensor observations. This information will also be required when we consider the fusion of multi-sensor input data. In that case we often include the sensor model within the general "background information" I.

We begin our development of the sensor model by distinguishing between the variable $\boldsymbol{\Theta}$ in which we are interested, and a sensor measurement $\mathbf{y}$. We directly observe N *raw* sensor measurements $\mathbf{y}_i, i \in \{1,2,\ldots,N\}$, while the variable of interest $\boldsymbol{\Theta}$ is not directly observed and must be *inferred.* In mathematical terms we interpret the task of inferring $\boldsymbol{\Theta}$ as estimating the *a posteriori* probability, $P(\boldsymbol{\Theta} = \boldsymbol{\theta}|\mathbf{y},I)$, where $\boldsymbol{\theta}$ represents the true value of the variable of interest $\boldsymbol{\Theta}$ and $\mathbf{y} = \left(\mathbf{y}_1^T, \mathbf{y}_2^T, \ldots, \mathbf{y}_N^T\right)^T$ denotes the vector of N sensor measurements [3]. This is in accordance with the Bayesian viewpoint (see Chapt. 9) adopted in this book and which assumes that all the available information concerning $\boldsymbol{\Theta}$ is contained in $p(\boldsymbol{\Theta} = \boldsymbol{\theta}|\mathbf{y},I)$ [4].

Fig. (2.7) shows a simple, but useful sensor model. Input to the model are three probability distributions:

A Priori pdf $\pi(\boldsymbol{\theta}|I)$. This is a continuous probability density function which describes our *a priori* beliefs about $\boldsymbol{\theta}$. In the absence of any further information we often model the distribution using a histogram of historical data, or we may construct it from *a priori* information we have concerning typical $\boldsymbol{\theta}$ values. For reasons of computational convenience we often assume that $\pi(\boldsymbol{\theta}|I)$ is a Gaussian distribution with a mean value $\boldsymbol{\mu}_0$ and a covariance matrix $\boldsymbol{\Sigma}_0$:

$$\pi(\boldsymbol{\theta}|I) = \mathcal{N}(\boldsymbol{\theta}|\boldsymbol{\mu}_0, \boldsymbol{\Sigma}_0) , \tag{2.3}$$

Sensor Reliability $\pi(\boldsymbol{\Lambda}|I)$. This is a *discrete* probability distribution which specifies the *a priori* reliability of the sensor. In the simplest model, there are two states: $\Lambda = \{\lambda_0, \lambda_1\}$ where λ_0 denotes fault-free operation and λ_1 denotes faulty operation, where ordinarily $\pi(\Lambda = \lambda_0|I) \approx 1$.

---

[3] We use bold letters for both the variable of interest $\boldsymbol{\Theta}$ and for the sensor measurements $\mathbf{y}_i, i \in \{1,2,\ldots,N\}$. This is to emphasize that, in general, $\boldsymbol{\Theta}$ and $\mathbf{y}_i$ may be multi-dimensional. Unless stated otherwise we shall also assume $\boldsymbol{\theta}$ and $\mathbf{y}_i, i \in \{1,2,\ldots,N\}$, are continuous. In this case, a probability which is a function of $\boldsymbol{\Theta}$ or $\mathbf{y}_i$ should be interpreted as a probability density function (pdf) or distribution. To emphasize the change in interpretation we shall denote probability density functions with a small p.

[4] For variables with continuous values we do not normally distinguish between the random variable and its value. Thus from now on we shall write the pdf $p(\boldsymbol{\Theta} = \boldsymbol{\theta}|I)$ as $p(\boldsymbol{\theta}|I)$. However, in the case of variables with discrete values, this may cause some confusion. For these variables we shall therefore continue to write $P(\boldsymbol{\Lambda} = \boldsymbol{\lambda}|I)$.

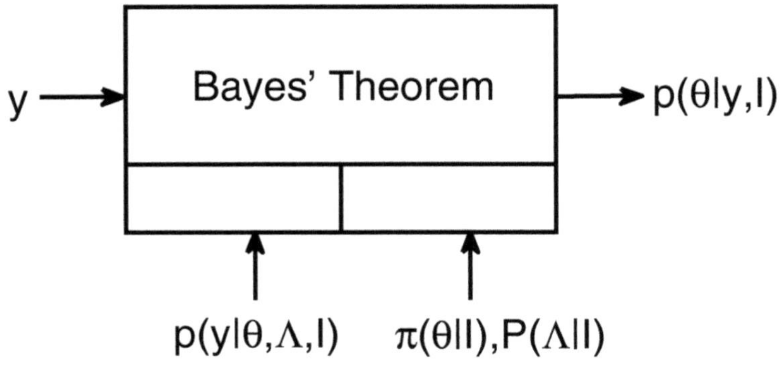

Fig. 2.7 Shows the faulty sensor model. Input to the model are three probabilities: the continuous *a priori* pdf, $\pi(\boldsymbol{\theta}|I)$, and likelihood function, $p(\mathbf{y}|\boldsymbol{\theta},I)$, and the discrete sensor reliability $\pi(\boldsymbol{\Lambda}|I)$. The model uses Bayes' theorem to calculate the joint probability distribution $p(\boldsymbol{\theta},\boldsymbol{\Lambda}|\mathbf{y},I) \sim p(\mathbf{y}|\boldsymbol{\theta},\boldsymbol{\Lambda},I)$ $\pi(\boldsymbol{\theta}|I)P(\boldsymbol{\Lambda}|I)$ and then eliminates $\boldsymbol{\Lambda}$ by marginalization. Output of the model is the *a posterior* pdf $p(\boldsymbol{\theta}|\mathbf{y},I) \sim \pi(\boldsymbol{\theta}|I)\int p(\mathbf{y}|\boldsymbol{\theta},\boldsymbol{\Lambda},I)P(\boldsymbol{\Lambda}|I)d\boldsymbol{\Lambda}$.

The status of the sensor may, of course, change from measurement to measurement. To emphasize this dependency we write the sensor status as $\boldsymbol{\Lambda} = (\Lambda_1,\Lambda_2,\ldots,\Lambda_N)^T$, where Λ_i denotes the status of the sensor when it makes the ith measurement $\mathbf{y}_i$.

Likelihood $p(\mathbf{y}|\boldsymbol{\theta},\boldsymbol{\Lambda},I)$. This is a continuous function which describes how the raw sensor measurements $\mathbf{y}$ depend on the true value $\boldsymbol{\theta}$, the background information I and the sensor status $\boldsymbol{\Lambda}$.

The following example illustrates the centralized fusion of multiple measurements when one or more of the sensor measurements may be spurious.

Example 2.9. Multi-Sensor Data Fusion with Spurious Data [16]. Given K scalar measurements $y_k, k \in \{1,2,\ldots,K\}$, of a given parameter $\Theta = \theta$, we assume a Gaussian sensor model:

$$p(y_k|\theta,\Lambda_k = \lambda_0,I) = \frac{1}{\sigma_k\sqrt{2\pi}}\exp\left[-\frac{1}{2}\left(\frac{\theta - y_k}{\sigma_k}\right)^2\right] .$$

Then we represent the probability that the measurement from the kth sensor S_k is *not* spurious given the true parameter value θ and given the sensor measurement y_k as follows:

$$p(\Lambda_k = \lambda_0|\theta,y_k,I) = \exp\left[-\left(\frac{\theta - y_k}{\alpha_k}\right)^2\right] ,$$

where the parameter α_k depends on the variances of the sensor models $\sigma_l, l \in \{1,2\ldots,K\}$, and the distance between the output of the kth sensor S_k with respect to the other sensors. The *a posteriori* pdf $p(\theta|y_1,y_2,\ldots,y_K,I)$ is given by

$$p(\theta|y_1,y_2,\ldots,y_K,I) = \frac{\pi(\theta|I)}{p(y_1,y_2,\ldots,y_K|I)} \prod_{k=1}^{K} \pi(\Lambda_k = \lambda_0|I) \times \frac{p(y_k|\theta,\Lambda_k = \lambda_0,I)}{p(\Lambda_k = \lambda_0|\theta,y_k,I)} ,$$

where the value of the parameter α_k is assumed to be given by

$$\alpha_k^2 = \beta_k^2 / \prod_{\substack{l=1 \\ l \neq k}}^{K} (y_k - y_l)^2 .$$

Substituting the expression for α_k into $p(\theta|y_1,y_2,\ldots,y_K,I)$ gives

$$p(\theta|y_1,y_2,\ldots,y_K,I) = \frac{\pi(\theta|I)}{p(y_1,y_2,\ldots,y_K,I)} \prod_{k=1}^{K} p(\Lambda_k = \lambda_0|I) \frac{1}{\sigma_k\sqrt{2\pi}} \times \exp\left[-(\theta - y_k)^2 \left(\frac{1}{2\sigma_k^2} - \prod_{\substack{l=1 \\ l \neq k}}^{K} (y_k - y_l)^2/\beta_k^2\right)\right] ,$$

where the value of the parameter β_k^2 is chosen to satisfy the following inequality

$$\beta_k^2 \geq 2\sigma_k^2 \prod_{\substack{l=1 \\ l \neq k}}^{K} (y_k - y_l)^2 .$$

By satisfying this inequality we ensure that the *a posteriori* probability distribution, $p(\theta|y_1,y_2,\ldots,y_K,I)$, is a Gaussian pdf with a single peak.

In the above example we have followed standard practice and used a Gaussian function for likelihood $p(\mathbf{y}|\boldsymbol{\theta},I)$. However, there are applications which require a different shaped pdf. For example in modeling an ultrasonic sensor we often use an asymmetric non-Gaussian likelihood function.

Example 2.10. Konolige Model for a Time-of-Flight Ultrasonic Sensor [13]. We consider a time-of-flight ultrasonic sensor (see Ex. 2.5). Suppose we obtain a range reading equal to R. Then the fault-free (i. e. $\Lambda = \lambda_0$) likelihood function is $p(r \circ R|r_0)$, where $r \circ R$ denotes that the *first* detected echo corresponds to a distance R ("$r = R$") and that no return less than R was received ("$r \nless R$"). Let $p(r = R|r_0)$ and $p(r \nless R|r_0)$ denote the conditional pdfs corresponding to $r = R$ and $r \nless R$, then the likelihood function for a time-of-flight ultrasonic sensor is

$$p(r \circ R|r_0) = p(r = R|r_0) \times p(r \nless R|r_0) ,$$

where

$$p(r \nless R|r_0) = 1 - \int_0^R p(r = x|r_0)dx .$$

To a good approximation,

$$p(r = R|r_0) \propto \frac{1}{\sqrt{2\pi\sigma^2}} \exp\left[-\frac{1}{2}\left(\frac{r - r_0}{\sigma}\right)^2\right] + F ,$$

where F is a small constant which takes into account multiple targets which may reflect the sonar beam, in addition to the target at r_0 (see Sect. 11.3.3).

In practice, we observe that the range error becomes proportionally larger and the probability of detection becomes proportionally smaller at increasing range. Incorporating this effects into the above likelihood function we obtain the Konolige likelihood function:

$$p(r \circ R|r_0) = \gamma\left[\frac{\alpha(r)}{\sqrt{2\pi\sigma^2(r)}} \exp\left[-\frac{1}{2}\left(\frac{r - r_0}{\sigma(r)}\right)^2\right] + F\right] \times \left(1 - \int_0^R p(r = x|r_0)dx\right) .$$

where γ is a normalization constant, $\alpha(r)$ describes the attenuation of the detection rate with increasing distance and $\sigma(r)$ describes the increase in the range variance with increasing distance.

2.8 Further Reading

General references on the use of smart sensors in multi-sensor data fusion are [6, 7, 12]. Sensor models are discussed in [4, 10]. Specific references on the sensor model for the time-of-flight ultrasonic sensors are [13, 17].

Problems

2.1. Use the concept of an interface file system (IFS) to explain the action of a brake pedal in an automobile (Ex. 2.3).

2.2. Compare and contrast the Real-time Service interface, the Diagnostic and management interface and the Configuration and Planning interface.

2.3. Measurement uncertainty may be broadly divided into random errors, systematic errors and spurious readings. Explain and define these errors.

2.4. The following are all systematic errors: calibration errors, loading errors, environmental errors and common representational errors. Compare and contrast these errors. Give examples of each type of error.

References

1. Boudjemaa, R., Forbes, A.B.: Parameter estimation methods for data fusion, National Physical laboratory Report No. CMSC 38–04 (2004)
2. Brajovic, V., Kanade, T.: When are the Smart Vision Sensors Smart? An example of an illumination-adaptive image sensor. Sensor Rev. 24, 156–166 (2004)
3. Celebi, M.E., Hwang, S., Iyatomi, H., Schaefer, G.: Robust border detection in dermoscopy images using threshold fusion. In: Proc. IEEE Int. Conf. Image Proc., pp. 2541–2544 (2010)
4. Durrant-Whyte, H.F.: Sensor models and multisensor integration. Int. J. Robot Res. 7, 97–113 (1988b)
5. Eidson, J.C., Lee, K.: Sharing a common sense of time. IEEE Instrument Meas. Mag., 26–32 (September 2003)
6. Elmenreich, W.: An introduction to sensor fusion. Institut fur Technische Informatik, Technischen Universitat Wien, Research Report 47/2001 (2002)
7. Elmenreich, W.: Sensor fusion in time-triggered systems PhD thesis, Institut fur Technische Informatik, Technischen Universitat Wien (2002)
8. Elmenreich, W., Haidinger, W., Kopetz, H.: Interface design for smart transducers. In: IEEE Instrument Meas. Tech. Conf. (IMTC), vol. 3, pp. 1643–1647 (2001)
9. Elmenreich, W., Haidinger, W., Kopetz, H., Losert, T., Obermaisser, R., Paulitsch, M., Trodhandl, C.: DSoS IST-1999-11585 Dependable systems of Systems. Initial Demonstration of Smart Sensor Case Study. A smart sensor LIF case study: autonomous mobile robot (2002)
10. Elmenreich, W., Pitzek, S.: The time-triggered sensor fusion model. In: Proc. 5th IEEE Int. Conf. Intell. Engng. Sys., Helsinki, Finland, pp. 297–300 (2001)
11. Elmenreich, W., Pitzek, S.: Smart transducers-principles, communications, and configuration. In: Proc. 7th IEEE Int. Conf. Intell. Engng. Sys., Egypt, Assuit, Luxor, pp. 510–515 (2003)
12. Fowler, K.R., Schmalzel, J.L.: Sensors: The first stage in the measurement chain. IEEE Instrument Meas. Mag., 60–65 (September 2004)
13. Konolige, K.: Improved occupancy grids for map building. Autonomous Robots 4, 351–367 (1997)
14. Kopetz, H., Bauer, G.: The time-triggered architecture. Proc. IEEE 91, 112–126 (2003)
15. Kopetz, H., Holzmann, M., Elmenreich, W.: A universal smart transducer interface: TTP/A. In: 3rd IEEE Int. Symp. Object-Oriented Real-Time Distributed Computing (2001)

16. Kumar, M., Garg, D.P., Zachery, R.A.: A method for the judicious fusion of inconsistent multiple sensor data. IEEE Sensors J. 7, 723–733 (2007)
17. O'Sullivan, S., Collins, J.J., Mansfield, M., Eaton, M., Haskett, D.: A quantitative evaluation of sonar models and mathematical update methods for map building with mobile robots. In: 9th Int. Symposium on Artificial Life and Robotics (AROB), Japan (2004)
18. Paulitsch, M.: Fault-tolerant clock synchronization for embedded distributed multi-cluster systems. PhD thesis, Institut fur Technische Informatik, Technischen Universitat Wien (2002)
19. Todini, E.: A Bayesian technique for conditioning radar precipitation estimates to rain-gauge measurements. Hydrology Earth Sys. Sci. 5, 187–199 (2001)

Chapter 3
Architecture

3.1 Introduction

In Chapt. 1 we introduced a formal framework in which we represented a multi-sensor data fusion system as a distributed system of autonomous modules. In this chapter we shall consider the architecture of a multi-sensor fusion system and, in particular, the architecture of the "data fusion block" (Sect. 1.4).

The modules in the data fusion block are commonly called *fusion nodes*. A communication system allows the transfer of information from one node to another node via an exchange of messages. An algorithmic description of the fusion block is provided by the software which is embedded in the nodes and which determines the behaviour of the block and coordinates its activities.

Although the data fusion block is represented as a distributed system we perceive it as a whole rather than as a collection of independent nodes. This phenomenum is known as *transparency* [17] and is concerned with the "concealment" of nodes in a distributed system. There are many different forms of transparency which reflect on the different motivations and goals of the distributed systems. Two important forms of transparency are:

Replication Transparency. Here we increase the reliability and performance of the system by using multiple copies of the same sensor. However the user, or the application program, has no knowledge of the duplicate sensors [17, 19].

Failure Transparency. Here we conceal failures in the hardware or software components and thereby allow the users and application programs to complete their tasks. However the user, or the application program, has no knowledge of the failures [17].

The following example illustrates a *fault-tolerant framework* using replication transparency.

Example 3.1. Fault-Tolerant Framework [2]. We consider the communication between a sensor node S and a fusion node F via an interface file system (IFS) (see Fig. 3.1). In general the environmental image produced by F is sensitive to errors in the value y received from S. We may reduce this sensitivity by using a fault-tolerant framework (Fig. 3.2) as follows. We replace the sensor S by a set of M redundant sensors $S_m, m \in \{1,2,\ldots,M\}$. If y_m denotes the y value produced by S_m, then we insert an extra component into the fusion node F in order to process the $y_m, m \in \{1,2,\ldots,M\}$, and produce a new y value which is less likely to be in error. Suppose y_{NEW} denotes the new y value, then the fusion algorithm remains unchanged: It operates on y_{NEW} value and has no knowledge of the y_m.

The fault-tolerance processing in Fig. 3.2 can be of various kinds. For example, in a *voting algorithm* we select one observation out of the multiple input observations (see Chapt. 14).

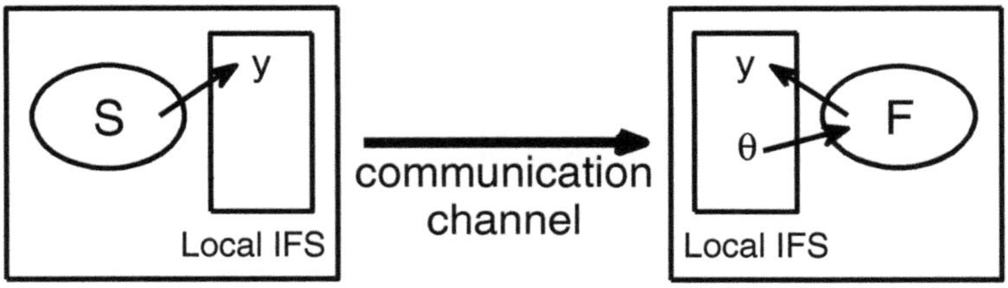

Fig. 3.1 Shows the communication and processing of data. If the sensor S provides a false value for y, then the fusion cell F will be affected by the false value and will in turn produce a wrong value for its output θ.

3.2 Fusion Node

We model the data fusion block as a distributed network of autonomous fusion cells or nodes. Each node receives one, or more, input *sensor observations*

$$O_i = \langle E_i, \mathbf{x}_i, t_i, \mathbf{y}_i, \Delta\mathbf{y}_i \rangle, \qquad i \in \{1,2,\ldots\}\,. \tag{3.1}$$

If necessary, the observations and any *a priori* knowledge, are converted to a common representational format, which are then fused together to produce new output observations. The output observation may, or may not, differ in position ($\mathbf{x}$), time (t), value ($\mathbf{y}$) and uncertainty ($\Delta\mathbf{y}$) from the input observation(s). However, the output observations are *always* assigned new entity names.

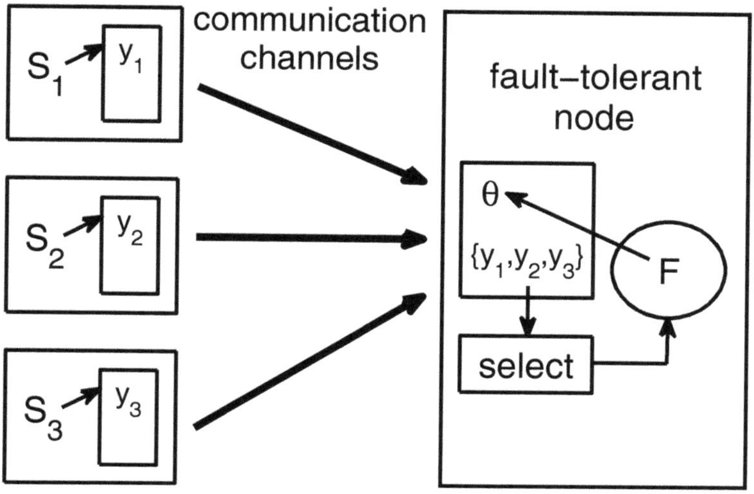

Fig. 3.2 Shows fault-tolerant communication and processing of data. As long as not more than one of the sensor nodes $S_m, m \in \{1,2,3\}$, provides a false value for y, then the fusion cell F will *not* be affected by the false value and will in turn produce a correct value for its output θ.

Example 3.2. Image Smoothing. We reconsider Ex. 2.6 and assume three input images $I_i, i \in \{1,2,3\}$, each characterized by an array of pixel gray-levels $\mathbf{G}_i$. The corresponding (censored) observations are $O_i = \langle I_i, \mathbf{X}_i, t_i, \mathbf{G}_i, *\rangle$. The fusion cell F performs pixel-by-pixel smoothing and generates an output $\widetilde{O}$, where $\widetilde{O}$ is a *new* image (entity name $\widetilde{I}$) characterized by an array of pixels with relative positions $\widetilde{\mathbf{X}}$ and gray-levels $\widetilde{\mathbf{G}}$:

$$\widetilde{O} = \langle \widetilde{I}, \widetilde{\mathbf{X}}, \widetilde{t}, \widetilde{\mathbf{G}}, *\rangle \ . \tag{3.2}$$

Fig. 3.3 shows a graphical representation of a single fusion cell or node [11, 23]. The node may receive three distinct types of input data or information:

Sensor Information. This includes data which comes *directly* from the sensors $S_m, m \in \{1,2,\ldots,M\}$, as well as input which has been produced by other fusion cells. Ordinarily, this constitutes most of the input data into the fusion cell.

Auxiliary Information. This includes additional data which is derived by specific processing of the sensor information.

External Knowledge. This includes additional data and consists of all the elements of the *a priori*, or external, knowledge.

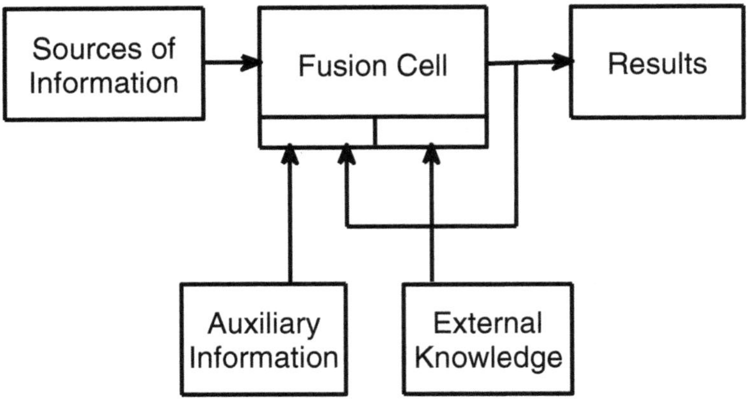

Fig. 3.3 Shows a single fusion node, its three types of input data or information: sources of information, auxiliary information and *a priori* external knowledge and its output results. In the figure we show feedback of the results as auxiliary information.

Example 3.3. Sensor Information vs. Auxiliary Information. We consider a fusion cell $\widetilde{F}$ which receives information from a physical sensor, S, and from another source-of-information A, in the form of data R. In classifying R we have two options: (1) We may regard A as a logical sensor in which case we regard R as sensor information, or (2) we may regard A as performing specific processing in which case we regard R as auxiliary information. We regard A as a logical sensor if R is of the same "type" as S or if the data rate of R is similar to that of S, otherwise we regard A as performing specific processing and R as auxiliary information.

3.2.1 Properties

The chief properties [7, 8] of the fusion node are:

Common Structure. Each node has the same structure. This allows designers to easily implement and modify all nodes.

Completeness. The node possesses all the information that it requires to perform its tasks. Some of this information, such as the sensor observations, is received through a communications channel.

Relevance. Each node only has information relevant to the functions it is required to perform.

Uncertainty. Observation error and noise mean that the information in a node is uncertain to a greater, or lesser, extent. To ensure a reasonable degree of operational reliability, we must take this uncertainty into account when processing the information in a node.

Communication. Individual nodes are only able to communicate with other nodes that are directly connected to it. A node is not able to send a message to, or receive a message from, another node if it is not directly connected to it.

Other properties which are sometimes important are

Self-Identification. This refers to the ability of the node to automatically identify itself. This facility is important in large networks containing very many nodes.

Self-Testing. Self-testing is the process of inspection and verification which ensures the node is functioning properly. In case of a malfunction the node is able to inform the network about its condition and thus avoid corrupting the system with wrong data.

3.3 Simple Fusion Networks

Fig. 3.3 is a graphical representation of the single fusion node or cell. It is the basic building block for all multi-sensor fusion networks. We now illustrate its versatility by using it to construct four different fusion networks.

3.3.1 Single Fusion Cell

The simplest multi-sensor fusion network consists of a single fusion cell. This cell may be used to fuse together raw sensor measurements. In the network we have the option of introducing external knowledge, in which case, the network represents a *supervised* process.

Example 3.4. Exploiting camera meta-data for scene classification [3]. Semantic scene classification based only on low-level vision cues (e. g. color, texture, etc.) has had limited success. To improve the classification performance we may use camera meta-data as follows:

Scene brightness. This includes exposure time, aperture, f-number, exposure time and shutter speed. Outdoor scenes tend to be brighter than indoor scenes, even under outcast skies. They therefore tend to have shorter exposure times and smaller aperture.

Flash. Because of the above lighting differences, camera flash is used on a much higher percentage of images of indoor scenes than of outdoor scenes.

Subject Distance. Most outdoor scenes and landscape images have a large subject distance.

Focal length. Zoom-in is more likely to be used for distant outdoor objects while zoom-out is more likely to be used for long-distance, outdoor scenes.

The above meta-data is auxiliary data. In addition, in order to exploit the meta-data, the fusion cell requires requires external data in the form of statistical tables.

3.3.2 Parallel Network

After the single fusion cell is the parallel network which represents the next higher level of complexity. In the parallel network (Fig 3.4) each fusion cell $F_m, m \in \{1,2,\ldots,M\}$, separately processes the input data it receives from the sensor S_m and delivers its results R_m to a fusion process $\widetilde{F}$ which generates the final result $\widetilde{R}$.

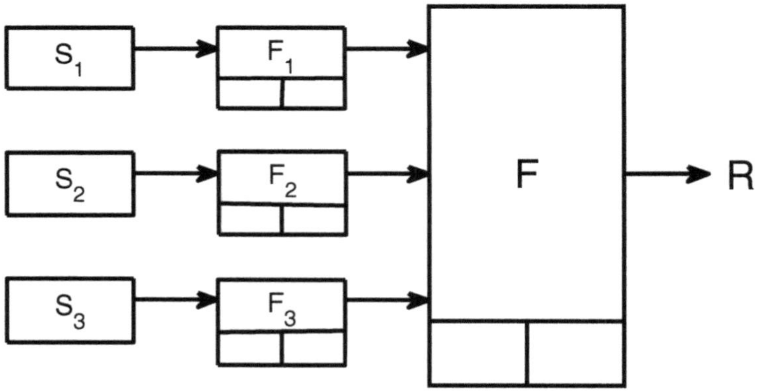

Fig. 3.4 Shows a parallel arrangement of three fusion cells $F_m, m \in \{1,2,3\}$. Each cell F_m acts as a virtual sensor which produces the input data R_m. The $R_m, m \in \{1,2,3\}$, are then fused together by $\widetilde{F}$.

Typically, the intermediate result $R_m, m \in \{1,2,\ldots,M\}$, are *redundant* and the fusion process $\widetilde{F}$ is used to increase reliability and accuracy of the final result $\widetilde{R}$.

Example 3.5. Tire Pressure Monitoring Device [18]. Maintaining the correct tire pressure for road vehicles is very important: incorrect pressure may adversely affect vehicle handling, fuel consumption and tire lifetime. The simplest way to monitor the tire pressure is to directly monitor the pressure in each tire using a separate pressure sensor mounted at the rim of each tire. This is, however, very expensive. The alternative is to indirectly measure the tire pressure using existing sensors. Two such sensors are:

Vibration Sensor S_1. This sensor relies on the fact that the rubber in the tire reacts like a spring when excited by road roughness. The idea is to monitor the resonance frequency which in turn is correlated with the tire pressure.

Wheel Radius Sensor S_2. This sensor relies on the fact that the tire pressure affects the rolling radius of the tire. The idea is to monitor the difference in wheel speeds, which should be close to zero when the tires are equally large.

The two fusion cells $F_m, m \in \{1,2\}$, carry information about the current tire inflation pressure in the form of two variables: (1) a flag, or *indicator*, I_m, which indicates whether, or not, the tire is under-inflated and (2) a confidence σ_m, which is associated with I_m. The two pairs of variables can be used *independently*: we may issue a warning that the tire is under-inflated if I_1 indicates under-inflation and σ_1 is greater than a given threshold T_1, *or* I_2 indicates under-inflation and σ_2 is greater than a given threshold T_2. In this case the output is the indicator I:

$$I = \begin{cases} 1 \text{ if } \max\big(I_1(\sigma_1 - T_1), I_2(\sigma_2 - T_2)\big) > 0\,, \\ 0 \text{ otherwise}\,, \end{cases}$$

where $I = 1$ indicates the presence of under-inflation and $I = 0$ indicates the absence of under-inflation.

We may, however, enhance both the robustness and the sensitivity of the system by *fusing* the two sets of measurements in a simple voting scheme (Fig. 3.5). In this case the output is the fused indicator I_F, where

$$I_F = \begin{cases} 1 \text{ if } \max\big(I_1(\sigma_1 - T_1), I_2(\sigma_2 - T_2), \frac{I_1+I_2}{2}(\frac{\sigma_1+\sigma_2}{2} - T_F)\big) > 0\,, \\ 0 \text{ otherwise}\,, \end{cases}$$

and $T_F < \min(T_1, T_2)$.

3.3.3 *Serial Network*

The serial, or cascaded, network (Fig. 3.6) contains M fusion cells, each cell $F_m, m \in \{1,2,\ldots,M-1\}$, receives one, or more, observations from the sensor S_m and it sends its results R_m to the next cell F_{m+1}. In a typical application, the intermediate results $R_m, m \in \{1,2,\ldots,M\}$, are *heterogeneous* and *complementary*.

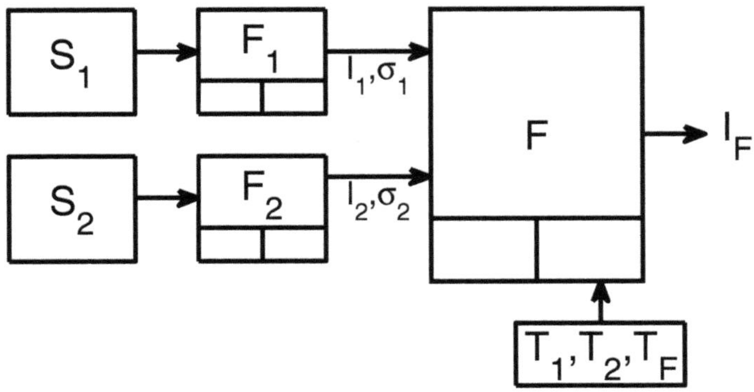

Fig. 3.5 Shows the operation of the new tire pressure monitoring system. The sensor S_1 sends measurements of the resonance frequency to F_1 which outputs the variables (I_1, σ_1). Similarly the sensor S_2 sends measurements of the wheel radii to F_2 which outputs the variables (I_2, σ_2). The cell F receives (I_1, σ_1) and (I_2, σ_2) as input data and generates the signal $\widetilde{I}$ as output. The thresholds $T_1, T_2, \widetilde{T}$ are calculated using a set of training samples D and are regarded as "external knowledge".

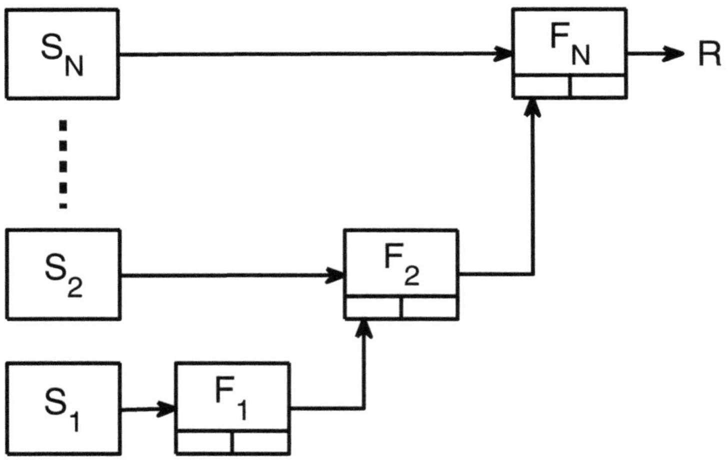

Fig. 3.6 Shows the serial arrangement of multiple fusion cells.

Example 3.6. Multi-Modal Biometric Identification Scheme [16]. Fig. 3.7 shows a multi-modal biometric identification scheme built using serial, or cascaded, architecture. The use of this architecture can improve the user convenience as well as allow fast and efficient searches in appropriate databases. For example, suppose, after processing the first k biometric features, we have sufficient confidence concerning the identity of the user. At this point, we may stop the process and the user will not be required to provide the remaining biometric features.

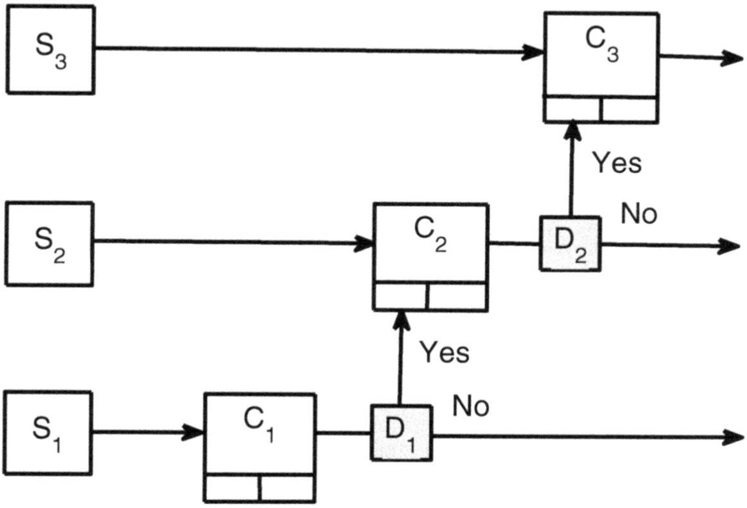

Fig. 3.7 Shows the serial arrangement of multiple classification cells $C_m, m \in \{1,2,3\}$, in a three-stage biometric identification scheme. Apart from the decision-boxes $D_m, m \in \{1,2\}$, the architecture is identical to that shown in Fig. 3.6. The function of the box D_m is to determine whether or not a biometric measurement from sensor S_{m+1} is required.

3.3.4 *Iterative Network*

Fig. 3.8 represents an iterative network in which we re-introduce the result R as auxiliary knowledge into the fusion cell F. This network is generally used in applications which possess dynamical data sources. In that case, S consists of data changing through time. This scheme integrates many applications such as temporal tracking of objects in an image sequence, or the navigation of a mobile robot. F could be, for example, a Kalman filter.

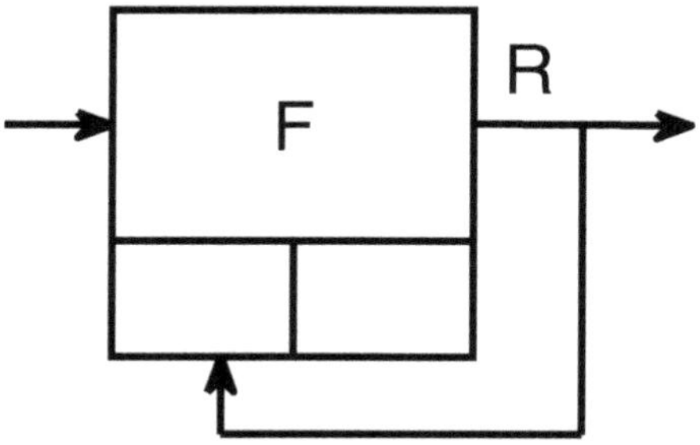

Fig. 3.8 Shows single fusion cell operating in an iterative mode. The result R is re-introduced as auxiliary knowledge into F.

*Example 3.7. Heart-Rate Estimation using a Kalman Filter** [20]. Analysis of heart rate variability requires the calculation of the mean heart rate which we adaptively estimate using a one-dimensional Kalman filter [4, 24]. Let $\mu_{k|k}$ denote the estimated mean heart rate at time step k given the sequence of measured heart-rates $y_{1:k} = (y_1, y_2, \ldots, y_k)$. Then we assume $\mu_{k|k}$ follows a random walk process:

$$\mu_{k|k} = \mu_{k-1|k-1} + v_{k-1} ,$$

where $v_{k-1} \sim \mathcal{N}(0, Q)$ represents the process noise at time step $k-1$. We suppose the measured heart-rate at time step k is y_k:

$$y_k = \mu_k + w_k ,$$

where $w_k \sim \mathcal{N}(0, R)$ represents the measurement noise. The Kalman filter estimates $\mu_{k|k}$ in a predictor/corrector algorithm: In the prediction phase we project the filtered heart rate $\mu_{k-1|k-1}$ and its covariance matrix $\Sigma_{k-1|k-1}$ to the next time step k. Let $\mu_{k|k-1}$ and $\Sigma_{k|k-1}$ be the corresponding *predicted* values, then

$$\begin{aligned} \mu_{k|k-1} &= \mu_{k-1|k-1} , \\ \Sigma_{k|k-1} &= Q + \Sigma_{k-1|k-1} . \end{aligned}$$

In the correction phase we fuse the predicted values with the measurement y_k made at the kth time step:

$$\begin{aligned} \mu_{k|k} &= \mu_{k|k-1} + K_k(y_k - \mu_{k|k-1}) , \\ \Sigma_{k|k} &= (I - K_k)\Sigma_{k|k-1} , \end{aligned}$$

where I is a unit matrix and

$$K_k = \Sigma_{k|k-1}(\Sigma_{k|k-1} + R)^{-1} .$$

3.4 Network Topology

Fundamentally the fusion nodes may be arranged in three different ways (topolgies), or architectures: *centralized*, *decentralized* and *hierarchical*. In Table 3.1 we list several multi-sensor data fusion applications and their architectures. We now briefly consider each of these architectures in turn.

Table 3.1 Multi-Sensor Data Fusion Applications: Their Networks and Architectures

Network	Application
Single (C)	Ex. 3.8 Satelite Imaging of the Earth.
Parallel/Serial (H)	Ex. 3.5 Tire Pressure Monitoring Device.
Serial (H)	Ex. 3.6 Multi-Modal Biometric Identification Scheme.
Iterative (C)	Ex. 3.7 Heart-Rate Estimation using a Kalman Filter.
Parallel (D)	Ex. 3.9 Covariance Intersection in a Decentralized Network
Parallel/Serial (H)	Ex. 3.11 Data Incest.
Parallel/Serial (H)	Ex. 3.12 Audio-Visual Speech Recognition System.
Parallel/Serial (H)	Ex. 3.13 Track-Track Fusion.
Parallel/Serial (H)	Ex. 8.22 Fusing Similarity Coefficients in a Virtual Screening Programme.
Single (C)	Ex. 10.6 Nonintrusive Speech Quality Estimation Using GMM's.
Single (C)	Ex. 12.1 INS/Radar Altimeter: A Hybrid Navigation Principle.
Single (C)	Ex. 12.3 Tracking a Satellite's Orbit Aound the Earth.
Single (C)	Ex. 12.6 Tracking Metrological Features.

The designations C, D and H refer, respectively, to the centralized, decentralized and hierarchical network topologies.

3.4.1 Centralized

In a centralized system, the sensor fusion unit is treated as a central processor or a node that collects all information from the different sensors (see Fig. 3.9). All decisions are made at this node and instructions, or task assignments, are given out to the respective sensors.

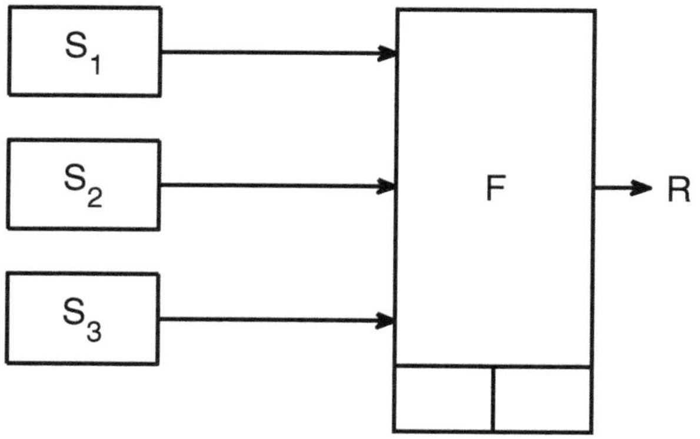

Fig. 3.9 Shows the operation of a centralized node F. The inputs to F are regarded as being produced by separate logical sensors, or sources-of-information, S_1, S_2, S_3.

Theoretically, a single fusion node has the best performance if all the sensors are accurately aligned[1]. However, small errors in the correct alignment of the sensors may cause substantial reduction in the performance of the fusion algorithm. In addition the centralized architecture suffers from the following drawbacks:

Communication. All sensors communicate directly with the centralized fusion node. This may cause a communication bottleneck.

Inflexibility. If the application changes, the centralized system may not easily adapt. For example, the central processor might not be powerful enough to handle the new application.

Vulnerability. If there is a failure in the central processor, or the central communications facility, the entire system fails.

Non-Modularity. If more sensors are added to a centralized system, some reprogramming is generally required: The communications bandwidth and computer power would need to increase (assuming a fixed total processing time constraint) in proportion to the amount of extra data.

Example 3.8. Satellite Imaging of the Earth [23]. Fig. 3.10 shows the mapping of the earth from different satellite images: MSI (multi-spectral image), RADAR, Geo (geographical map) and HRI (high-resolution image). For this application the fusion method is a classifier (see Chapt. 13). The output is a segmented image showing the distribution of the different classes.

3.4.2 Decentralized

In a decentralized system, the sensor measurements, or information, is fused locally using a set of local fusion nodes rather than by using a single central fusion node. The main advantage of a decentralized architecture is the lack of sensitivity regarding the correct alignment of the sensors. Other advantages of a decentralized architecture are:

Communication. The architecture is relatively immune to communication and signal processing bottlenecks. This is because the information flow between nodes is reduced, and the data processing load is distributed.

Scalable. The decentralized architecture is scalable in structure without being constrained by centralized computational bottleneck or communication bandwidth limitations.

Robustness. The decentralized architecture is robust against loss of sensing nodes and any changes in the network.

Modular Design. The decentralized architecture is modular in the design and implementation of the fusion nodes.

---

[1] Technically. a centralized architecture requires a single common representational format in which all sensors are accurately located. See Chapt. 4.

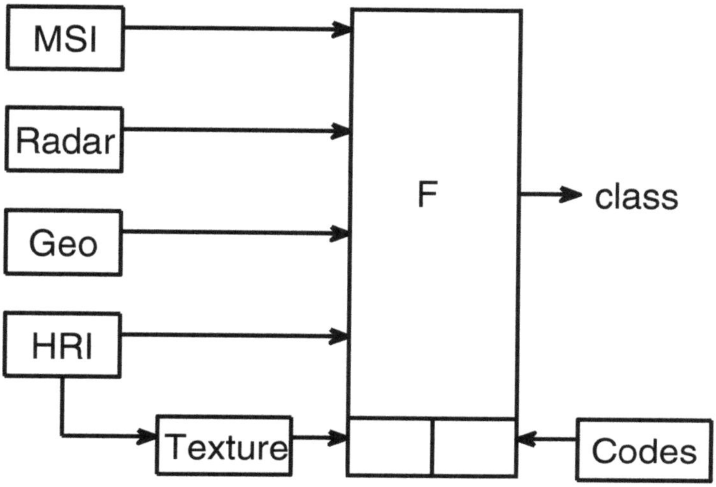

Fig. 3.10 Shows the mapping of a landscape using Earth observation satellite images. The fusion, or classification, is performed using a centralized architecture. The texture information is produced by a specific processing of the HRI (high-resolution image) and is passed onto the fusion cell F as auxiliary information. The classifier uses a set of pre-calculated codes which are considered to be external knowledge.

However, the decentralized network suffers from effects of *redundant* information [22]. For example, suppose a node A receives a piece of information from a node B, the topology of the network may be such that B is passing along the information it originally received from A. Thus if A were to fuse the "new" information from B with the "old" information it already has under the assumption of independence, then it would *wrongly* decrease the uncertainty in the fused information.

Theoretically we could solve the problem of redundant information by keeping track of all the information in the network. This is not, however, practical. Instead, fusion algorithms have been developed which are robust against redundant information. The method of *covariance intersection* [5, 12] is one such algorithm and is illustrated in the following example.

Example 3.9. Covariance Intersection in a Decentralized Network [5]. We consider a decentralized network in which a given node F receives the *same* piece of information from m different sources of information $S_m, m \in \{1,2,\ldots,M\}$. We suppose the piece of information sent by S_m is a scalar measurement y_m, and an uncertainty σ_m^2. If we regard the measurements and uncertainties as independent, then the corresponding fused maximum likelihood measurement value and uncertainty (see Chapt. 10) are given by:

$$\tilde{y}_{ML} = \sum_{m=1}^{M} \frac{y_m}{\sigma_m^2} / \sum_{m=1}^{M} \frac{1}{\sigma_m^2} ,$$

$$\tilde{\sigma}_{ML}^2 = 1 / \sum_{m=1}^{M} \frac{1}{\sigma_m^2} .$$

Since the sources S_m all supply the same piece of information, then

$$y_1 = y = y_2 = \ldots = y_M ,$$

$$\sigma_1 = \sigma = \sigma_2 = \ldots = \sigma_M ,$$

and thus

$$\tilde{y}_{ML} = y , \quad \tilde{\sigma}_{ML}^2 = \sigma^2 / M .$$

The $1/M$ reduction in uncertainty is clearly incorrect: it occurs because the node F assumes that each $(y_m, \sigma_m^2), m \in \{1, 2, \ldots, M\}$, represents an *independent* piece of information.

In the method of covariance intersection we solve this problem by fusing the measurements $y_m, m \in \{1, 1, \ldots, M\}$, using a *weighted* maximum likelihood estimate. The weights Ω_m, are chosen such that $\Omega_m \geq 0$ and $\sum_{m=1}^{M} \Omega_m = 1$. In this case, the corresponding fused estimate and uncertainty are

$$\tilde{y}_{CI} = \sum_{m=1}^{M} \Omega_m \frac{y_m}{\sigma_m^2} / \sum_{m=1}^{M} \frac{\Omega_m}{\sigma_m^2} ,$$

$$\tilde{\sigma}_{CI}^2 = 1 / \sum_{m=1}^{M} \frac{\Omega_m}{\sigma_m^2} .$$

If we assume the Ω_m are all equal to $1/M$, then the covariance intersection estimates are:

$$\tilde{y}_{CI} = y , \quad \tilde{\sigma}_{CI}^2 = \sigma^2 .$$

In this case, we observe that by using the method of covariance intersection we no longer obtain a reduction in uncertainty.

In the general case we have M multi-dimensional measurements $\mathbf{y}_m, m \in \{1, 2, \ldots, M\}$, with covariance matrices $\boldsymbol{\Sigma}_m$. The covariance intersection estimate is

$$\tilde{\mathbf{y}}_{CI} = \sum_{m=1}^{M} \Omega_m \tilde{\boldsymbol{\Sigma}}_{CI} \boldsymbol{\Sigma}_m^{-1} \mathbf{y}_m , \tag{3.3}$$

where

$$\widetilde{\Sigma}_{CI}^{-1} = \sum_{m=1}^{M} \Omega_m \Sigma_m^{-1} . \tag{3.4}$$

Example 3.10. Covariance Intersection of Two 2D Observations [12]. Fig. 3.11 shows two 2D observations $O_1 = (\mathbf{y}_1, \boldsymbol{\Sigma}_1)$ and $O_2 = (\mathbf{y}_2, \boldsymbol{\Sigma}_2)$ and their covariance intersection $\widetilde{O}_{CI} = (\widetilde{\mathbf{y}}_{CI}, \widetilde{\boldsymbol{\Sigma}}_{CI})$.

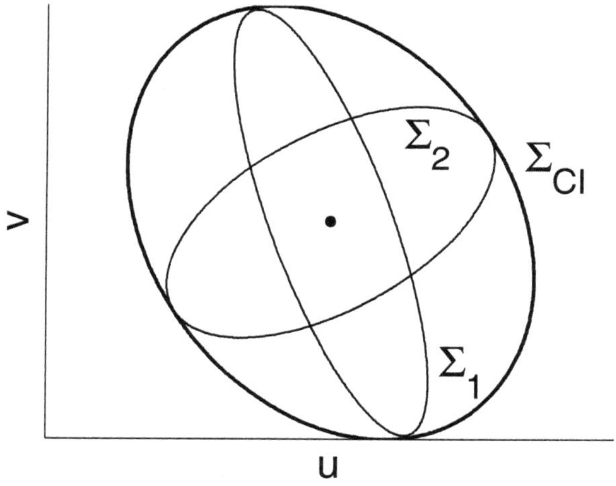

Fig. 3.11 Shows the covariance of two coincident sensor observations $O_1 = (\mathbf{y}_1, \boldsymbol{\Sigma}_1)$ and $O_2 = (\mathbf{y}_2, \boldsymbol{\Sigma}_2)$ in two dimensions (u, v). The position of the observations is shown by a small filled circle. The covariances, $\boldsymbol{\Sigma}_1$ and $\boldsymbol{\Sigma}_2$, of the two observations are two ellipses drawn with thin solid lines. The covariance of the covariance intersection, $\widetilde{\boldsymbol{\Sigma}}_{CI}$, is the ellipse drawn with a thick solid line.

3.4.3 Hierarchical

This design can be thought of as a hybrid architecture in which we mix together the centralized and decentralized systems. The hierarchical architecture combines the advantages of the centralized and decentralized architectures without some of their disadvantages. For example, the performance of the hierarchical architecture is relatively insensitive to any errors in the correct alignment of the sensors. However, the hierarchical network may still suffer from the effects of redundant information, or *data incest*, as the following example shows.

Example 3.11. Data Incest [14]. Three mobile sensors A,B,C send information concerning a target T to a fusion cell F (Fig. 3.12). C is always out of range of F and it can only relay its information to F via A or B. However C is not always in range of A and B. To increase the chances of its information reaching F, C broadcasts its information to *both* A and B. Assuming F fuses the pieces of information it receives as if they were independent, then data incest occurs when the information from C reaches both A and B and is used twice over by F. Table 3.2 shows the *a posteriori* probability density function (pdf) that F may calculate assuming the transmission of information from C to A (via the channel $C \rightarrow A$) and from C to B (via the channel $C \rightarrow B$) are successful or not. Data incest only occurs when when both $C \rightarrow A$ and $C \rightarrow B$ are successful and the corresponding *a posteriori* pdf is $p(T|A,B,C,C)$.

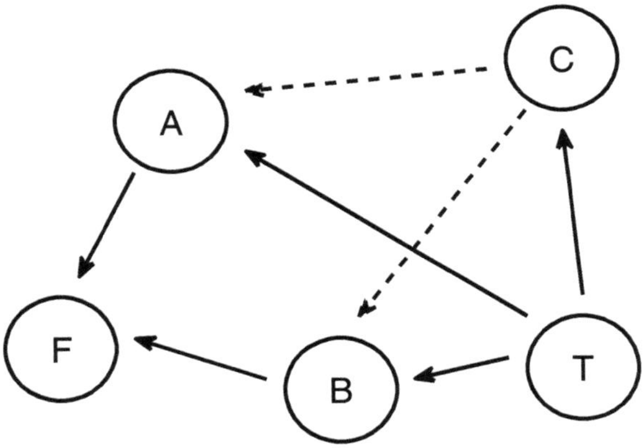

Fig. 3.12 Illustrates the possibility of data incest when the sensor C successfully sends its information regarding the target T to A (via communication channel $C \rightarrow A$) and B (via communication channel $C \rightarrow B$) which in turn forward the information onto the fusion cell F. *Note*: Unreliable communication channels are drawn with dashed arrows.

Table 3.2 *A Posteriori* pdf at F.

$C \rightarrow A$	$C \rightarrow B$	*A Posteriori* pdf
No	No	$p(T\|A,B)$
Yes	No	$p(T\|A,B,C)$
No	Yes	$p(T\|A,B,C)$
Yes	Yes	$p(T\|A,B,C,C)$

In the hierarchical architecture there are often several hierarchical levels where the top level contains a single centralized fusion node and last level is made up of several decentralized (local) fusion nodes. Each of the local fusion nodes may receive inputs from a small group of sensors or from an individual sensor. The next two examples illustrate the implementation of a fusion system using a hierarchical architecture.

Example 3.12. Audio-Visual Speech Recognition System [15]. Fig. 3.13 shows the implementation of the audio-visual speech recognition system using a hierarchical architecture (see Ex. 1.13).

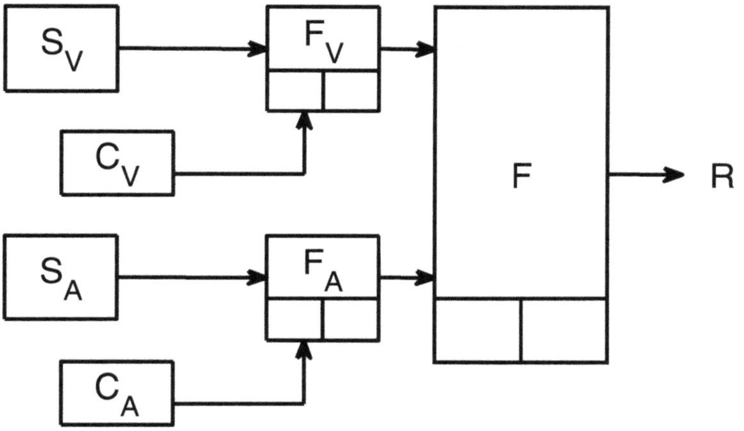

Fig. 3.13 Shows the hierarchical architecture for an audio-visual speech recognition system. The network consists of a single central fusion node F which receives input from an audio fusion node F_A and a visual fusion node F_V. These nodes receive, respectively, *raw sensor data* from the audio and visual sensors S_A and S_V and *auxiliary data* from the secondary classifiers C_A and C_V.

Example 3.13. Track-to-Track Fusion [1]. Fig. 3.14 shows the implementation of the track-to-track fusion system using a hierarchical architecture (see Sect. 12.6.2).

3.5 Software

Matlab code for the covariance intersection is given in [12]. However, for fast computation of the covariance intersection we recommend the algorithm in [10].

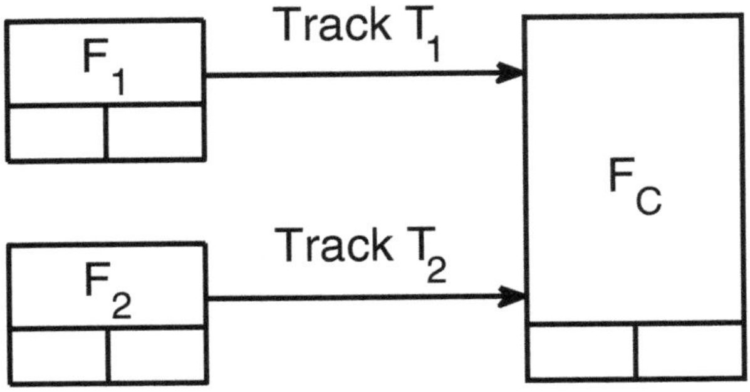

Fig. 3.14 Shows the hierarchical architecture for a track-to-track fusion system. The network consists of two local fusion nodes F_1 and F_2 which form local tracks T_1 and T_2 using the observations received from the sensors S_1 and S_2. The local tracks are sent to the central fusion node F_C where they are fused together into a single track T_C.

3.6 Further Reading

General references on the different architectures used in multi-sensor data fusion are: [6, 9, 11, 13, 21]. Covariance intersection is now widely used in decentralized and hierarchical systems. Recent references on this topic are [12, 22].

Problems

3.1. Explain the concept of replication transparency in a decentralized multi-sensor data fusion network.

3.2. Explain the concept of failure transparency in a decentralized multi-sensor data fusion network. Compare and contrast it to replication transparency.

3.3. List the relative advantages and disadvantages of the centralized, decentralized and hierarchical architectures.

3.4. Give an example of a multi-sensor data fusion system which uses a parallel network. What are its advantages and its disadvantages?

3.5. Give an example of a multi-sensor data fusion system which uses a serial network (Sect. 3.3.3). What are its advantages and its disadvantages?

3.6. Give an example of a multi-sensor data fusion system which uses an iterative network.

3.7. Define data incest. Explain why it does not affect multi-sensor data fusion systems employing centralized architectures.

3.8. Define the method of covariance intersection. Explain how it solves the method of data incest. What are is its disadvantages?

3.9. Suggest an alternative approach to prevent data incest.

References

1. Bar-Shalom, Y., Campo, L.: The effect of the common process noise on the two sensor fused track covariance. IEEE Trans. Aero Elec. Syst. 22, 803–805 (1986)
2. Bauer, G.: Transparent fault tolerance in a time-triggered architecture. PhD thesis. Institut fur Technische Informatik, Technischen Universitat Wien (2001)
3. Boutell, M., Luo, J.: Beyond pixels: exploiting camera metadata for photo classification. In: Proc. Comp. Vis. Patt. Recogn. CVPR 2004 (2004)
4. Brown, R.G., Hwang, P.Y.C.: Introduction to random signal analysis and Kalman filtering. John Wiley and Sons (1997)
5. Chen, L., Arambel, P.O., Mehra, R.K.: Estimation under unknown correlation: covariance intersection revisited. IEEE Trans. Automatic Control 47, 1879–1882 (2002)
6. Delvai, Eisenmann, U., Elmenreich, W.: A generic architecture for integrated smart transducers. In: Proc. 13th Int. Conf. Field Programmable Logic and Applications, Lisbon, Portugal (2003)
7. Elmenreich, W.: An introduction to sensor fusion. Institut fur Technische Informatik, Technischen Universitat Wien, Research Report 47/2001 (2001)
8. Elmenreich, W.: Sensor fusion in time-triggered systems PhD thesis, Institut fur Technische Informatik, Technischen Universitat Wien (2002)
9. Elmenreich, W., Pitzek, S.: The time-triggered sensor fusion model. In: Proceedings of the 5th IEEE Int Conf. Intell. Engng. Sys., Helsinki, Finland, pp. 297–300 (2001)
10. Franken, D., Hupper, A.: Improved fast covariance intersection for distributed data fusion. In: Proc. Inform. Fusion (2005)
11. Houzelle, S., Giraudon, G.: Contribution to multisensor fusion formalization. Robotics Autonomous Sys. 13, 69–75 (1994)
12. Julier, S., Uhlmann, J.: General decentralized data fusion with covariance intersection (CI). In: Hall, D., Llians, J. (eds.) Handbook of Multidimensional Data Fusion, ch. 12. CRC Press (2001)
13. Kopetz, H., Bauer, G.: The time-triggered architecture. Proc. IEEE 91, 112–126 (2003)
14. McLaughlin, S., Krishnamurthy, V., Challa, S.: Managing data incest in a distributed sensor network. In: Proc. IEEE Int. Conf. Accoust, Speech Signal Proc., vol. 5, pp. 269–272 (2003)
15. Movellan, J.R., Mineiro, P.: Robust sensor fusion: analysis and applications to audio-visual speech recognition. Mach. Learn. 32, 85–100 (1998)
16. Nandakumar, K.: Integration of multiple cues in biometric systems. MSc thesis, Michagan State University (2005)
17. Paulitsch, M.: Fault-tolerant clock synchronization for embedded distributed multi-cluster systems. PhD thesis, Institut fur Technische Informatik, Technischen Universitat Wien (2002)
18. Persson, N., Gustafsson, F., Drevo, M.: Indirect tire pressure monitoring using sensor fusion. In: Proc. SAE 2002, Detroit, Report 2002-01-1250, Department of Electrical Engineering, Linkoping University, Sweden (2002)
19. Poledna, S.: Replica determinism and flexible scheduling in hard real-time dependable systems. In: Research Report Nr. 21/97 November 1997, Institut fur Technische Informatik, Technishe Universitat Wien, Austria (1997)
20. Schlogl, A., Fortin, J., Habenbacher, W., Akay, M.: Adaptive mean and trend removal of heart rate variability using Kalman filtering. In: Proc. 23rd Int. Conf. IEEE Engng. Med. Bio. Soc. (2001)

21. Trodhandl, C.: Architectural requirements for TP/A nodes. MSc. thesis, Institut Technische Informatik, Technischen Universitat Wien (2002)
22. Uhlmann, J.K.: Covariance consistency methods for fault-tolerant distributed data fusion. Inform. Fusion 4, 201–215 (2003)
23. Wald, L.: Data Fusion: Definitions and Architectures. Les Presses de l'Ecole des Mines, Paris (2002)
24. Welch, G., Bishop, G.: (2001) An introduction to the Kalman filter. Notes to accompany Siggraph Course 8, (2001), http://www.cs.unc.edu/welch+

Chapter 4
Common Representational Format

4.1 Introduction

The subject of this chapter is the *common representational format*. Conversion of all sensor observations to a common format is a basic requirement for all multi-sensor data fusion systems. The reason for this is that only after conversion to a common format are the sensor observations compatible and sensor fusion may be performed. The following example, taken from the field of brain research, illustrates the concept of a common representational format.

Example 4.1. A Standardized Brain Atlas [56]. In order to compare different brains and, to facilitate comparisons on a voxel-by-voxel basis, we use a standardized anatomically-based coordinate system or *brain atlas*. The idea is that, in the new coordinate system, all brains have the same orientation and size. The transformation to this coordinate system also gives us the means to enhance weak, or noisy, signals by averaging the transformed images. The standardized brain atlas allows us to catalogue the anatomical, metabolic, electrophysiological, and chemical architecture of different brains into the same coordinate systems.

Table 4.1 list several different types of brain atlas which are in common use today.

In (2.1) we introduced the following notation for a sensor observation:

$$O = \langle E, \mathbf{x}, t, \mathbf{y}, \Delta\mathbf{y} \rangle \;, \qquad (4.1)$$

where E is the entity name which designates the physical property underlying the observation and $\mathbf{x}$, t, $\mathbf{y}$ and $\Delta\mathbf{y}$ are, respectively, the spatial position, the time, the value and the uncertainty of the physical property as measured by the sensor. Corresponding to the four types of information - entity, spatial, temporal and value

Table 4.1 Different Brain Atlases

Name	Description
Adaptable Brain Atlas	This is a brain atlas which can be individualized to the anatomy of new subjects. This allows the automated labeling of structures in new patients' scans.
Probabilistic Brain Atlas	This is a population-based atlas in which the atlas contains detailed quantitative information on cross-subject variations in brain structure and function. The information stored in the atlas is used to identify anomalies and label structures in new patients [38].
Disease-Specific Brain Atlas	This is a brain atlas which relates to the anatomy and physiology of patients suffering from a specific disease.
Dynamic ($4D$) Brain Atlas	This is an atlas which incorporates probabilistic information on the structural changes which take place during brain development, tumor growth or degenerative disease processes.

- the process of converting the sensor observations into a common representational format involves the following four functions:

Spatial Alignment. Transformation of the local spatial positions x to a common coordinate system. The process involves geo-referencing the location and field-of-view of each sensor and is considered in Chapt. 5.

Temporal Alignment. Transformation of the local times t to a common time axis. In many applications, is performed using a dynamic time warping algorithm and is considered in Chapt. 6.

Semantic Alignment. Conversion of the multiple inputs so they refer to the same objects or phenomena. In many applications the process is performed using an assignment algorithm and is considered in Chapt. 7.

Radiometric Normalization. Calibration or normalization of the sensor values $\mathbf{y}$ and their uncertainties $\Delta\mathbf{y}$ to a common scale. The process is often performed using a statistical analysis and is considered in Chapt. 8.

In many multi-sensor data fusion applications, the construction of the common coordinate system is the primary fusion algorithm. Two examples are given below. Additional examples are listed in Table 4.2. The first example describes the fusion of K direction bearing measurements by creating a common Cartesian coordinate system (Fig. 4.2).

Example 4.2. Bayesian Triangulation of Direction-Bearing Measurements [44]. Let $\phi_m, m \in \{1,2,\ldots,M\}$, denote, respectively, the *true* bearings of the object O. Then

$$\phi_m = \tan^{-1}\frac{x - X_m}{y - Y_m} ,$$

where the *measured* bearings are distorted by random noise w_m:

Table 4.2 Applications in which Construction of the Common Coordinate System is the Primary Fusion Algorithm

Class		Application
DaI-DaO	Ex. 4.1	A Standardized Brain Atlas: A Common Representational Format for Brain Research.
	Ex. 4.3	Image Fusion of Visible and Thermal Images for Fruit Detection
	Ex. 4.4	A Distributed Surveillance System.
	Ex. 4.9	Myocardial Imaging.
	Ex. 4.14	Object recognition based on photometric color invariants.
	Ex. 4.15	Debiased Cartesian Coordinates.
	Ex. 5.14	Spatial Alignment in Cardiac MR Images.
	Ex. 5.18	Mosaic Fingerprint Image.
	Ex. 6.1	Cardiac MR Image Sequences.
DaI-FeO	Ex. 4.2	Bayesian Triangulation of Direction-Bearing Measurements.
	Ex. 4.18	Chemical Sensor Analysis: Principal Discriminant Analysis (PDA) for Small-Sample Size Problems.
	Ex. 4.20	PCA and LDA Fusion for Face Recognition.
	Ex. 5.17	Multi-Modal Cardiac Imaging.
	Ex. 6.2	Visualization of the Dysphagia (Disorders Associated with Swallowing).

The designations DaI-DaO, DaI-FeO refer, respectively, to the Dasarathy input/output classifications: "Data Input-Data Output" and "Data Input-Feature Output" (Sect. 1.3.3).

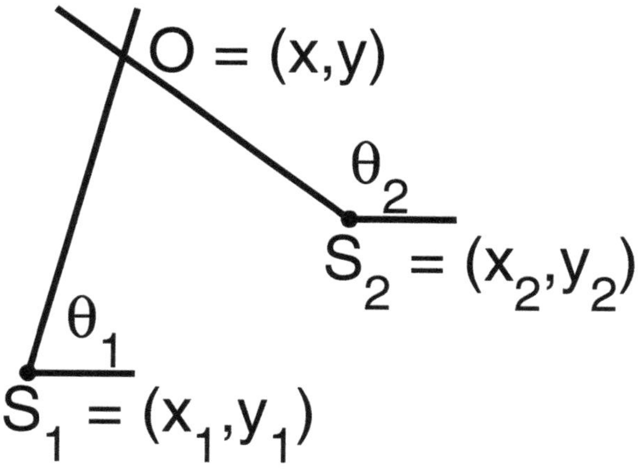

Fig. 4.1 Shows the triangulation of two sensors S_1 and S_2. The sensors are located, respectively, and at (X_1, Y_1) and (X_2, Y_2), and only measure the bearings θ_1 and θ_2 to the object O. Fusion of the two sensor measurements gives an estimate $(\hat{x}, \hat{y})$ of the targets true location (x, y)

$$\theta_m = \phi_m + w_m .$$

We assume the w_m are independent and identically distributed (iid) according to a zero-mean Gaussian distribution, $w_m \sim \mathcal{N}(0, \sigma_m^2)$. Given the bearings $\phi = (\phi_1, \phi_2, \ldots, \phi_M)^T$, the *a posteriori* probability density $p(\theta|\phi, I)$ is

$$\begin{aligned} p(\theta|\phi,I) &= \prod_{m=1}^{M} \frac{1}{\sigma_m\sqrt{2\pi}} \exp\left[-\left(\frac{\theta_m - \phi_m}{\sigma_m\sqrt{2}}\right)^2\right] , \\ &= \exp\left[-\sum_{m=1}^{M} \frac{1}{2\sigma_m^2}\left(\theta_m - \tan^{-1}\frac{x - X_m}{y - Y_m}\right)^2\right] / \prod_{m=1}^{M} \sigma_m\sqrt{2\pi} , \end{aligned}$$

where $\theta = (\theta_1, \theta_2, \ldots, \theta_M)^T$. Inverting this equation using Bayes' theorem gives

$$p(x,y|\theta) \sim \pi(x,y|I) \exp\left[-\sum_{m=1}^{M} \frac{1}{2\sigma_m^2}\left(\theta_m - \tan^{-1}\frac{x - X_m}{y - Y_m}\right)^2\right] ,$$

where $\pi(x,y|I)$ is an *a priori* probability density which we postulate for each position (x,y) given the background information I. The estimated location of the object O is given by the mean values:

$$\begin{aligned} \hat{x} &= \int x p(x,y|\theta) dx dy , \\ \hat{y} &= \int y p(x,y|\theta) dx dy . \end{aligned}$$

The second example describes the fusion of visible and thermal images into a common representational format for fruit detection.

Example 4.3. Image Fusion of Visble and Thermal Images for Fruit Detection [5]. The fusion of a thermal image and a visible image of an orange canopy scene gives an improved fruit detection performance. After spatial alignment the images are converted to a common radiometric gray-scale as follows. The thermal image $T(i,j)$ is converted to an 8-bit gray-scale using the following equation:

$$I_T(i,j) = \frac{255 \times (T(i,j) - T_{\min})}{T_{\max} - T_{\min}} ,$$

and the RGB color image is converted to an 8-bit gray-scale image using the following equation:

$$I_V(i,j) = \frac{255 \times R(i,j)}{R(i,J) + B(i,j) + G(i,j)} .$$

4.2 Spatial-temporal Transformation

Let T denote a transformation which maps the position $\mathbf{x}$ and time t of a sensor observation O to a position $\mathbf{x}'$, and a time, t', in a common representational format:

$$(\mathbf{x}', t') = T(\mathbf{x}, t) . \tag{4.2}$$

In general, T has the following form:

$$T(\mathbf{x}, t) = (T_{\mathbf{x}}(\mathbf{x}, t), T_t(\mathbf{x}, t)) , \tag{4.3}$$

where both the spatial transformation $T_{\mathbf{x}}(\mathbf{x}, t)$ and the temporal transformation $T_t(\mathbf{x}, t)$ are functions of $\mathbf{x}$ and t. However, in many multi-sensor fusion applications we approximate (4.3) with a "decoupled" spatial-temporal transformation:

$$T(\mathbf{x}, t) = (T_{\mathbf{x}}(\mathbf{x}), T_t(t)) . \tag{4.4}$$

Even when (4.4) is not very accurate we often find it convenient to use it as a first-order approximation. In some applications (4.4) is true by definition e. g. when the system under observation is not changing in time. The following example illustrates the construction of a common representational format for an environment which is essentially static and in which the sensors are all of the same type. In this case temporal alignment, semantic alignment and radiometric calibration functions are not required and the construction of the common representational format reduces to the construction of a common spatial coordinate system.

Example 4.4. A Distributed Surveillance System [62]. The demand for surveillance activities for safety and security purposes has received particular attention for remote sensing in transportation applications (such as airports, maritime environments, railways, motorways) and in public places (such as banks, supermarkets, department stores and parking lots). Such systems typically consist of a number of video-based television cameras located in multiple locations. Consider a sequence of M narrow field-of-view "spot" images $I_m, m \in \{1, 2, \ldots, M\}$, taken of a wide surveillance area.

We establish a common coordinate system by building a panoramic or "mosaic" image I^* from the sequence of images I_m (Fig. 4.2). For each image I_m, we find a geometric transformation T_m which maps the local "camera-centered" coordinate system of I_m to the common "object-centered" coordinate system of I^*. We then form the mosaic image I^* by "stitching" or "compositing" together the transformed images $T_m(I_m)$. In this case, the aim of a stitching algorithm [51, 52, 73] is to produce a visually plausible mosaic image I^* in which, geometrically and photometrically, I^* is as similar as possible to the input images $T_m(I_m)$ and the seams between the stitched images are invisible. In Table 4.3 we list the two types of stitching algorithms which are commonly used nowadays.

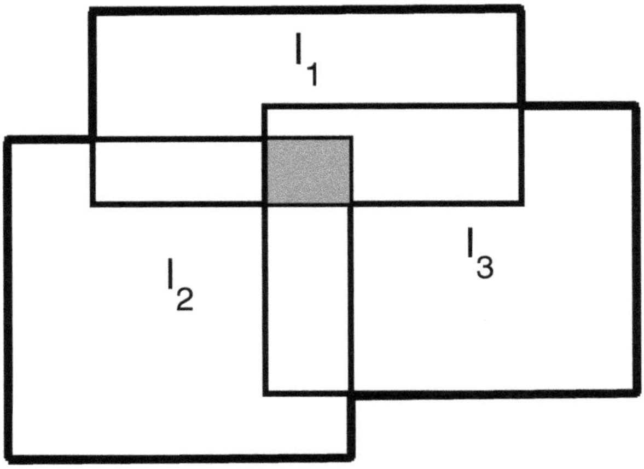

Fig. 4.2 Shows the surveillance of a wide-area site with a sequence of "spot" images $I_m, m \in \{1,2,3\}$. Each spot image I_m is transformed to a common coordinate system using a transformation T_m. The region defined by the heavy border is the mosaic image I^*. The shaded area denotes a region where the transformed images $T_1(I_1), T_2(I_2)$ and $T_3(I_3)$ overlap. In this region the mosaic image I^* is formed by "stitching" $T_1(I_1), T_2(I_2)$ and $T_3(I_3)$.

Table 4.3 Stitching Algorithms

Algorithm	Description
Optimal Seam Algorithms	These algorithms search for a curve in the overlap region on which the differences between the $T_m(I_m)$ are minimized. Then each image is copied to the corresponding side of the seam [51].
Transition Smoothing Algorithms	These algorithms minimize seam artifacts by smoothing the transition region (a small region which is spatially near to the seam) [73].

The following example describes the transition smoothing, or feathering, of two input images I_1 and I_2.

Example 4.5. Feathering [73]. In feathering the mosaic image I^* is a weighted combination of the input images I_1, I_2, where the weighting coefficients vary as a function of the distance from the seam. In general, feathering works well as long as there is no significant misalignment. However, when the misalignments are significant, the mosaic image displays artifacts such as double edges. A modification of feathering which is less sensitive to misalignment errors, is to stitch the derivatives of the input images instead of the images themselves. Let $\partial I_1/\partial x$, $\partial I_1/\partial y$, $\partial I_2/\partial x$ and $\partial I_2/\partial y$ be the derivatives of the input images. If F_x and F_y denote the derivative images formed by feathering $\partial I_1/\partial x$ and $\partial I_2/\partial x$ and $\partial I_1/\partial y$ and $\partial I_2/\partial y$, then we choose the final mosaic image I^* to be the image whose derivatives $\partial I^*/\partial x$ and $\partial I^*/\partial y$ are closest to F_x and F_y.

4.3 Geographical Information System

An important example of a common representational format is a *Geographical Information System*. In a Geographic Information System (GIS) we combine multiple images of the earth obtained from many different sensors and maps, including demographic and infrastructure maps, into a common coordinate system.

Example 4.6. Detection and Tracking of Moving Objects in Geo-Coordinates [31]. The goal of video surveillance is to identify and track all relevant moving objects in a scene. This involves detecting moving objects, tracking them while they are visible and re-acquiring the objects once they emerge from occlusion. To track the objects from a moving camera, we need to describe the motion of the objects in a common coordinate system. In [31] the authors use absolute geo-coordinates, i. e. longitude and latitude, from a staellite image or map as the common coordinate system. By spatially aligning the images from the moving camera to a satellite image we generate the absolute geo-location of the moving objects. An important advantage of tracking in geo-coordinates is we obtain physically meaningful measurements of object motion.

In building a GIS we often need to convert a set of sensor measurements $y(\mathbf{u}_i), i \in \{1,2,\ldots,N\}$, made at specified spatial locations $\mathbf{u}_i$, into a continuous map. One way of doing this is known as *Kriging* [8, 20, 47], where the term Kriging refers to a family of best (in the least squares sense) linear unbiased estimators (BLUE) (see Table 4.4). We interpret the $y_i = y(\mathbf{u}_i)$ as the outcome of a random field $y(\mathbf{u})$. If $y_i, i \in \{1,2\ldots,N\}$, denotes the sensor measurements which are located in the neighborhood of a given point $\mathbf{u}_0$, then the estimated sensor value at $\mathbf{u}_0$, is

Table 4.4 Kriging Estimators

Estimator	Description
Simple	Assumes a known mean value $\mu(\mathbf{u}_0) = \mu_0$ in the local neighbourhood of $\mathbf{u}_0$.
Ordinary	Assumes a constant but unknown mean value $\mu(\mathbf{u}_0) = \mu_0$ in the local neighbourhood of $\mathbf{u}_0$.
Universal	Assumes the mean value $\mu(\mathbf{u})$ varies smoothly in neighbourhood of $\mathbf{u}_0$. $\mu(\mathbf{u}_0)$ is modelled as a linear combination of known basis functions $f_l(\mathbf{u})$: $\mu(\mathbf{u}_0) = \sum_{l=1}^{L} \beta_l f_l(\mathbf{u}_0)$, where the parameters $\beta_l, l \in \{1,2,\ldots,L\}$ are found by solving the appropriate Kriging equations.
Indicator	Uses a threshold to create binary data. The ordinary Kriging procedure is then applied to the binary data. The covariance matrix used in the Kriging eqs. is calculated on the binary data.
Disjunctive	Uses multiple thresholds to create multiple binary input data. Cokriging procedure is then applied to the multiple binary data.
Co-Kriging	Combines the primary input measurements $y(\mathbf{u}_i)$ with K auxiliary variables $z_1(\mathbf{u}), z_2(\mathbf{u}), \ldots, z_K(\mathbf{u})$. Often used when the auxiliary variables are known at a relatively few places $\mathbf{u}_j$.
Kriging with External Drift	Similar to Co-Kriging except the auxiliary variables are known at all $\mathbf{u}$ including the required location $\mathbf{u}_0$.

$$\hat{y}(\mathbf{u}_0) = \mu(\mathbf{u}_0) + \sum_{i=1}^{N} \lambda_i (y_i - \mu(\mathbf{u}_0)) , \tag{4.5}$$

where $\mu(\mathbf{u}_0)$ is the mean value at $\mathbf{u}_0$ and the weights $\lambda_i, i \in \{1,2,\ldots,N\}$, are chosen to give the best unbiased solution. Mathematically, $\mu(\mathbf{u}_0)$ and λ_i, are solutions of a system of linear equations known as the *Kriging equations*.

Example 4.7. Equations for Ordinary Kriging [19]. In ordinary Kriging the mean $\mu(\mathbf{u})$ is equal to an unknown constant μ_0 in the local region of $\mathbf{u}_0$. In this case, the estimated sensor value at $\mathbf{u}_0$ is

$$\hat{y}(\mathbf{u}_0) = \mu_0 + \sum_{i=1}^{N} \lambda_i (y_i - \mu_0) .$$

The unknown mean μ_0 and the weights λ_i are given by the following matrix equation:

$$\begin{pmatrix} \lambda_1 \\ \vdots \\ \lambda_N \\ \mu_0 \end{pmatrix} = \begin{pmatrix} \Sigma(\mathbf{u}_1,\mathbf{u}_1) & \ldots & \Sigma(\mathbf{u}_N,\mathbf{u}_1) & 1 \\ \vdots & \ddots & \vdots & \vdots \\ \Sigma(\mathbf{u}_1,\mathbf{u}_N) & \ldots & \Sigma(\mathbf{u}_N,\mathbf{u}_N) & 1 \\ 1 & \ldots & 1 & 0 \end{pmatrix}^{-1} \begin{pmatrix} \Sigma(\mathbf{u},\mathbf{u}_1) \\ \vdots \\ \Sigma(\mathbf{u},\mathbf{u}_N) \\ 1 \end{pmatrix} ,$$

where $\Sigma(\mathbf{u}_i,\mathbf{u}_j)$ is the spatial covariance,

$$\Sigma(\mathbf{u}_i,\mathbf{u}_j) = E\big((y(\mathbf{u}_i) - \mu(\mathbf{u}_i))(y(\mathbf{u}_j) - \mu(\mathbf{u}_j))^T\big) ,$$

and $\mu(\mathbf{u}_i)$ is the mean sensor value in the neighbourhood of $\mathbf{u}_i$.

In ordinary Kriging, the weights λ_i sum to one. In this case, we may write $\hat{y}(\mathbf{u}_0)$ more simply as:

$$\hat{y}(\mathbf{u}_0) = \sum_{i=1}^{N} \lambda_i y_i \,.$$

Apart from assuming a given model for the spatial mean $\mu(\mathbf{u})$, the only information required by the Kriging equations is the spatial covariance function $\Sigma(\mathbf{u}_i, \mathbf{u}_j)$.

4.3.1 Spatial Covariance Function

Conventionally we learn the spatial covariance, $\Sigma(\mathbf{u}_i, \mathbf{u}_j)$, from the input measurements y_i subject to any *a priori* knowledge and given constraints. Some of the constraints are *necessary*. For example, we require $\Sigma(\mathbf{u}_i, \mathbf{u}_j)$ to be *positive definite*. Mathematically, this constraint requires $\sum_i \sum_j a_i a_j \Sigma(\mathbf{u}_i, \mathbf{u}_j) \geq 0$ for all $\mathbf{u}_i$, $\mathbf{u}_j$ and real numbers a_i, a_j. Other constraints may *not* be necessary, but we shall, nevertheless, require them to be true. For example, we shall require $\Sigma(\mathbf{u}_i, \mathbf{u}_j)$ to be *second-order stationary*. Mathematically, this constraint requires $\mu(\mathbf{u})$ to be the same for all $\mathbf{u}$ and $\Sigma(\mathbf{u}_i, \mathbf{u}_j)$ to depend only on the difference $\mathbf{h} = (\mathbf{u}_i - \mathbf{u}_j)$. One way of satisfying these constraints is to fit [1] the experimentally measured covariance function, $\tilde{\Sigma}(\mathbf{h})$, to a *parametric* curve $\Sigma(\mathbf{h}|\boldsymbol{\theta})$, which is known to be positive definite and second-order stationary. See Table 4.5 for a list of parametric univariate covariance functions which are positive-definite and second-order stationary.

Let $N(\mathbf{h})$ be the number of pairs of measurements $(y(\mathbf{u}_i), y(\mathbf{u}_j))$ whose separation $(\mathbf{u}_i - \mathbf{u}_j)$ is close to $\mathbf{h}$. Then an experimentally measured estimate of $\Sigma(\mathbf{h})$ is

$$\tilde{\Sigma}(\mathbf{h}) \approx \frac{1}{N(\mathbf{h})} \sum_{(\mathbf{u}_i - \mathbf{u}_j) \approx \mathbf{h}} (y(\mathbf{u}_i) - \mu)(y(\mathbf{u}_j) - \mu)^T \,, \tag{4.6}$$

where $\mu = \frac{1}{N} \sum_{i=1}^{N} y(\mathbf{u}_i)$ is an estimate of the mean value of y [2].

---

[1] See Ex. 10.8 for an example of the least square fitting of the experimental measured covariance $\tilde{\Sigma}(\mathbf{h})$.

[2] Eq. (4.6) is sensitive to outliers (see Chapt. 11). If outliers are thought to be present, an alternative "robust"estimate should be used. Two such estimates are:

$$\begin{aligned} \tilde{\Sigma}(\mathbf{h}) &= \left(\sum a_{ij}\right)^4 / (0.457N(\mathbf{h}) + 0.494) \,, \\ \tilde{\Sigma}(\mathbf{h}) &= \left(\mathrm{med}\{a_{ij}\}\right)^4 / 0.457 \,, \end{aligned}$$

where $a_{ij} = \sqrt{|(y(\mathbf{u}_i) - \mu)(y(\mathbf{u}_j) - \mu)^T|}$ and the summation and the median are taken over all pairs of points $(\mathbf{u}_i, \mathbf{u}_j)$ for which $(\mathbf{u}_i - \mathbf{u}_j) \approx \mathbf{h}$.

Table 4.5 Positive Definite Second-Order Stationary Univariate Covariance Functions

Name	$\Sigma(h\|\boldsymbol{\theta})$
Linear	$\alpha-\beta h.$
Spherical	$\alpha-\beta\min(\frac{3h}{2H}-\frac{1}{2}(\frac{h}{H})^3,1).$
Exponential	$\alpha-\beta(1-\exp-(\frac{h}{H})).$
Gaussian	$\alpha-\beta(1-\exp-(\frac{h}{H})^2).$
Exponential Power	$\alpha-\beta(1-\exp-(\|\frac{h}{H}\|^p)).$

The formulas in the table only apply for $h>0$. By definition $\Sigma(h=0|\boldsymbol{\theta})=1$, where $\boldsymbol{\theta}$ denotes the parameters $\{\alpha,\beta,H,p\}$ which are commonly referred to as nuggett (α), sill (β) and practical range (H).

The following example illustrates how we may use $\widetilde{\boldsymbol{\Sigma}}(\mathbf{h})$ to estimate the standard deviation of the noise in an input image.

Example 4.8. Standard Deviation in an Input Image [48]. Given an input image $I(\mathbf{u})$ we calculate the spatial covariance matrix $\widetilde{\boldsymbol{\Sigma}}(h)$ for different values of $h=|\mathbf{h}|$. We fit $\widetilde{\boldsymbol{\Sigma}}(h)$ to a parametric function $f(A,B,\ldots)$. The square root of the best fit at $h\to 0$ gives us the standard deviation of the noise in the image. Recommended functions for $f(A,B,\ldots)$ include the Gaussian function:

$$f(A,B,C)=A+B(1-\exp(-h/C)^2)\ ,$$

and the exponential function:

$$f(A,B,C)=A+B(1-\exp(-h/C))\ .$$

4.4 Common Representational Format

It is clear from the above that the common representational format plays a crucial role in a multi-sensor data fusion system. In fact in most cases the choice of common representational format will govern what type of fusion algorithm is used. The following example illustrates this situation when a human being performs the multi-sensor data fusion.

Example 4.9. Myocardial Imaging [2]. Polar maps, or "bull's-eye" images, are a standard way of displaying myocardial functions and are well established in clinical settings. They are constructed by combining images from multiple planes so that information about the entire myocardium can be displayed in a single image. Polar maps can be compared to a three-dimensional cone-shaped heart activity image projected onto a single plane. Each image plane forms a ring in the polar map. Although the rings may be divided into an arbitrary number of sectors, in practice, a clinician uses four (anterior, lateral, inferior and septal) or six (anterior, anterior-lateral, inferior-lateral, inferior, inferior-septal and anterior-septal) sectors for his visual interpretation of the image. Fig. 4.3 shows the polar image representation of a left ventricle.

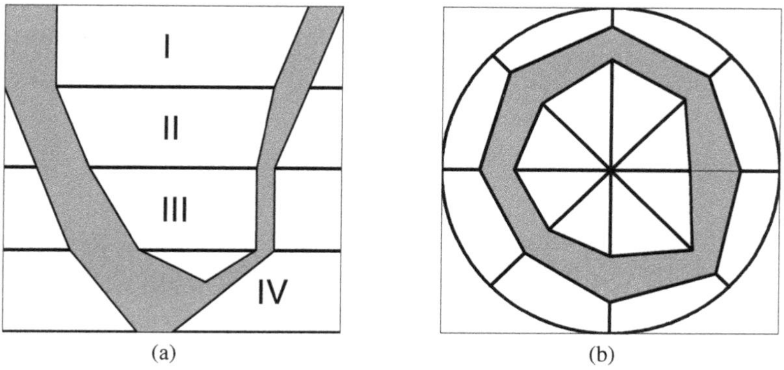

Fig. 4.3 Shows a polar image of a left ventricle. **(a)** Shows the the left ventricle divided into four slices: I (Basal), II (Mid-Basal), III (Mid-Apical) and IV (Apical). **(b)** Shows a given slice divided into 8 sectors.

For object recognition we require a common representational format which is insensitive to certain variations that can occur to an object. For example, in comparing shapes for object recognition, we require a common representational format which is invariant to translation, rotation and scaling and is insensitive to variations due to articulation, occlusion and noise (see Fig. 4.4)

The census transform and the local binary pattern (LBP) is a common representational format which is widely used in face recognition and texture classification.

Example 4.10. Census Transform and Local Binary Pattern (LBP) [59]. Let G be an gray-scale input image. At each pixel (i,j) we define a set of N neighbouring pixels which lie equally spaced on a radius R centered at (i,j). Each neighbouring pixel is labeled $\phi_n = 0$ if the center pixel has higher gray-scale value than the neighbouring pixel ($g_c > g_n$); otherwise $\phi_n = 1$. The census

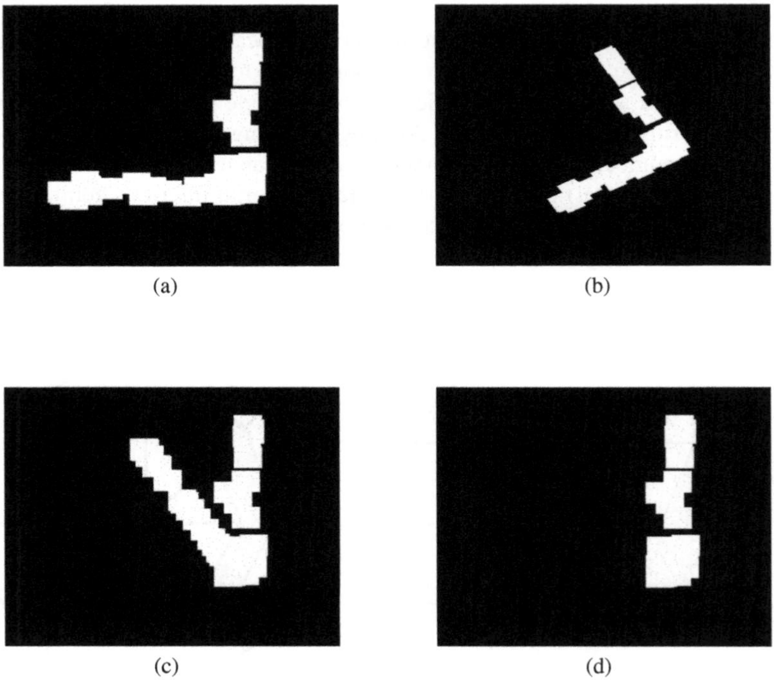

Fig. 4.4 **(a)** Shows the original image. **(b)** Shows the rotation and scaling of the original image.**(c)** Shows the articulation of the original image.**(d)** Shows the occlusion of the original image.

transform for the center pixel (i, j) is then given by an N-dimensional binary vector $\mathbf{V}(i, j)$:

$$\mathbf{V}(i, j) = (\phi_1, \phi_2, \ldots, \phi_N)^T .$$

The corresponding LBP label is found by converting $\mathbf{V}(i, j)$ into an integer in the range $[0, 2^N)$:

$$LBP(i, j) = \sum_{n=1}^{N} \phi_n 2^{(n-1)} .$$

Variations of the LBP include:

Rotationally invariant LBP.

$$LBP_{\mathrm{ROT}}(i, j) = \min_{0 \leq k \leq N} \left(\sum_{n=1}^{N} \phi_n 2^{[(n+k-1) \mod N]} \right) .$$

Dominant LBP [30]. In the dominant LBP we calculate the frequency of occurrence of all the rotationally invariant LBP's. We then select a small number of the most frequently occurring patterns.

Probabilistic LBP [53].

$$LBP_{\mathrm{PROB}}(i,j) = \sum_{n=1}^{N} p_n \phi_n 2^{n-1} / \sum_{n=1}^{N} p_n \ ,$$

where

$$p_n = \begin{cases} 1 - \frac{1}{2}\exp\left[-\left|\frac{g_n - g_c}{\sigma g_c}\right|^2\right] & \text{if } g_c \geq g_n \ , \\ \frac{1}{2}\exp\left[-\left|\frac{g_n - g_c}{\sigma g_c}\right|^2\right] & \text{otherwise} \ , \end{cases}$$

and σ is a scaling parameter (standard deviation).

Detecting and tracking humans in images is a challenging task owing to their variable appearance and wide range of poses that they can adopt. For this purpose a histogram of oriented gradients (HOG) [9] is often used . This is a common representational format which is robust against local geometric and photometric transformations.

Example 4.11. Histogram of Oriented Gradients [9]. The image is divided into small spatial regions ("cells"). In each cell we calculate a local histogram of gradient directions, or edge orientations, over the pixels in the cell. A HOG feature vector $\mathbf{f}(i,j)$ is created at (i,j) by concatenating the histograms of several cells which close to (i,j).

Variations of HOG include:

Co-occurrence HOG (CoHOG) [65]. CoHOG are histograms which are constructed from *pairs* of gradient orientations.

HOG-LBP [64]. In the HOG-LBP operator, in each cell we construct a conventional histogram of oriented gradients (HOG) and a histogram of LBP patterns. The HOG-LBP vector is formed by concatenating the two histograms which is found to give excellent performance for human detection.

Among the precursors of the HOG is the SIFT operator. The SIFT operator is designed to extract a small number of distinct invariant features. These features are then used to perform reliable matching between different views of an object or of a scene.

Example 4.12. Scale Invariant Feature Transform (SIFT) [36, 39]. SIFT features are invariant to image scale and rotation and provide robust matching across a substantial range of affine distortion changes due to changes in camera viewpoint, occlusion of noise and changes in illumination. In addition, the features are highly distinctive. The SIFT descriptor is computed by partitioning the image region surrounding each detected key-point into a 4×4 grid of cells, and computing a HOG of 8 bins in each cell. For each key-point we form a 128-long SIFT feature vector by concatenating the 16 HOG's. Traditionally, key-points in two different images are matched between images by minimizing the Euclidean distance between SIFT vectors. A ratio test, comparing the Euclidean distances between the best and the second best match for a given key-point, is used as a measure of match quality. Note, however, that alternative distance measures may be used. For example, [45] recommends using the earth mover's distance (see Ex. 8.5)

Note. Traditionally the pixel gradients used in the SIFT operator are calculated with pixel differences. However a significant improvement in key-point distinctiveness may be obtained if, instead, the pixel gradients are calculated using a smooth derivative filter [42]

Another key-point operator is SURF [1]. This has comparable performance to SIFT but is much faster.

The following example illustrate the shape context and the inner distance which may be regarded as the counterpart of the SIFT operators when comparing two-dimensional shapes and contours.

Example 4.13. Shape Context and Inner Distance [4, 33]. Let $\mathbf{z}_i = (x_i, y_i), i \in \{1, 2, \ldots, N\}$, denote a set of N sample points on a given contour (Fig. 4.5). Then the shape context $\mathbf{S}_i$ of the point $\mathbf{z}_i$ describes the distance and orientation of the points $\mathbf{z}_j, j \neq i$, relative to $\mathbf{z}_i$. Let r_{ij} and θ_{ij} denote the distance and orientation of the point $\mathbf{z}_j$ relative to $\mathbf{z}_i$. We divide the (r, θ) space into L vertical columns $\Theta_l, l \in \{1, 2, \ldots, L\}$, and K horizontal rows $R_k, k \in \{1, 2, \ldots, K\}$. If $h(k, l)$ is the number of points $\mathbf{z}_j, j \neq i$, for which r_{ij} lies in the kth row and θ_{ij} lies in the lth column, then

$$\mathbf{S}_i = \begin{pmatrix} h(1,1) & h(1,2) & \ldots & h(1,K) \\ h(2,1) & h(2,2) & \ldots & h(2,K) \\ \vdots & \vdots & \ddots & \vdots \\ h(K,1) & h(K,2) & \ldots & h(K,L) \end{pmatrix} .$$

The shape context is invariant to translation and rotation and approximately invariant to occlusion and to noise. In addition, the shape context may be made scale invariant by dividing the distances r_{ij} through with the average distance

$$\bar{r} = \frac{1}{N(N-1)} \sum_{i=1}^{N} \sum_{\substack{j=1 \\ j \neq i}}^{N} r_{ij} .$$

For articulated objects the inner distance [33] is often used instead of the shape context. This is identical to the shape context except we replace r_{ij} with $\widetilde{r}_{ij}$, where $\widetilde{r}_{ij}$ is the length of the shortest line which joins $\mathbf{z}_i$ and $\mathbf{z}_j$ and which lies within the given contour.

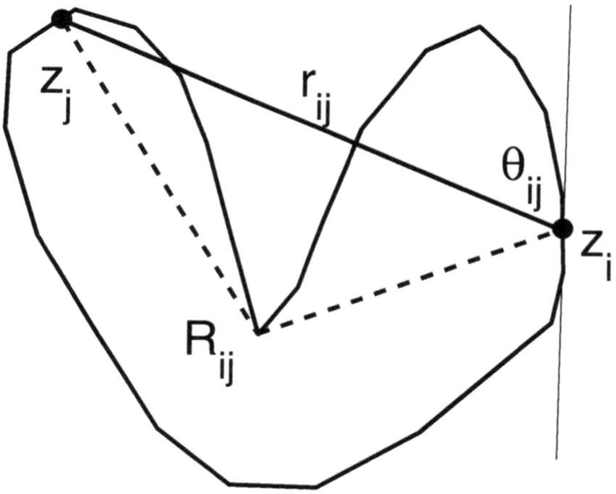

Fig. 4.5 Shows a closed contour defined by N points $\mathbf{z}_i, i \in \{1,2,\ldots,N\}$. In the figure we show the Euclidean distance r_{ij} and the inner distance $\widetilde{r}_{kj}$ of the point $\mathbf{z}_j$ relative to $\mathbf{z}_i$.

Sometimes what is important is how the sensor measurement uncertainty propagates in a given common representational format. The following two examples illustrate these concerns.

Example 4.14. Object Recognition Based on Photometric Color Invariants [17]. A simple and effective scheme for three-dimensional object recognition is to represent and match images on the basis of color histograms. For effective object recognition we should use a color space which is invariant to changes in viewing direction, object orientation and illumination.

In Table 4.6 we list several color spaces which are commonly used for this purpose. We observe that measurement uncertainty is propagated differently in each space: the normalized rg space is unstable around $R = G = B = 0$ ($\sigma_r, \sigma_g \to \infty$) and hue H is unstable around $R = G = B$ ($\sigma_H \to \infty$) while the opponent color space o_1, o_2 is relatively stable at all RGB values [17].

Table 4.6 Photometric Invariant Color Space

Color Space	Definition	Uncertainty
Normalized rg	$r = R/(R+G+B)$ $g = G/R+G+B)$	$\sigma_r = \sqrt{R^2(\sigma_B^2 + \sigma_G^2) + (G+B)^2\sigma_R^2/}$ $(R+G+B)^2$, $\sigma_g = \sqrt{G^2(\sigma_B^2 + \sigma_R^2) + (R+B)^2\sigma_G^2/}$ $(R+G+B)^2$.
Opponent $o_1 o_2$	$o_1 = (R-G)/2$ $o_2 = (2B-R-G)/4$	$\sigma_{o_1} = \sqrt{\sigma_G^2 + \sigma_R^2/2}$, $\sigma_{o_2} = \sqrt{4\sigma_B^2 + \sigma_G^2 + \sigma_R^2/4}$.
Hue H	$H = \tan^{-1}(\sqrt{3}(G-B)/$ $(2R-G-B))$	$\sigma_H = \frac{1}{2}\sqrt{3}[(-2BR\sigma_G^2 + R^2\sigma_B^2\sigma_G^2 + G^2(\sigma_B^2 + \sigma_R^2)/\ \Delta) + (\sigma_G^2 + \sigma_R^2) - 2G(R\sigma_B^2 + B\sigma_R^2)/\Delta)]$, where $\Delta = B^2 + G^2 - GR + R^2 - B(G+R)^2$.

Example 4.15. Debiased Cartesian Coordinates [14]. We consider tracking an object O in two-dimensions using a range/bearing sensor measurements (r_i, θ_i). Let (r_0, θ_0) denote the true range/bearing of the object O, then we assume

$$r_i = r_0 + \delta r\ ,$$
$$\theta_i = \theta_0 + \delta\theta\ ,$$

where $\delta r \sim \mathcal{N}(0, \sigma_r^2)$ and $\delta\theta \sim \mathcal{N}(0, \sigma_\theta^2)$ are iid random measurement errors.

For applications which involve target tracking (see Chapt. 12) it is convenient to use a Cartesian coordinate system (x, y) since this defines an inertial reference frame. If (x_i, y_i) are the Cartesian coordinates corresponding to (r_i, θ_i), then formally we may write

$$x_i = r_i \cos\theta_i\ ,$$
$$y_i = r_i \sin\theta_i\ .$$

However, $E(x_i) \neq 0$ and $E(y_i) \neq 0$. This means x_i and y_i are *not* unbiased estimates of the true Cartesian coordinates $x_0 = r_0 \cos\theta_0$ and $y_0 = r_0 \sin\theta_0$. To correct for these bias effects we use modified, or "debiased", Cartesian coordinates $(\widetilde{x}_i, \widetilde{y}_i)$ [14]. For range/bearing measurements, the bias is multiplicative and the corrected coordinates are found by dividing through by a correction factor λ:

$$(\widetilde{x}_i, \widetilde{y}_i) \approx \left(\frac{x_i}{\lambda}, \frac{y_i}{\lambda}\right) ,$$

where $\lambda = \exp(\sigma_\theta^2/2)$.

4.5 Subspace Methods

In many multi-sensor data fusion applications our paramount concern is to keep the computational load and/or the storage requirements low. This may be achieved by using a *low-dimensional* common representational format. One way of producing such a format is to apply a dimension-reducing, or subspace, technique to the raw input data.

Example 4.16. Common Representational Format for a Vehicle Sound Signature Application [66]. All moving vehicles make some kind of noise in which vehicles of the same kind and working in similar conditions generate similar noises. This fact is used to detect and classify vehicles moving in a given surveillance region: Each vehicle is classified by its power spectrum in the frequency range $\sim [5, 6500]$ Hz. If each spectrum is sampled every ~ 5 Hz, the corresponding common representational format has $N \sim 1200$ dimensions. However, by using a principal component analysis subspace technique we may reduce the number of dimensions from $N \sim 1200$ to $L \sim 30$.

Table 4.7 lists some of the principal subspace techniques which are commonly used for dimension-reducing purposes.

4.5.1 Principal Component Analysis

Principal component analysis (PCA) is an *unsupervised* technique which builds a low-dimensional commmon representational format using a set of input vectors $\mathbf{y}_i, i \in \{1, 2, \ldots, N\}$. The following example explains the mathematical framework which underpins principal component analysis.

Table 4.7 Subspace Techniques

Technique	Description
Principal Component Analysis (PCA)	Linear transformation chosen so the projected components have maximum variance [27].
Linear Discriminant Analysis (LDA)	Linear transformation for $K \geq 2$ classes. Transformation is chosen so the projected components for each class are maximally separated from the projected components of the other classes [71].
Independent Component Analysis (ICA)	Linear transformation chosen so the projected components have maximized independence [25].
Non-Negative Matrix Factorization (NMF)	Finds factors with non-negative elements.
Canonical Correlation Analysis (CCA)	For K classes defines K linear transformations [21]. For $K=2$ the transformations are chosen so the projected components of each class are maximally correlated.
Informative Sample Subspace (ISS)	Uses a maximal mutual information criterion to search a labeled training data set D directly for the subspaces projection base vectors [61].

Example 4.17. Principal Component Analysis [27]. The aim of principal component analysis (PCA) is to find a L-dimensional *linear* projection that best represents the input data in a least squares sense. Let the input data be a set of M-dimensional input vectors $\mathbf{y}_i, i \in \{1,2,\ldots,N\}$, where $\mathbf{y}_i = (y_i^{(1)}, y_i^{(2)}, \ldots, y_i^{(M)})^T$. If we define a reduced L-dimensional space using a set of orthonormal axes $\mathbf{u}_l, l \in \{1,2,\ldots,L\}$, then

$$\boldsymbol{\theta}_i = \mathbf{U}^T(\mathbf{y}_i - \boldsymbol{\mu}) ,$$

is a L-dimensional representation of $\mathbf{y}_i$, where $\boldsymbol{\theta}_i = \left(\theta_i^{(1)}, \theta_i^{(2)}, \ldots, \theta_i^{(L)}\right)^T$ and $\mathbf{U} = (\mathbf{u}_1, \mathbf{u}_2 \ldots, \mathbf{u}_L)$. *Note*: Given $\boldsymbol{\theta}_i$, the optimal reconstruction of $\mathbf{y}_i$ is $\tilde{\mathbf{y}}_i = \mathbf{U}\boldsymbol{\theta}_i + \boldsymbol{\mu} = \mathbf{U}\mathbf{U}^T(\mathbf{y}_i - \boldsymbol{\mu}) + \boldsymbol{\mu}$.

Mathematically, the orthonormal axes $\mathbf{u}_l, l \in \{1,2,\ldots,L\}$, are given by the L dominant eigenvectors of the sample covariance matrix $\boldsymbol{\Sigma}$, i. e.

$$\boldsymbol{\Sigma}\mathbf{u}_l = \lambda_l \mathbf{u}_l ,$$

where

$$\boldsymbol{\Sigma} = \frac{1}{N}\sum_{i=1}^{N}(\mathbf{y}_i - \boldsymbol{\mu})(\mathbf{y}_i - \boldsymbol{\mu})^T ,$$

$$\boldsymbol{\mu} = \frac{1}{N}\sum_{i=1}^{N}\mathbf{y}_i .$$

Fig. 4.6 shows the PCA of a set of two-dimensional input measurements $\mathbf{y}_i, i \in \{1,2,\ldots,N\}$.

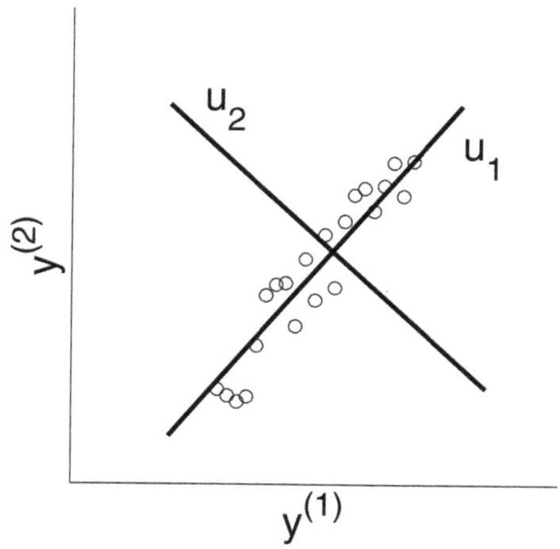

Fig. 4.6 Shows the two-dimensional input measurements $\mathbf{y}_i = (y_i^{(1)}, y_i^{(2)}), i \in \{1,2,\ldots,N\}$. The PCA representation consists of vectors $\mathbf{u}_1$ and $\mathbf{u}_2$, where the dominant vector is $\mathbf{u}_1$.

It is common practice to select the principal components with the largest eigenvalues [27]. Recently Thomaz and Giraldi [55] suggested selecting the *most discriminant* principal components. For this purpose they used the LDA algorithm (section 4.5.2).

In many multi-sensor data fusion applications it is convenient to use a *probabilistic* common representational format in which we transform the measurements $\mathbf{y}_i$ into L-dimensional *a posteriori* probabilities $p(\boldsymbol{\theta}_i|\mathbf{y}_i,I)$. To carry out this transformation we use a probabilistic PCA (see Sect. 10.6.3).

In Table 4.8 we list some of the PCA variants which are in common use.

Table 4.8 Principal Component Analysis and its Variants

Technique	Description
Principal Component Analysis (PCA)	The basic PCA algorithm. PCA maximizes the variance of all linear projections [27].
Probabilistic PCA (PPCA)	Places PCA in a probabilistic framework. See Sect. 10.6.3.
Weighted PCA	Each input measurement $\mathbf{y}_i$ is given a different weight before PCA is performed [49].
Incremental PCA	Given an existing PCA and a new measurement $\mathbf{y}$, incremental PCA calculates the new PCA without recalculating the covariance matrices [49].
Robust PCA	Performs PCA using robust techniques. See Chapt. 11.
Kernel PCA	Generalizes the PCA to non-linear transformations.
2D-PCA	Performs PCA on two-dimensional input matrices. Useful in image processing and computer visual applications [7, 68, 70].

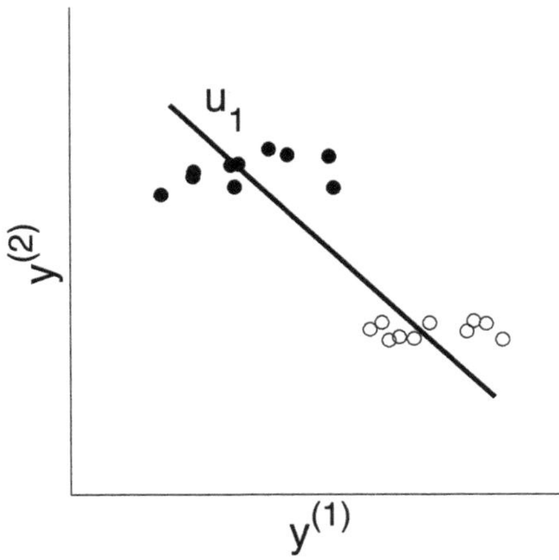

Fig. 4.7 Shows the two-dimensional input measurements $\mathbf{y}_i, i \in \{1,2,\ldots,N\}$, with their class labels. Measurements which belong to class c_1 are denoted by an open circle and measurements which belong to class c_2 are denoted by a closed circle. For a two-class problem the LDA representation has a single vector labeled $\mathbf{u}_1$ which joins $\boldsymbol{\mu}_1$ and $\boldsymbol{\mu}_2$.

4.5.2 Linear Discriminant Analysis

The method of *linear discriminant analysis* (LDA) is a *supervised* technique in which we suppose each input measurement $\mathbf{y}_i$ is associated with a given class $C = c_k, k \in \{1,2,\ldots,K\}$. The aim of LDA is to find a L-dimensional subspace $\mathbf{U} = (\mathbf{u}_1,\mathbf{u}_2,\ldots,\mathbf{u}_L)$ in which the different classes are maximally separated (Fig. 4.7). If $\boldsymbol{\theta}_i = \big(\theta_i^{(1)},\theta_i^{(2)},\ldots,\theta_i^{(L)}\big)^T$ denotes the L-dimensional LDA representation of $\mathbf{y}_i$, then

$$\boldsymbol{\theta}_i = \mathbf{U}^T\mathbf{y}_i \; , \tag{4.7}$$

where $\mathbf{u}_l$ is a solution of the eigenvector equation:

$$\mathbf{H}\mathbf{u}_l = \lambda_l\mathbf{u}_l \; . \tag{4.8}$$

In (4.8), $\mathbf{H} = \boldsymbol{\Sigma}_W^{-1}\boldsymbol{\Sigma}_B$, where $\boldsymbol{\Sigma}_B$ and $\boldsymbol{\Sigma}_W$ are two covariance matrices which measure, respectively, the scatter of the $\mathbf{y}_i$ "between-classes" and "within-classes". Mathematically, the covariance matrices are defined as follows:

$$\boldsymbol{\Sigma}_B = \frac{1}{N}\sum_{k=1}^{K} n_k(\boldsymbol{\mu}_k - \boldsymbol{\mu}_G)(\boldsymbol{\mu}_k - \boldsymbol{\mu}_G)^T \; , \tag{4.9}$$

$$\boldsymbol{\Sigma}_W = \frac{1}{N}\sum_{k=1}^{K} n_k\boldsymbol{\Sigma}_k \; , \tag{4.10}$$

where n_k is the number of input measurements which belong to class c_k, and $\boldsymbol{\mu}_k$ and $\boldsymbol{\Sigma}_k$ are, respectively, the mean vector and covariance matrix for the kth class, and

$$\mu_G = \frac{1}{N}\sum_{k=1}^{K} n_k \mu_k \;, \tag{4.11}$$

is the global mean vector.

The decomposition of **H** in (4.8) is only possible if the matrix is non-singular. This requires the number of training samples, N, to be larger than L, i. e. to be larger than $K-1$. Even if $N > L$, the LDA has a tendency to "overfit" when $N \lesssim 10L$ (see Ex. 9.8) [24, 63, 72]. One way to prevent overfitting is to regularize the solution by performing the eigenvalue decomposition of a regularized matrix, $\widetilde{\mathbf{H}}$, as illustrated by the following example.

Example 4.18. Chemical Sensor Analysis: Principal Discriminant Analysis (PDA) for Small-Sample Size Problems [63]. The PDA is based on the eigenvalue of the regularized matrix, $\widetilde{\mathbf{H}}$:

$$\widetilde{\mathbf{H}} = (1-\varepsilon)\boldsymbol{\Sigma}_W^{-1}\boldsymbol{\Sigma}_B + \varepsilon\boldsymbol{\Sigma}_G \;,$$

where $\boldsymbol{\Sigma}_G = E\big((\mathbf{x}-\mu_G)(\mathbf{x}-\mu_G)^T\big) = \boldsymbol{\Sigma}_W + \boldsymbol{\Sigma}_B$ is the global covariance matrix and $\varepsilon \in [0,1]$ is a regularization parameter. For $\varepsilon > 0$ the LDA solution is regularized by incorporating information on $\boldsymbol{\Sigma}_G$. *Note*. The PDA also increases the number of non-zero eigenvalues beyond the upper limit set by LDA.

In (4.9) the rank of $\boldsymbol{\Sigma}_B$ is bounded from above to $K-1$. This, in turn, implies $L \leq (K-1)$. In many multi-sensor data fusion applications we require a richer, i. e. a higher dimensional, common representational format. We may obtain such a representation by using the recursive LDA algorithm [67].

Example 4.19. Recursive LDA Algorithm [67]. In the recursive LDA the feature vectors are obtained recursively one at a time as follows. The first feature vector $\mathbf{u}_1$ is obtained as in the conventional LDA algorithm:

$$\boldsymbol{\Sigma}_W^{-1}\boldsymbol{\Sigma}_B\mathbf{u} = \lambda\mathbf{u}_1 \;.$$

Before the second feature vector $\mathbf{u}_2$ is computed, the information represented by the first feature vector $\mathbf{u}_1$ is discarded from all the input vectors $\mathbf{y}_i$. Mathematically, this is formulated as:

$$\mathbf{y}_i^{(2)} = \mathbf{y}_i - (\mathbf{u}_1^T\mathbf{y}_i)\mathbf{u}_1 \;.$$

The second feature vector $\mathbf{u}_2$ is then calculated using the conventional LDA algorithm:

$$(\boldsymbol{\Sigma}_W^{(2)})-1\boldsymbol{\Sigma}_B^{(2)}\mathbf{u} = \lambda \mathbf{u}_1 ,$$

where $\boldsymbol{\Sigma}_B^{(2)}$ and $\boldsymbol{\Sigma}_W^{(2)}$ denote the "between-classes" and "within-classes" co-variance matrices of $\mathbf{y}_i^{(2)}$. Mathematically, this may be written as

$$(\mathbf{W}_2^T\mathbf{W}_2)^{-1}\mathbf{W}_2^T\mathbf{B}_2\mathbf{u}_2 = \lambda \mathbf{u}_2 ,$$

where $\mathbf{W}_2 = (\boldsymbol{\Sigma}_W^{(2)}\mathbf{u}_1^T)^T$ and $\mathbf{B}_2 = (\boldsymbol{\Sigma}_B^{(2)}\mathbf{0})^T$.

Similarly, the kth feature vector $\mathbf{u}_k$ is the solution with the largest eigenvalue of the eigenvector equation:

$$(\mathbf{W}_k^T\mathbf{W}_k)^{-1}\mathbf{W}_k^T\mathbf{B}_k\mathbf{u}_k = \lambda \mathbf{u}_k ,$$

where $\mathbf{W}_k = (\boldsymbol{\Sigma}_W^{(k)}\mathbf{u}_1^T \ldots \mathbf{u}_{k-1}^T)^T$, $\mathbf{B}_k = (\boldsymbol{\Sigma}_B^{(k)}\mathbf{0}\ldots\mathbf{0})^T$ and the within-class scatter matrix $\boldsymbol{\Sigma}_W^{(k)}$ and the between-class scatter matrix $\boldsymbol{\Sigma}_B^{(k)}$ are calculated from the preprocessed samples $\mathbf{y}_i^{(k)}$ in which all the information represented by previous feature vectors $\mathbf{u}_1,\mathbf{u}_2,\ldots,\mathbf{u}_{k-1}$ are eliminated using the following equation:

$$\mathbf{y}_i^{(k)} = \mathbf{y}_i^{(k-1)} - (\mathbf{u}_{k-1}^T\mathbf{y}_i^{(k-1)})\mathbf{u}_{k-1}, \quad i = 1,2,\ldots,N .$$

Outliers and atypical observations might have an undue influence on the results obtained, since the LDA is based on non-robust estimates of the population parameters. If outliers are thought to be present, the population parameters should be calculated using robust techniques (see Sect. 11.5).

In Table 4.9 we list some of the LDA variants which are in common use.

The following example illustrates the fusion of PCA and LDA images for face recognition.

Example 4.20. Fusion of PCA and LDA Images for Face Recognition [37]. Suppose $\mathbf{y}$ denotes a test face image which is to be identified by matching it against a training set of face images $\mathbf{Y}_n, n \in \{1,2,\ldots,N\}$. We project $\mathbf{y}$ and $\mathbf{Y}_n, n \in \{1,2,\ldots,N\}$, on the K-dimensional PCA and LDA subspaces. Let $\boldsymbol{\theta} = \big(\theta(1),\theta(2),\ldots,\theta(K)\big)^T$ and $\boldsymbol{\Theta}_n = \big(\Theta_n(1),\Theta_n(2),\ldots,\Theta_n(K)\big)^T$ and $\boldsymbol{\phi} = \big(\phi(1),\phi(2),\ldots,\phi(K)\big)^T$ and $\boldsymbol{\Phi}_n = \big(\Phi_n(1),\Phi_n(2),\ldots,\Phi_n(K)\big)^T$ denote the corresponding projected face patterns. Let $\widetilde{d}_n$ and $\widetilde{D}_n$ denote, respectively, the *normalized* Euclidean distance between $\boldsymbol{\theta}$ and $\boldsymbol{\Theta}_n$ and between $\boldsymbol{\phi}$ and $\boldsymbol{\Phi}_n$:

Table 4.9 LDA and its Variants

Name	**H**
Original LDA	$\boldsymbol{\Sigma}_W^{-1}\boldsymbol{\Sigma}_B$.
PCA+LDA	$(\mathbf{V}^T\boldsymbol{\Sigma}_W\mathbf{V})^{-1}(\mathbf{V}^T\boldsymbol{\Sigma}_B\mathbf{V})$, where $\mathbf{V}$ denotes the principal component matrix [3]. A variant of the PCA+LDA algorithm is to replace the PCA with a kernel PCA.
PDA	$(\boldsymbol{\Sigma}+\lambda\boldsymbol{\Omega})^{-1}\boldsymbol{\Sigma}_B$, where $\boldsymbol{\Omega}$ is an appropriate regularization matrix [22].
Heteroscedastic LDA	$\boldsymbol{\Sigma}_W^{-1}\widetilde{\boldsymbol{\Sigma}}_B$. For a two-class problem, $\widetilde{\boldsymbol{\Sigma}}_B=\boldsymbol{\Sigma}_W^{1/2}[\boldsymbol{\Sigma}_W^{-1/2}(\boldsymbol{\mu}^{(1)}-\boldsymbol{\mu}^{(2)})(\boldsymbol{\mu}^{(1)}-\boldsymbol{\mu}^{(2)})^T\boldsymbol{\Sigma}_W^{-1/2}-2\ln(\boldsymbol{\Sigma}_W^{-1/2}\boldsymbol{\Sigma}^{(1)}\boldsymbol{\Sigma}_W^{-1/2})-2\ln(\boldsymbol{\Sigma}_W^{-1/2}\boldsymbol{\Sigma}^{(2)}\boldsymbol{\Sigma}_W^{-1/2})]\boldsymbol{\Sigma}_W^{1/2}$ [35].
Generalized LDA	$\boldsymbol{\Sigma}_G^{+}\boldsymbol{\Sigma}_B$, where $\boldsymbol{\Sigma}_G^{+}$ denotes the pseudo-inverse of $\boldsymbol{\Sigma}_G$ [26].
Maximum Margin Criterion	$\boldsymbol{\Sigma}_B-\boldsymbol{\Sigma}_W$ [29, 34].
Uncorrelated LDA	LDA in which the extracted features are statistically uncorrelated [69].
Recursive LDA	LDA in which the feature vectors are obtained recursively one at a time [67].
Maximum Uncertainty LDA	$\widetilde{\boldsymbol{\Sigma}}_W^{-1}\boldsymbol{\Sigma}_B$, where $\widetilde{\boldsymbol{\Sigma}}_W=(N-K)\boldsymbol{\phi}\widetilde{\boldsymbol{\Lambda}}\boldsymbol{\phi}^T$; $\boldsymbol{\phi}$ and $\boldsymbol{\Lambda}=\mathrm{diag}(\lambda_1,\lambda_2,\ldots,\lambda_L)$ are, respectively, the eigenvectors and eigenvalues of $\boldsymbol{\Sigma}_W/(N-K)$; $\widetilde{\boldsymbol{\Lambda}}=\mathrm{diag}(\max(\lambda_1,\bar{\lambda}),\max(\lambda_2,\bar{\lambda}),\ldots,\max(\lambda_L,\bar{\lambda}))$ and $\bar{\lambda}=\sum_{l=1}^{L}\lambda_l/L$ [54].
2D-LDA	LDA on two-dimensional input matrices. Useful in image processing and computer vision applications [28].

$$\widetilde{d}_n=\left(d_n-\min_m(d_m)\right)/\left(\max_m(d_m)-\min_m(d_m)\right),$$
$$\widetilde{D}_n=\left(D_n-\min_m(D_m)\right)/\left(\max_m(D_m)-\min_m(D_m)\right),$$

where

$$d_n=\sum_{k=1}^{K}\left(\theta(k)-\Theta_n(k)\right)^2,$$
$$D_n=\sum_{k=1}^{K}\left(\phi(k)-\Phi_n(k)\right)^2.$$

We then fuse the normalized distances $\widetilde{d}_n$ and $\widetilde{D}_n$ using a simple arithmetic mean: $F_n=\frac{1}{2}(\widetilde{d}_n+\widetilde{D}_n)$. We use the distance F_n to classify the test face as belonging to the n^*th training face, where

$$n^*=\arg\min_n(F_n).$$

4.6 Multiple Training Sets

A recent development in multi-sensor data fusion is *ensemble learning* (see Chapt. 14) in which we employ an ensemble, or collection, of *multiple* functions, or classifiers, $S_m, m \in \{1,2,\ldots,M\}$, where each function S_m is learnt on its own training set D_m. Given a common training set D we may generate an ensemble of training sets, $D_m, m \in \{1,2,\ldots,M\}$, which share the same common representational format by simply sub-sampling D.

Example 4.21. Bagging: Multiple Training Sets Sharing a Common Representational Format. Let D denote a training set of N measurements $\mathbf{y}_i, i \in \{1,2,\ldots,N\}$. In *bagging*, we generate multiple training sets $D_m, m \in \{1,2,\ldots, M\}$, from a common training set D. Each training set D_m is formed by *bootstrapping* D, i. e. randomly sampling D with replacement. The sampling process is made with a uniform probability of selection.

The following matlab code creates an ensemble of bootstrapped training sets D_m.

Example 4.22. Matlab Code for Creating an Ensemble of Bootstrapped Training Sets D_m

Input

Training set D containing N measurements $y(i), i \in \{1,2,\ldots,N\}$.
Number of bootstrapped training sets M.

Output

Bootstrapped training sets $Boot(m).D, m \in \{1,2,\ldots,M\}$.

Code

```
for m = 1 : M
    for i = 1 : N
        k = min(floor(ran(1,1) * N) + 1, N);
        Boot(m).D(i) = y(k);
    end
end
```

In Table 4.10 we list some methods for sampling the common training set D. An important sampling method is boosting which is described in detail in Chapt. 14.

Sometimes we require each training set D_m to have its own common representational format. This is a case of *multiple common representational formats*. Given a common training set D, we may generate an ensemble of training sets $D_m, m \in \{1,2,\ldots,M\}$, where each D_m has a different common representational format, by applying a subspace technique to D and then sub-sampling (with, or without, replacement) the result. In Table 4.11 we list some variants of this technique.

Table 4.10 Methods for Sampling a Common Training Set D

Method	Description
Cross-Validation	Partition the common training set D into M disjoint slices (similar to that used in cross-validation). Each classifier S_m is trained on a training set D_m, where D_m is the common training set D *less* the examples in the m slice.
Bagging (Bootstrap)	Perturb D by randomly sampling D with replacement. The sampling is made with a uniform probability random selection procedure. The entire procedure is repeated M times to create M different, although overlapping, training sets D_m. Each D_m contains N samples. On average each perturbed training set will have 63.2% of the samples in D, the rest being duplicates.
Wagging	Similar to bagging except that instead of sampling D using a uniform probability we use a non-uniform probability. The non-uniform probabilities are calculated by stochastically assigning a different weight to each sample in D and then normalizing the weights to one. Each training set D_m is generated using a different set of non-uniform probabilities.
Boosting	We use the classification results obtained with the mth classifier, S_m, to learn D_{m+1}. The classifier S_m is itself learnt on D_m. The training set D_{m+1} is created by re-sampling D such that samples which are misclassified by S_m have a higher chance of being chosen than samples which were correctly classified by S_m.
Multi-Boosting	Create a set of M data sets $D_m, m \in \{1,2,\ldots,M\}$, by boosting a common training set D. Each boosted training set D_m is then bagged, or wagged, to generate M' new training sets $\{D_m^{(1)}, D_m^{(2)}, \ldots, D_m^{(M')}\}$.

Example 4.23. Input Decimated Samples [60]. Figure 4.8 shows the application of the method of input decimated samples on a common training set D which contains N two-dimensional samples $\mathbf{y}_i, i \in \{1,2,\ldots,N\}$.

The following example illustrates a simple pattern recognition algorithm using multiple training sets generated using random projections.

Example 4.24. Pattern Recognition Via Multiple Random Projection [15]. Classical K-nearest neighbor (K-NN) works as follows: Given a test target $\mathbf{x}$ we classify it by comparing it with a set of train targets $\mathbf{y}_n, n \in \{1,2,\ldots,N\}$, contained in a database. Each target $\mathbf{y}_n$ has a label l_n associated with it which denotes the class to which it belongs. We select the K train targets which are closest to $\mathbf{x}$. Let S_k denote the number of these train targets which belong to the class k, then we give $\mathbf{x}$ a label l where

$$l = \arg\max_k (S_k) \; .$$

Table 4.11 Methods for Generating Multiple Common Representational Formats

Name	Description
Random Subspace Method (RSM)	Form training sets $D_m, m \in \{1,2,\ldots,M\}$, by randomly selecting which dimensions to retain [23].
Random Projections	Form a random $L \times L'$ matrix $\mathbf{R}$, where L' is the number of dimensions which are to be retained. Each element in $\mathbf{R}$ is random number chosen from $\mathcal{N}(0,1)$ distribution and normalizing the columns to one. We apply $\mathbf{R}$ to each sample $\mathbf{y}_i, i \in \{1,2,\ldots,N\}$, and thus obtain a new (L'-dimensional) sample $\mathbf{y}'_i = \mathbf{R}^T \mathbf{y}_i$ [13].
Input Decimated Samples	Divide samples in D according to their class variable. For each class, $C = c_k$, find the corresponding PCA transformation $\mathbf{U}$. Generate D_k by applying $\mathbf{U}$, or only the dominant columns in $\mathbf{U}$ (i. e. only eigenvectors with large eigenvalues), to all samples in D [60]. *Note*: In Input Decimated Sample the number of training sets is limited to the number of classes K.
Rotation Forest	Similar to the method of Input Decimated Samples. However, in Rotation Forest the number of D_m is not limited to the number of classes present [46].
Skurichina-Duin	Find PCA transformation $\mathbf{U}$ on the common training set D. Then form the D_m as follows: D_1 is formed by applying the first K columns of $\mathbf{U}$ to D, D_2 is formed by applying the next K columns of $\mathbf{U}$ to D, etc., where K is a suitably chosen number [50].

In the table we assume that each training sample $\mathbf{y}_i$ in D is an L-dimensional vector: $\mathbf{y}_i = (y_i^{(1)}, y_i^{(2)}, \ldots, y_i^{(L)})^T$.

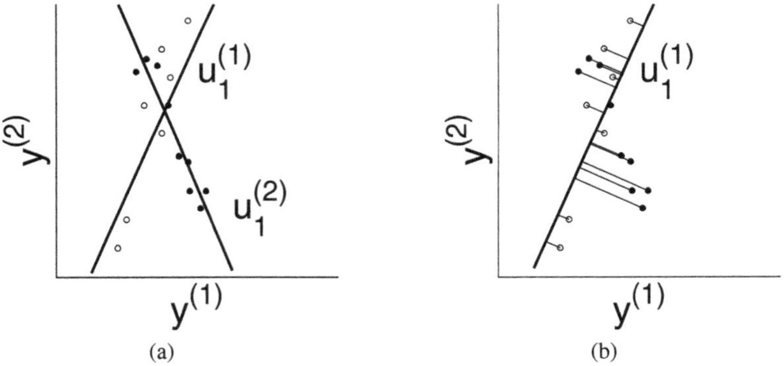

Fig. 4.8 **(a)** Shows a set of two-dimensional measurements $\mathbf{y}_i = (y_i^{(1)}, y_i^{(2)})^T, i \in \{1,2,\ldots,N\}$, which belong to a common training set D. Measurements $i \in \{1,2,\ldots,n\}$, belong to class $C = c_1$ (and are shown by open circles) and measurements $i \in \{n+1, n+2, \ldots, N\}$, belong to class $C = c_2$ (and are shown by closed circles). Overlaying the $\{\mathbf{y}_i\}$ are the dominant principal axis $\mathbf{u}_1^{(k)}$ for each class $C = c_k$. **(b)** Shows the creation of a training set D_1 by projecting the samples $\mathbf{y}_i, i \in \{1,2,\ldots,N\}$, onto the leading axis $\mathbf{u}_1^{(1)}$.

In [15] we apply a subset of R randomly chosen direction $\theta_r, r \in \{1,2,\ldots,R\}$. We project the test target $\mathbf{x}$ and the train targets $\mathbf{y}_n, n \in \{1,2,\ldots,N\}$, along each θ_r and perform the K-NN classification on the projected signals. Suppose for θ_r the K-NN classification is l_r. Then we combine the $l_r, r \in \{1,2,\ldots,R\}$, to give a final label l^*, where l^* is the label with the most votes.

4.7 Software

DACE. A matlab toolbox for performing Kriging. Authors: Soren N. Lophaven, Hans Brun Nielsen, Jacob Sondergaard. The toolbox contains m-files for performing simple, ordinary and universal Kriging.

FASTICA. A matlab toolbox for performing ICA. Authors: Hugo Gavert, Jarmo Hurri, Jaakko Sarela, Aapo Hyvarinen.

LIBRA. A matlab toolbox for performing robust statistics (see Chapt. 11). Authors: Sabine Verboven, Mia Hubert. The toolbox contains m-files on various robust subspace techniques.

MATLAB STATISTICAL TOOLBOX. The mathworks matlab statistical toolbox. The toolbox contains m-files on various subspace techniques.

NETLAB. A matlab toobox for performing neural network pattern recognition. Author: Ian Nabney. The toolbox contains m-files on various subspace techniques including PPCA.

STPRTOOL. A matlab toolbox for performing statistical pattern recognition. Authors: Vojtech Franc, Vaclav Hlavac.

4.8 Further Reading

The question of what constitutes an appropriate common representational format has been intensely investigated for mobile robots by Thrun [57, 58], Durrant-Whyte [11, 12] and others. For target tracking the choice of representational format has been discussed in [6, 41] and specific references on debiased Cartesian coordinates are given in [10, 14]. Recent articles on the polar maps, or Bull's-eye images in coronary artery disease include [32, 43]. The subject of invariant color spaces has been intensely investigated in recent years. Two modern references which provide many pointers to the literature on this subject are: [16, 18]. Two comprehensive references on Kriging are [8, 47]. Modern reviews of the brain atlas are given in [38, 56]. A large literature exists on the different subspace methods. Pointers to the literature are given in Table 4.7.

Problems

4.1. Develop the algorithm for Bayesian triangulation (Ex. 4.2)

4.2. In Ex. 4.15 we considered the conversion of range-bearing measurements to Cartesian coordinates. Show the conventional conversion $x = r\cos(\theta)$ and $y = r\sin(\theta)$ gives biased estimates. Use a simple example to show the coordinates $x_{DB} = r\cos(\theta)/\lambda$ and $y_{DB} = r\sin(\theta)/\lambda$ are less biased.

4.3. Describe the main characteristics of the PCA and LDA subspaces.

4.4. Describe a nearest neighbor face recognition scheme using PCA.

4.5. Describe a nearest neighbor face recognition scheme using LDA.

4.6. Describe the main characteristics of the linear subspace techniques PCA, LDA, ICA, NMF and CCA.

4.7. Explain the main differences between bagging, wagging, boosting and multi-boosting.

References

1. Bay, H., Tuytelaars, T., Van Gool, L.: SURF: Speeded-Up Robust Features. In: Leonardis, A., Bischof, H., Pinz, A. (eds.) ECCV 2006. LNCS, vol. 3951, pp. 404–417. Springer, Heidelberg (2006)
2. Behloul, F., Lelieveldt, B.P.E., Boudraa, A., Janier, M., Revel, D., Reiber, J.H.C.: Neuro-fuzzy systems for computer-aided myocardial viability assessment. IEEE Trans. Med. Imag. 20, 1302–1313 (2001)
3. Belhumeur, P.N., Hespanha, J., Kriegman, D.J.: Eigenfaces vs. Fisherfaces: Recognition using class specific linear projection. IEEE Trans. Patt. Anal. Mach. Intell. 19, 711–720 (1997)
4. Belongie, S., Malik, J., Puzicha, J.: Shape matching and object recognition using shape context. IEEE Trans. Patt. Anal. Mach. Intell. 24, 509–522 (2002)
5. Bulanon, D.M., Burks, T.F., Alchanatis, V.: Image fusion of visible and thermal images for fruit detection. Biosystems Engng. 103, 12–22 (2009)
6. Blackman, S.S., Popoli, R.F.: Design and analysis of modern tracking Systems. Artech House, Norwood (1999)
7. Chen, S.C., Zhu, Y.L., Zhang, D.Q., Yang, J.Y.: Feature extraction approaches based on matrix pattern: MatPCA and MatFLDA. Patt. Recogn. Lett. 26, 1157–1167 (2005)
8. Cressie, N.A.C.: Statistics for spatial data. John Wiley and Sons (1993)
9. Dalal, N., Triggs, B.: Histograms of oriented gradients for human detection. In: Int. Conf. Comp Vis Patt. Recogn, CVPR 2005 (2005)
10. Duan, Z., Han, C., Li, X.R.: Comments on " Unbiased converted measurements for tracking". IEEE Trans. Aero Elect. Sys. 40, 1374–1376 (2004)
11. Durrant-Whyte, H.F.: Consistent integration and propagation of disparate sensor observations. Int. J. Robotics Res. 6, 3–24 (1987)
12. Durrant-Whyte, H.F.: Sensor models and multisensor integration. Int. J. Robotics Res. 7, 97–113 (1988)
13. Fern, X.Z., Brodley, C.E.: Random projection for high dimensional data clustering: a cluster ensemble approach. In: Proc. 20th Int. Conf. Mach. Learn., pp. 186–193 (2003)

14. Fletcher, F.K., Kershaw, D.J.: Performance analysis of unbiased and classical measurement conversion techniques. IEEE Trans. Aero Elect. Sys. 38, 1441–1444 (2002)
15. Fraiman, D., Justel, A., Svarc, M.: Pattern recognition via projection-based KNN rules. Comp. Stat. Data Anal. 54, 1390–1403 (2010)
16. Gevers, T., De Weijer, J., van Stockman, H.: Color feature detection. In: Lukac, R., Plataniotis, N. (eds.) Color Image Processing: Emerging Applications. CRC Press (2006)
17. Gevers, T., Stockman, H.: Robust histogram construction from color invariants for object recognition. IEEE Trans. Patt. Anal. Mach. Intell. 25, 113–118 (2004)
18. Geusebroek, J.-M., van den Boomgaard, R., Smeulders, A.W.M., Geerts, H.: Color invariance. IEEE Trans. Patt. Anal. Mach. Intell. 23, 1338–1350 (2001)
19. Goovaerts, P.: Geostatistics in soil science: state-of-the-art and perspectives. Geoderma 89, 1–45 (1999)
20. Goovaerts, P.: Geostatistical approaches for incorporating elevation into the spatial interpolation of rainfall. Hydrology 228, 113–129 (2000)
21. Hardoon, D.R., Szedmak, S., Shaw-Taylor, J.: Canonical correlation analysis: an overview with application to learning methods. Neural Comp. 16, 2639–2664 (2004)
22. Hastie, T., Tibshirani, R.: Penalized discriminant analysis. Ann. Stat. 23, 73–102 (1995)
23. Ho, T.K.: The random subspace method for constructing decision forests. IEEE Trans. Patt. Anal. Mach. Intell. 20, 832–844 (1998)
24. Howland, P., Wang, J., Park, H.: Solving the small sample size problem in face recognition using generalized discriminant analysis. Patt. Recogn. 39, 277–287 (2006)
25. Hyvarinen, A., Karhunen, A., Oja, E.: Independent Component Analysis. John Wiley and Sons (2001)
26. Ji, S., Ye, J.: A unified framework for generalized linear discriminant analysis. IEEE Conf. Comp. Vis. Patt. Recogn. CVPR (2008)
27. Jolliffe, I.T.: Principal Component Analysis. Springer, Heidelberg (1986)
28. Li, M., Yuan, B.: 2D-LDA: A novel statistical linear discriminant analysis for image matrix. Patt. Recogn. Lett. 26, 527–532 (2005)
29. Li, H., Zhang, K., Jiang, T.: Efficient and Robust Feature Extraction by Maximum Margin Criterion. IEEE Trans. Neural Networks (2006)
30. Liao, S., Law, W.K., Chung, A.C.S.: Dominant local binary patterns for texture classification. IEEE Trans. Image Proc. 18, 1107–1118 (2009)
31. Lin, Y., Yu, Q., Medioni, G.: Efficient detection and tracking of moving objects in geo-coordinates. Mach. Vis. Appl. 22, 505–520 (2011)
32. Lindhal, D., Palmer, J., Pettersson, J., White, T., Lundin, A., Edenbrandt, L.: Scintigraphic diagnosis of coronary artery disease: myocardial bull's-eye images contain the important information. Clinical Physiology 18, 554–561 (1998)
33. Ling, H., Jacobs, D.W.: Shape classification using the inner distance. IEEE Trans. Patt. Analy. Mach. Intell. 29, 286–299 (2007)
34. Liu, J., Chen, S., Tan, X.: A study on three linear discriminant analysis based methods in small sample size problems. Patt. Recogn. 41, 102–111 (2008)
35. Loog, M., Duin, R.P.W.: Linear dimensionality reduction via a heteroscedastic extension of LDA: The Chernoff criterion. IEEE Trans. Patt. Anal. Mach. Intell. 26, 732–739 (2004)
36. Lowe, D.: Distinctive image features from scale-invariant keypoints. Int. J. Comp. Vision 20, 91–110 (2004)
37. Marcialis, G.L., Roli, F.: Decision-level fusion of PCA and LDA-band recognition algorithms. Int. J. Imag. Graphics 6, 293–311 (2006)
38. Mazziotta, J., Toga, A., Fox, P., Lancaster, J., Zilles, K., Woods, R., Paus, T., Simpson, G., Pike, B., Holmes, C., Collins, L., Thompson, P., MacDonald, D., Iacoboni, M., Schormann, T., Amunts, K., Palomero-Gallagher, N., Geyer, S., Parsons, L., Narr, K., Kabani, N., Le Goualher, G., Feidler, J., Smith, K., Boomsma, D., Pol, H.H., Cannon, T., Kawashima, R., Mazoyer, B.: A four-dimensional probabilistic atlas of the human brain. Am. Med. Inform. Assoc. 8, 401–430 (2001)
39. Mikolajczyk, K., Schmid, C.: A performance evaluation of local descriptors. IEEE Trans. Patt. Anal. Mach. Intell. 27, 1615–1630 (2005)

40. Miller, M.D., Drummond, O.E.: Coordinate transformation bias in target tracking. In: Proc. SPIE Conf. Sig. Data Proc. Small Targets, vol. 3809, pp. 409–424 (1999)
41. Moore, J.R., Blair, W.D.: Practical aspects of multisensor tracking. In: Bar-Shalom, Y., Blair, W.D. (eds.) Multitarget-Multisensor Tracking: Applications and Advances, vol. III, pp. 1–76. Artech House (2000)
42. Morena, P., Bernardino, Santos-Victor, J.: Improving the SIFT descriptor with smooth derivative filters. Patt. Recogn. Lett. 30, 18–26 (2009)
43. Moro, C.M.C., Moura, L., Robilotta, C.C.: Improving reliability of Bull's Eye method. Computers in Cardiology, 485–487 (1994)
44. Nunn, W.R.: Position finding with prior knowledge of covariance parameters. IEEE Trans. Aero. Elect. Sys. 15, 204–208 (1979)
45. Pele, O., Werman, M.: A Linear Time Histogram Metric for Improved SIFT Matching. In: Forsyth, D., Torr, P., Zisserman, A. (eds.) ECCV 2008, Part III. LNCS, vol. 5304, pp. 495–508. Springer, Heidelberg (2008)
46. Rodriguez, J.J., Kuncheva, L.I., Alonso, C.J.: Rotation forest: A new classifier ensemble method. IEEE Trans. Patt. Anal. Mach. Intell. 28, 1619–1630 (2006)
47. Schabenberger, O., Gotway, C.A.: Statistical Methods for Spatial Data Analysis. Chapman and Hall (2005)
48. Sanchez-Brea, L.M., Bernabeu, E.: On the standard deviation in charge-coupled device cameras: A variogram-based technique for non-uniform images. Elect. Imag. 11, 121–126 (2002)
49. Skocaj, D.: Robust subspace approaches to visual learning and recognition. PhD thesis, University of Ljubljana (2003)
50. Skurichina, M., Duin, R.P.W.: Combining Feature Subsets in Feature Selection. In: Oza, N.C., Polikar, R., Kittler, J., Roli, F. (eds.) MCS 2005. LNCS, vol. 3541, pp. 165–175. Springer, Heidelberg (2005)
51. Soille, P.: Morphological image compositing. IEEE Trans. Patt. Anal. Mach. Intell. 28, 673–683 (2006)
52. Szeliski, R.: Computer Vision: Algorithms and Applications. Springer, Heidelberg (2011)
53. Tan, N., Huang, L., Liu, C.: In: Proc. Int. Conf. Image Proc. (ICIP), pp. 1237–1240 (2009)
54. Thomaz, C.E., Gillies, D.F.: A maximum uncertainty LDA-based approach for limited sample size problems - with application to face recognition. In: 18th Brazilian Symp. Comp. Graph Imag. Proc. SIG-GRAPI 2005, pp. 89–96 (2005)
55. Thomaz, C.E., Giraldi, G.A.: A new ranking method for principal component analysis and its application to face image analysis. Im. Vis. Comp. 28, 902–913 (2010)
56. Thompson, P.M., Mega, M.S., Narr, K.L., Sowell, E.R., Blanton, R.E., Toga, A.W.: Brain image analysis and atlas construction. In: Handbook of Medical Imaging. Medical Image Processing and Analysis, vol. 2. SPIE Press (2000)
57. Thrun, S.: Learning metric-topological maps for indoor mobile robot navigation. Art. Intell. 99, 21–71 (1998)
58. Thrun, S.: Learning occupancy grids with forward sensor models. Autonomous Robots 15, 111–127 (2003)
59. Topi, M.: The local binary pattern approach to texture analysis - extensions and application. PhD thesis, University of Oulu (2003)
60. Tumer, K., Oza, N.C.: Input decimated ensembles. Patt. Anal. Appl. 6, 65–77 (2003)
61. Qiu, G., Fang, J.: Classification inan informative sample subspace. Patt. Recogn. 41, 949–960 (2008)
62. Valera, M., Velastin, S.A.: Intelligent distributed surveillance systems: a review. In: IEE Proc. - Vis. Imag. Sig. Proc., vol. 152, pp. 192–204 (2005)
63. Wang, M., Perera, A., Gutierrez-Osuna, R.: Principal discriminant analysis for small-sample-size problems: application to chemical sensing. In: Proc. 3rd IEEE Conf. Sensors, Vienna, Austria (2004)
64. Wang, X., Han, T., Yan, S.: An HOG-LBP human detector with partial occlusion handling. In: Int. Conf. Comp. Vis. ICCV 2009 (2009)
65. Watanabe, T., Ito, S., Yokoi, K.: Co-occurrence histograms of oriented gradients for human detection. IPSJ Trans. Comp. Vis. Appl. 2, 39–47 (2010)

66. Wu, H., Siegel, M., Khosla, P.: Vehicle sound signature recognition by frequency vector principal component analysis. IEEE Trans. Instrument Meas. 48, 1005–1009 (1999)
67. Xiang, C., Fan, X.A., Lee, T.H.: Face recognition using recursive Fisher linear discriminant. IEEE Trans. Image. Proc. 15, 2097–2105 (2006)
68. Yang, J., Zhang, D., Frangi, A.F., Yang, J.Y.: Two-dimensional PCA: A new approach to appearance-based face representation and recognition. IEEE Trans. Patt. Anal. Mach. Intell. 26, 131–137 (2004)
69. Ye, J., Janardan, R., Li, Q., Park, H.: Feature reduction via generalized uncorrelated linear discriminant analysis. IEEE Trans. Knowledge Data Engng. 18, 1312–1322 (2006)
70. Zhang, D., Zhou, Z.-H., Chen, S.: Diagonal principal component analysis for face recognition. Patt. Recogn. 39, 140–142
71. Zhang, P., Peng, J., Riedel: Discriminant analysis: a unified approach. In: Proc. 5th Int. Conf. Data Mining (ICDM 2005), Houston, Texas (2005)
72. Zheng, W., Zhao, L., Zou, C.: An efficient algorithm to solve the small sample size problem for LDA. Patt. Recogn. 37, 1077–1079 (2004)
73. Zomet, A., Levin, A., Peleg, S., Weiss, Y.: Seamless image stitching by minimizing false edges. IEEE Trans. Image Process. 15, 969–977 (2006)

Chapter 5
Spatial Alignment

5.1 Introduction

The subject of this chapter is *spatial alignment*. This is the conversion of local spatial positions to a common coordinate system and forms the first stage in the formation of a common representational format. To keep our discussion focused we shall limit ourselves to two-dimensional (x,y) image sensors. In this case the process of spatial alignment is more commonly referred to as *image registration*.

In many multi-sensor data fusion applications spatial alignment is the primary fusion algorithm. In Table 5.1 we list some of these applications together with the classification of the type of fusion algorithm involved.

Table 5.1 Applications in which Spatial Alignment is the Primary Fusion Algorithm

Class		Application
DaI-DaO	Ex. 4.1	A Standardized Brain Atlas: A Common Representational Format for Brain Research.
	Ex. 4.4	A Distributed Surveillance System.
	Ex. 4.9	Myocardial Imaging.
	Ex. 5.3	Background subtraction in video surveillance.
	Ex. 5.9	Optical flow for driver fatigue detection.
	Ex. 5.11	Demosaicing a Bayer Image.
	Ex. 5.14	Spatial Alignment in Cardiac MR Images.
	Ex. 5.15	Fingerprint warping using thin-plate spline.
	Ex. 5.18	Mosaic Fingerprint Image.

The designation DaI-DaO refers to the Dasarathy input/output classification: "Data Input-Data Output" (Sect. 1.3.3).

5.2 Image Registration

Let I_1 and I_2 denote two input images, where $I_m(x,y), m \in \{1,2\}$, denotes the gray-level of I_m at the pixel (x,y). The registration process is defined as finding the transformation T which "optimally" maps coordinates in I_1 to coordinates in I_2. Thus, given a pixel (x,y) in I_1, the corresponding location in I_2 is $(x',y') = T(x,y)$. Often a transformed point (x',y') does not fall on a pixel in I_2. In this case, strictly speaking, we have no gray-level for the point (x',y'). To handle such situations we resample/interpolate the test image so that a gray-level is defined for all points (x',y').

A popular technique for multi-sensor spatial alignment is to use an information theoretic topic known as "mutual information". The basic concept behind this approach is to find a transformation, which when applied to the test signal, will maximize the mutual information [1].

The success of mutual information image registration lies in its inherent simplicity. It makes few assumptions regarding the relationship that exists between different signals. It only assumes a *statistical* dependence. The idea is that although different sensors may produce very different signals, since they are imaging the same underlying scene, there will exist some inherent mutual information between the signals.

Example 5.1. Medical Image Registration: A Thought Experiment [37]. Consider a pair of medical images from different modalities (e. g. Magnetic Resonance and Computerized Tomography) that are in perfect alignment. At many corresponding spatial locations in the two images, the associated gray-levels will exhibit a consistent mapping. In other words, there will exist some kind of global relation, or mapping, between the gray-levels of the reference image with the gray-levels of the test image. Whatever the mapping is (one-to-one, one-to-many, many-to-one or many-to-many) the statistical dependency between the images is strong and the mutual information measure has a high value.

Suppose now that we move out of registration by gradually deforming the test image. The mapping between the gray-levels of the two images will degrade, i. e. will become less consistent, and the value of the mutual information will decrease. In the extreme case, the two images are completely independent of one another, and the mutual information will take a value of zero which signifies that it is not possible to predict any part of one image from the information stored in the other.

---

[1] The mutual information $MI(u,v)$ between two random variables u, v is defined formally in Sect. 5.3. For the moment, we may regard $MI(u,v)$ as a generalization of the correlation coefficient $\rho(u,v)$ which can be used to detect nonlinear dependencies.

5.3 Mutual Information*

The mutual information between two random variables u and v is given by

$$MI(u,v) = H(u) + H(v) - H(u,v) , \tag{5.1}$$

where $H(u)$ and $H(v)$ are the marginal entropies of the two variables u and v and $H(u,v)$ is their joint entropy. By rewriting (5.1) as

$$MI(u,v) = H(u) - H(u|v) = H(v) - H(v|u) , \tag{5.2}$$

we see that we may interpret MI as representing the reduction in the uncertainty of u when v is known, or equivalently, as representing the reduction in the uncertainty in v when u is known.

In the continuous domain, the (differential) entropy of a random variable u and the joint (differential) entropy of two random variables u and v are defined as

$$H(u) = -\int_{-\infty}^{\infty} p_U(u)\log_2 p_U(u)du , \tag{5.3}$$

$$H(u,v) = -\int_{-\infty}^{\infty} p_{UV}(u,v)\log_2 p_{UV}(u,v)dudv , \tag{5.4}$$

where $p_U(u)$ and $p_{UV}(u,v)$ are, respectively, the marginal and joint probability density functions of the random variables u and v. In this case the corresponding formula for the mutual information is

$$\begin{aligned} MI(u,v) &= -\int_{-\infty}^{\infty} p_U(u)\log_2(p_U(u))du - \int_{-\infty}^{\infty} p_V(v)\log_2(p_V(v))dv \\ &\quad + \int_{-\infty}^{\infty} p_{UV}(u,v)\log_2(p_{UV}(u,v))dudv , \\ &= \int\int p(u,v)\log_2 \frac{p_{UV}(u,v)}{p_U(u)p_V(v)}dudv . \end{aligned} \tag{5.5}$$

In the case of two input images I_1 and I_2, u and v represent, respectively, the gray-levels $I_1(i,j)$ and $I_2(i,j)$ of the same pixel (i,j).

5.3.1 *Histogram Estimation*

Given N simultaneous measurements $(u_i,v_i), i \in \{1,2,\ldots,N\}$, the most straightforward, and widely used, approach to calculate the mutual information $MI(u,v)$ is to estimate the marginal probability densities $p_U(u)$ and $p_V(v)$ and the joint probability density $p_{UV}(u,v)$ using a histogram technique. We divide the (u,v) space into K rows and L columns. The intersection of the rows and columns define a set of rectangular bins $B_{k,l}, k \in \{1,2,\ldots,K\}, l \in \{1,2,\ldots,L\}$. If n_{kl} is the number of simultaneous measurements which lie in the k,lth bin B_{kl}, then the mutual information is defined as

$$MI(u,v) \approx \log_2 N + \frac{1}{N}\sum_{kl} n_{kl} \log_2 \frac{n_{kl}}{N_k^c + N_l^r} + \Delta MI \ , \tag{5.6}$$

where $N_k^r = \sum_{l=1}^{L} n_{kl}$ and $N_l^c = \sum_{k=1}^{K} n_{kl}$ are, respectively, the number of measurement pairs which lie in the kth row and lth column. ΔMI is a *systematic* error correction which arises because we are using a finite number of measurements. If the bins are of equal size, then to first order

$$\Delta MI(u,v) = \frac{M^B - M^c - M^r + 1}{2KL} \ , \tag{5.7}$$

where M^B, M^c and M^r are, respectively, the number of non-empty bins, columns and rows.

In (5.6) the number of histogram bins, K and L, is crucial: If K and L are too small the estimated probability densities are accurate but do not follow the shape of the true density curves. On the other hand, if K and L are too large the estimated probability densities follow the shape of the true density curves but are inaccurate. Table 5.2 lists several formulas which are used to estimate the optimum number of bins in a regular histogram.

Table 5.2 Methods for Calculating Optimum Number Bins in a Regular Histogram

Name	Formula
Sturges	$K_{\text{STURGES}} = 1 + \log_2(N)$, where N is the number of data points.
Scott	$K_{\text{SCOTT}} = RN^{1/3}/(3.49s)$, where R is the range of the data and s is the sample standard deviation of the input data [31].
Knuth	$K_{\text{KNUTH}} = \arg\max_K \big(N \ln K + \ln\Gamma(K/2) - K \ln\Gamma(N + (K/2)) - \sum_{k=1}^{K} \ln\Gamma(n_k + (1/2))\big)$, where n_k is the number of data points which fall in the kth bin and Γ is the gamma function [17].
Birge and Rosenhole	See Ex.8.3.

Instead of using histograms with fixed bin widths, we may improve the MI estimate by adaptively partitioning the (u,v) space. For example, we may construct a hierarchy of partitions which recursively divide the (u,v) space into smaller and smaller bins [7, 8]. The idea is that regions in the (u,v) space, where pairs of measurements (u_i,v_i) are uniformly distributed, cease to contribute further to the estimated mutual information under further partitioning. For these regions, there is therefore no point in subdividing them further. The result is that dense regions containing many simultaneous measurements are covered with small bins, while regions containing few, or no, simultaneous measurements are covered with large bins. Although these estimators are much better than estimators using fixed bin size, they still suffer from systematic errors. For a full discussion concerning these errors see [8].

The following example illustrates the principles underlying the implementation of a variable partitioning MI algorithm.

Example 5.2. Darbellay-Tichavsky Variable Partitioning Algorithm [7]. The Darbellay-Tichavsky algorithm works as follows. Suppose at some stage in the above recursive procedure we have a rectangular bin B_{kl} which contains $n_{kl} > 4$ pairs of measurements. The corresponding horizontal row and vertical column contain, respectively, $N_k^r = \sum_{l=1}^{L} n_{kl}$ and $N_l^c = \sum_{k=1}^{K} n_{kl}$ measurement pairs. Tentatively, we divide B_{kl} into four rectangular sub-bins $B_{kl}^{(i)}, i \in \{1,2,3,4\}$, containing $n_{kl}^{(1)}, n_{kl}^{(2)}, n_{kl}^{(3)}$ and $n_{kl}^{(4)}$ measurement pairs. The corresponding vertical columns contain $a = (n_{kl}^{(1)} + n_{kl}^{(4)})$ and $b = (n_{kl}^{(2)} + n_{kl}^{(3)})$ measurement pairs and the corresponding horizontal rows contain $c = (n_{kl}^{(1)} + n_{kl}^{(2)})$ and $d = (n_{kl}^{(3)} + n_{kl}^{(4)})$ measurement pairs. The division into four sub-bins is chosen such that $|a-b|$ and $|c-d|$ are as close to zero as possible. If the resulting sub-bins satisfy the equality:

$$n_{kl}^{(1)} \approx n_{kl}^{(2)} \approx n_{kl}^{(3)} \approx n_{kl}^{(4)} \approx \frac{n_{kl}}{4} ,$$

then the division of B_{kl} is *not* carried out and the bin remains unchanged. If we use a 5% significance test to measure the above approximate equalities, then the above equation becomes

$$\frac{4}{n_{kl}} \sum_{i=1}^{4} \left(n_{kl}^{(i)} - \frac{n_{kl}}{4} \right)^2 < 7.81 .$$

5.3.2 Kernel Density Estimation

Instead of using discrete bins we may use continuous bins, or *kernels*,to calculate the probability densities $p_U(u)$, $p_V(v)$ and $p_{UV}(u,v)$ which appear in (5.5). This is known as kernel, or Parzen-window, density estimation [32, 34] and is a generalization of histogram binning. The estimated probability density $p_U(u)$ is given by

$$p_U(u) = \frac{1}{NH} \sum_{i=1}^{N} K\left(\frac{u - u_i}{H} \right) , \tag{5.8}$$

where the input data consists of N measurements $u_i, i \in \{1,2,\ldots,N\}$, and $\int K(u)du = 1$. In general the density estimate $p(u)$ is more sensitive to the choice of the bandwidth H and less sensitive to the choice of the kernel $K(u)$. For this

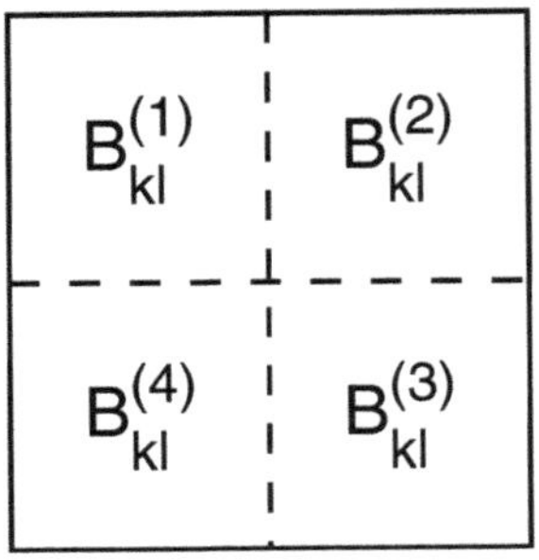

Fig. 5.1 Shows the division of the bin B_{kl} into four rectangles $B_{kl}^{(i)}, i \in \{1,2,3,4\}$.

reason we often let $K(u) = \mathcal{N}(0,1)$, where $\mathcal{N}(0,1)$ denotes a Gaussian function with zero-mean and unit standard deviation. Table 5.3 lists several schemes which are commonly used to calculate the optimal bandwidth H.

Table 5.3 Methods for Calculating Optimum One-Dimensional Bandwidth H

Name	Description
Rule-of-Thumb	Suppose the input data $y_i, i \in \{1,2,\ldots,N\}$, is generated by a given parametric density function, e. g. a Gaussian function. In this case $H = 1.06\hat{\sigma}N^{-1/5}$, where $\hat{\sigma}$ is the sample standard deviation. Robust versions of this bandwidth are available: $H = 1.06\min(\hat{\sigma}, \hat{Q}/1.34)N^{-1/5}$ and $H = 1.06\hat{s}N^{-1/5}$, where $\hat{Q}$ is the sample interquartile distance and $\hat{s} = \text{med}_j\lvert y_j - \text{med}_i y_i \rvert$.
Cross-Validation (CV)	Use a CV procedure to directly minimize the MISE or the AMISE (see sect. 13.6). CV variants include least square, biased and smoothed CV [16].
Plug-in	Minimize the AMISE using a second bandwidth known as the *pilot* bandwidth L. In the *solve-the-equation plug-in* method we write L as a function of the kernel bandwidth H [16].
Cross-entropy	See [3].
Boosting	See Sect. 14.10.

MISE is the mean integrated square error and is defined as $\text{MISE}(p, \hat{p}_H) = \int \left(p(y) - \hat{p}_H(y)\right)^2 dy$, where $\hat{p}_H(y)$ is the kernel approximation to $p(y)$. AMISE denotes the asymptotic MISE and represents a large number approximation of the MISE.

Apart from its use in MI calculations, accurate density estimation plays an important role in many multi-sensor data fusion algorithms. The following example illustrates one such application

Example 5.3. Background Subtraction in Video Surveillance [9]. In video surveillance systems, stationary cameras are typically used to monitor activity at a given site. Since the cameras are stationary, the detection of moving objects can be achieved by comparing each incoming image with a representation of the background scene. This process is called *background subtraction.* Let $x_1, x_2, \ldots, x_N$ be N consecutive gray-level values for a given pixel. Suppose the pixel has a gray-level x_t at time t, we estimate the probability that x_t belongs to the background:

$$p(x_t|x_1, x_2, \ldots, x_N, I) = \frac{1}{NH} \sum_{i=1}^{N} K\left(\frac{x_t - x_i}{H}\right) ,$$

where $\int K(u)du = 1$.

The pixel is considered to be in the foreground if $p(x_t|x_1, x_2, \ldots, x_N, I) < t$, where t is a given threshold.

To estimate the kernel bandwidth H Ref. [9] suggest the following equation:

$$H = \frac{\text{med}(|x_1 - x_2|, |x_2 - x_3|, \ldots, |x_{N-1} - x_N|)}{\sqrt{2} \times 0.68} .$$

The motivation for this equation is that although the intensity of a given pixel may vary in time, differences between consecutive intensity values are generally small, and only a few pairs are expected to exhibit a jump in intensity. By using the median operation our estimate should not be affected by the few jumps in intensity.

5.3.3 *Regional Mutual Information*

A drawback to the use of mutual information is that it does not take into account local relationships which may exist in the input data. indexMutual information!neighbourhood mutual information The following example explains a simple way to incorporate these relationships into the MI.

Example 5.4. Neighbourhood Mutual Information [30]. In the neighborhood MI we simply replace the random variables u and v in (5.5) with two-dimensional random vectors $\mathbf{u} = (u^{(1)}, u^{(2)})^T$ and $\mathbf{v} = (v^{(1)}, v^{(2)})^T$. In the case of two input images I_1 and I_2, $u^{(1)}$ and $v^{(1)}$ represent, respectively, the gray-levels $I_1(i,j)$ and $I_2(i,j)$ at a pixel (i,j) and $u^{(2)}$ and $v^{(2)}$ represent, respectively, the mean gray-levels $\bar{I}_1(i,j)$ and $\bar{I}_2(i,j)$ at the same pixel (i,j).

To incorporate more spatial relationships we require random vectors **u** and **v** with more dimensions. However, in practice, the number of dimensions is limited since the number of data points required for an accurate estimate of the marginal and joint probability densities increases exponentially with the number of dimensions. One way to overcome this problem (known as the curse of dimensionality) is known as the Regional MI algorithm which is described in the following example.

Example 5.5. Regional Mutual Information [30]. In regional MI we include all the pixels which are in a $(2r+1)\times(2r+1)$ local neighborhood of (i,j). By assuming the distribution of the points is approximately normal, we can *analytically* calculate the mutual information MI as follows:

$$MI = \ln\big((2\pi)^{D/2}\det(\boldsymbol{\Sigma}_A)^{1/2}\big) + \ln\big((2\pi)^{D/2}\det(\boldsymbol{\Sigma}_B)^{1/2}\big) - \ln\big((2\pi)^{D}\det(\boldsymbol{\Sigma}_J)^{1/2}\big) ,$$

where $\boldsymbol{\Sigma}_A$ is the covariance matrix of the image A and its $D=(2r+1)^2$ local neighbors, $\boldsymbol{\Sigma}_B$ is the covariance matrix of the image B and its D local neighbors, and $\boldsymbol{\Sigma}_J$ is the joint covariance matrix of the images A and B and their $2D$ local neighbors.

The main steps in calculating the regional MI are:

1. Create a stack, $\boldsymbol{\Phi}$, of $D=(2r+1)^2$ images of A and her corresponding neighbors and D images of B and her corresponding neighbors (Fig. 5.2). If the original image is of size $(m\times n)$, then $\boldsymbol{\Phi}$ is a $2D\times(m-2r)(n-2r)$ matrix
2. Let $\overline{\boldsymbol{\Phi}}$ denotes the mean of $\boldsymbol{\Phi}$. We subtract $\overline{\boldsymbol{\Phi}}$ from each point in $\boldsymbol{\Phi}$ and denote the result as $\widetilde{\boldsymbol{\Phi}}$.
3. Calculate the covariance $\boldsymbol{\Sigma}_J$ of $\widetilde{\boldsymbol{\Phi}}$, where Σ_J is a $2D\times 2D$ matrix given by
$$\boldsymbol{\Sigma}_J = \frac{1}{(m-2r)(n-2r)}\widetilde{\boldsymbol{\Phi}}\widetilde{\boldsymbol{\Phi}}^T .$$
4. The joint entropy of A and B and their corresponding neighbors is $H(A,B)$, where
$$H(A,B) = \ln\big((2\pi)^{D}\det(\boldsymbol{\Sigma}_J)^{1/2}\big) .$$
5. Similar calculations give the marginal entropy of A and its neighbors and the marginal entropy of B and its neighbors:
$$H(A) = \ln\big((2\pi)^{D/2}\det(\boldsymbol{\Sigma}_A)^{1/2}\big) ,$$
$$H(B) = \ln\big((2\pi)^{D/2}\det(\boldsymbol{\Sigma}_B)^{1/2}\big) ,$$
where $\boldsymbol{\Sigma}_A$ and $\boldsymbol{\Sigma}_B$ are, respectively, the $D\times D$ sub-matrices in the top-left and bottom-right corners of $\boldsymbol{\Sigma}_J$.

6. The (regional) mutual information is then given by (5.1):

$$MI = H(A) + H(B) - H(A,B) .$$

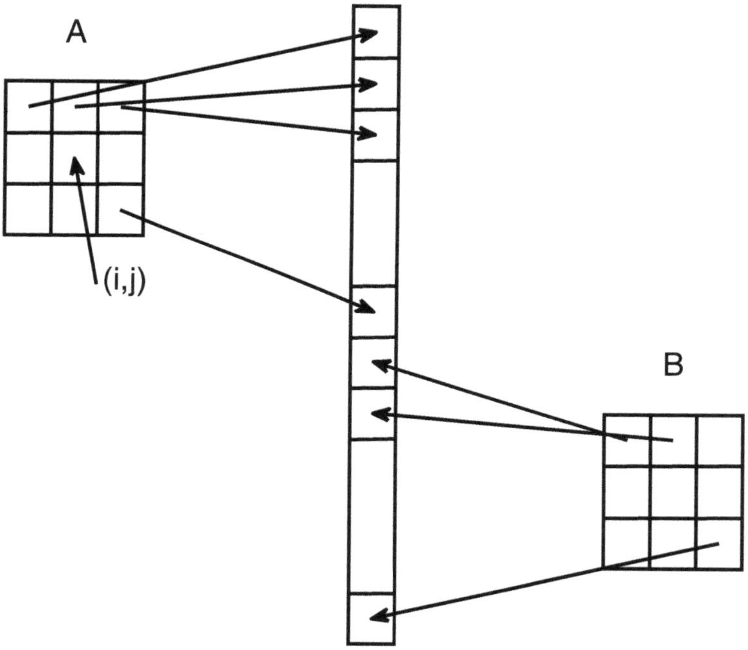

Fig. 5.2 Shows the creation of a stack Φ of $2D = 2(2r+1)^2$ images for the pixel (i, j). In the figure we assume $r = 1$, i. e. $D = (2r+1)^2 = 9$. The first D images are derived from the image A and the second D images are derived from the image B.

5.4 Optical Flow *

Optical flow [11, 13] is a two-dimensional field of velocities associated with the variation of brightness in the input image. This suggests the constant brightness assumption that the brightness of an image at any point on the object is invariant under motion.

Let $\mathbf{z} = (u, v)$ be the position of an image of some point P of the scene at time t and $\mathbf{w} = (u, v)$ be the projection of the velocity vector of this point onto the image plane. Then after time δt the image of the point P will move to a new point $\mathbf{z} + \delta \mathbf{z}$. The brightness assumption implies

$$I(\mathbf{z},t) = I(\mathbf{z}+\mathbf{w}\delta t, t+\delta t) , \tag{5.9}$$

where I is the brightness in the image plane. Using a Taylor series expansion:

$$I(\mathbf{z}+\mathbf{w}\delta t, t+\delta) \approx I(\mathbf{z},t) + \frac{\partial I(\mathbf{z},t)}{\partial x} u\delta t + \frac{\partial I(\mathbf{z},t)}{\partial y} v\delta t + \frac{\partial I(\mathbf{z},t)}{\partial t} \delta t . \tag{5.10}$$

Combining (5.9) and (5.10) gives

$$\frac{\partial I}{\partial x} u\delta t + \frac{\partial I}{\partial y} v\delta t + \frac{\partial I}{\partial t} \delta t \approx 0 . \tag{5.11}$$

Assuming an infinitesimal time interval we obtain the *optical flow equation*:

$$\frac{\partial I}{\partial x} u + \frac{\partial I}{\partial y} v + \frac{\partial I}{\partial t} = 0 . \tag{5.12}$$

Eq. (5.12) involves two unknowns at each point of the image plane: the optical flow components u and v. In order to determine we must make additional constraints. The simplest constraint is to assume the optical flow is constant in a small window. The resulting equations are known as the Lucas-Kanade equations.

Example 5.6. Lucas-Kanade Equations [11, 13]. Consider a small image patch containing K pixels. Then the residual error for the image patch is

$$E = \sum_k \left(\frac{\partial I(k)}{\partial x} u + \frac{\partial I(k)}{\partial y} v + \frac{\partial I(k)}{\partial t} \right)^2 ,$$

which we rewrite as

$$E = \sum_k (uI_x(k) + vI_y(k) + I_t)^2 ,$$

where $I_x(k) = \partial I(k)/\partial x$, $I_y(k) = \partial I(k)/\partial y$ and $I_t(k) = \partial I(k)/\partial t$. Minimizing E leads to the Lucas-Kanade equation which is identical to an ordinary least square solution:

$$\begin{pmatrix} u \\ v \end{pmatrix} = \begin{pmatrix} \sum_k I_x(k)I_x(k) & \sum_k I_x(k)I_y(k) \\ \sum_k I_x(k)I_y(k) & \sum_k I_y(k)I_y(k) \end{pmatrix}^{-1} \begin{pmatrix} -\sum_k I_x(k)I_t(k) \\ -\sum_k I_y(k)I_t(k) \end{pmatrix} .$$

The Lucas-Kanade equation is found to yield a biased estimate of the flow. The reason for this is the implicit assumption that the estimated derivatives I_x, I_y and I_t are noise-free. If we use an additive Gaussian noise model for the errors then the maximum likelihood estimate of the optical flow is the smallest eigenvector of the matrix M, where

$$M = \begin{pmatrix} \overline{I_x I_x} & \overline{I_x I_y} & \overline{I_x I_t} \\ \overline{I_x I_y} & \overline{I_y I_y} & \overline{I_y I_t} \\ \overline{I_x I_t} & \overline{I_y I_t} & \overline{I_t I_t} \end{pmatrix} .$$

For RGB color images we may simply apply (5.12) to each color plane. A more robust approach is to transform the RGB color space to the Hue-Saturation-Value (HSV) color space, in which the hue channel (H) is invariant under multiplicative illumination changes including shadow, shading, highlights and specularities. In this case, we may write the optical flow for H by replacing I in (5.12) with H.

Example 5.7. Constant Color Optical Flow [1]. In addition to the hue channel, the saturation channel is invariant with respect to shadow and shading. In this case, we may write two optical flow equations: one for H and one for S:

$$\frac{\partial H}{\partial x}u + \frac{\partial H}{\partial y}v + \frac{\partial H}{\partial t} = 0 ,$$

$$\frac{\partial S}{\partial x}u + \frac{\partial S}{\partial y}v + \frac{\partial S}{\partial t} = 0 .$$

where

$$H = \tan^{-1}\left(\frac{\sqrt{3}(G-B)}{2R-G-B}\right) ,$$

$$S = 1 - \frac{3}{R+G+B}\min(R,G,B) .$$

The following example illustrates how optical flow may be used to help create a common representational format.

Example 5.8. Converting a Smiling Face to a Neutral Face [14, 27]. The human face displays variety of expressions, e. g. smile, sorrow, surprise etc. These expressions lead to a significant change in the appearance of a facial image, which in turn, leads to a reduction in recognition rate of a face recognition system which in general have been trained on neutral faces and which represent the "common representational format". We may alleviate the reduction in performance by using optical flow to warp the test image to a neutral face i. e. to the common representational format [14].

Optical flow may also be used as a classifier in its own right as the following example shows.

Example 5.9. Optical flow for driver fatigue detection [33]. A major cause of road accidents is due to the lack of concentration in a driver due to fatigue. A strong indicator of driver fatigue is "percentage eye closure" $\mathscr{P}$, i. e. the percentage of the time the eye is closed. In [33] the movement of the upper eyelid is tracked using image velocities based on optical flow. Eye closures and eye openings are detected which are, in turn, used to estimate $\mathscr{P}$.

5.5 Feature-Based Image Registration

In feature-based image registration we represent the salient structures in the two input images as compact geometrical entities, for examples, points, curves or surfaces. In recent years the SIFT point operator (see Ex. 4.12) has been used for this purpose. After finding the correspondence between the two sets of features we are able to calculate the corresponding spatial transformation, or mapping, $T(x,y)$. The following example illustrates feature-based registration when the features are two sets of key-points.

Example 5.10. Point-based Image Registration [20, 41]. We separately apply the SIFT operator to the two input images A and B. Each key-point has associated with it a 128 long descriptor. The key points in A are matched with those in B by minimizing the distance between the descriptors. A ratio test, comparing the distances between the best and the second best match for a given key-point is used as a measure of match quality. We select the N key-point pairs which have the highest match quality. Given these N key-points we then calculate the corresponding spatial transformation.

5.6 Resample/Interpolation

As we pointed out in Sect. 5.2, the resample/interpolation process is a crucial part of any registration scheme because it uniquely defines the behaviour of the transformation T when the samples do not fall on a pixel. Two detailed reviews on modern image interpolation methods are [19, 36]. In image interpolation we reconstruct a two-dimensional continuous image $I(x,y)$ from its discrete pixel values $I(i,j)$. Thus the amplitude at the position (x,y) must be estimated from its discrete neighbours. This is often modeled as a convolution of the discrete image samples with a continuous two-dimensional impulse response $h(x,y)$:

$$I(x,y) = \sum_i \sum_j I(i,j) h(x-i, y-j) . \tag{5.13}$$

Usually, symmetrical and separable interpolation kernels are used to reduce the computational complexity

$$h(x,y) = h_x(x) h_y(y) , \tag{5.14}$$

where $h_x(x) = h_x(-x)$ and $h_y(y) = h_y(-y)$. In Table 5.4 we list several popular kernel functions $h(x)$ used for image interpolation [19].

Table 5.4 Kernel Functions $h(x)$ for Image Interpolation.

Name	$h(x)$
Nearest Neighbour	1 if $0 \le \|x\| < \frac{1}{2}$; otherwise 0.
Linear	$1 - \|x\|$ if $0 \le \|x\| < 1$; otherwise 0.
Quadratic	$\begin{cases} -2\|x\|^2 + \frac{1}{4} & \text{if } 0 \le \|x\| < \frac{1}{2} , \\ \|x\|^2 - \frac{5}{2}\|x\| + \frac{3}{8} & \text{if } \frac{1}{2} \le \|x\| < \frac{3}{2} , \\ 0 & \text{otherwise.} \end{cases}$
Cubic ($N = 4$)	$\begin{cases} (a+2)\|x\|^3 - (a+3)\|x\|^2 + 1 & \text{if } 0 \le \|x\| < 1 , \\ a\|x\|^3 - 5a\|x\|^2 + 8a\|x\| - 4a & \text{if } 1 \le \|x\| < 2 , \\ 0 & \text{otherwise ,} \end{cases}$ where a can take the values $a = -1/2, -2/3, -3/4, -1, -4/3$.

The above formulas in the table assume x and y are given in units of the sampling interval.

Apart from the fixed coefficient kernels listed in Table 5.4 we may also use the spatial Kriging estimators listed in Table 4.4.

The following example illustrates the use of linear interpolation in image demosaicing.

Example 5.11. Demosaicing a Bayer Image [22]. In a typical digital camera, the colors of the scene are captured by a single CCD or CMOS sensor array, where for each pixel the sensor detects a particular color channel, for example, red, green or blue. This kind of sensor is called a color filter array (CFA). The most popular CFA pattern was introduced by Bayer and it samples the green band using a quincunx grid, while red and blue are obtained by a rectangular grid as shown in Fig. 5.3. In this way, the density of the green samples is twice than that of the red and blue channels. Due to the subsampling of the color components, an interpolation step is required in order to reconstruct a full color representation of the image. This is called *demosaicing*. A high performance demosaicing algorithm [22] works by reconstructing two possible full color images H and V. Then at each pixel (m,n) we make a decision choosing $H(m,n)$ or $V(m,n)$ as the optimal reconstructed pixel value.

In reconstructing H we first estimate each missing green sample by interpolating along the horizontal lines or rows. Mathematically, each missing green sample is estimated as follows:

$$\hat{G}^H_{m,n} = \begin{cases} \frac{1}{2}(G_{m,n-1} + G_{m,n+1}) + \\ \quad \frac{1}{4}(2R_{m,n} - R_{m,n-2} - R_{m,n+2}) & \text{if } m \text{ is odd}, \\ \frac{1}{2}(G_{m,n-1} + G_{m,n+1}) + \\ \quad \frac{1}{4}(2B_{m,n} - B_{m,n-2} - B_{m,n+2}) & \text{if } m \text{ is even}, \end{cases}$$

where the superscript H denotes that the estimated green sample is made by interpolating along the horizontal lines in the input picture and $G_{m,n}, R_{m,n}$ and $B_{m,n}$ denote, respectively, the green, red and blue channel gray-levels of the pixel (m,n).

After interpolation of the missing green samples, the missing red and blue samples are estimated using a linear interpolation of the color differences $R - \hat{G}^H$ and $B - \hat{G}^H$. The final result is a complete reconstructured full-color image H. The second image, V, is reconstructed in exactly the same way except the interpolation is performed along the vertical lines, or columns. Finally, at each pixel (m,n) we make a decision and choose $H(m,n)$ or $V(m,n)$ as the reconstructured pixel value.

R	G	R	G
G	B	G	B
R	G	R	G
G	B	G	B

Fig. 5.3 Shows the Bayer filter wherein for each pixel the sensor detects only one color channel: red (R), green (G) or blue (B).

In the case of image registration using mutual information, special attention must be given to the process of resampling/interpolation, because the process introduces new gray-levels which may interfere with the computation of the MI [26]. For example, in applications involving remote sensing, nearest neighbour interpolation is often found to perform better than either linear, or cubic convolution, interpolation [6].

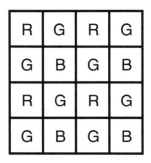

Example 5.12. Entropy Changes and Interpolation [37]. Consider a 10×10 binary image centered on a step function. The linear interpolation of this image after a very small sub-pixel translation across the edge introduces 10 new members with a third gray-level. Before translation the entropy of the image is $H = 1$. After translation the entropy is $H = 1.03 + 0.33$, the last value being the contribution of the new gray-level which represents 10% of all pixels.

5.7 Pairwise Transformation

The (pairwise) transformation $T : (x', y') = T(x, y)$ is a mathematical relationship that maps a spatial location (x, y) in one image to a new location, (x', y'), in another image. It arises in many image analysis problems, whether we wish to remove optical distortions introduced by a camera or a particular viewing perspective, or to register an image with a given map or template, or simply to align two input images. The choice of transformation is always a compromise between a smooth distortion and a distortion which achieves a good match. One way to ensure smoothness is to assume a parametric form of low-order for the transformation [12, 40] such as that given in Table 5.5. In most applications, the transformation T is chosen on the grounds of mathematical convenience. However, sometimes, we may have information regarding the physical processes which govern the formation of the pictures. In this case, we may use physical arguments in order to *derive* the transformation T.

Table 5.5 Spatial Transformations $T : (x', y') = T(x, y)$

Name	Formula
Translation	$x' = x + a_1, y' = y + a_2$.
Similarity	$x' = a_1x + a_2y + a_3$, $y' = -a_2x + a_1y + a_4$.
Affine	$x' = a_1x + a_2y + a_3$, $y' = a_4x + a_5y + a_6$.
Perspective	$x' = (a_1x + a_2y + a_3)/(a_7x + a_8y + 1)$,
	$y' = (a_4x + a_5y + a_6)/(a_7x + a_8y + 1)$.
Polynomial	$x' = \sum a_{ij}x^iy^j$, $y' = \sum b_{ij}x^iy^j$.

Example 5.13. Transformation for Retinal Images [4, 5]. Let T denote the pairwise transformation between a pair of retinal images. If we model the retina as a quadratic surface and assume a rigid weak-perspective transformation between the two images, then the corresponding transformation is

$$x' = a_1x^2 + a_2xy + a_3y^2 + a_4x + a_5y + a_6 ,$$
$$y' = a_7x^2 + a_8xy + a_9y^2 + a_{10}x + a_{11}y + a_{12} .$$

The assumptions made in deriving this transformation may be justified on physical grounds [4, 5]. Alternatively we may "derive" the transformation by performing a Taylor expansion of the general polynomial transformation in x and y and only retaining terms upto, and including, second order.

In some multi-sensor data fusion applications, the images undergo local deformations. In this case, we cannot describe the alignment of two images using a single low-order transformation. In this case we often use instead a "non-rigid" transformation, which we write as a sum of a low-order global transformation $T_G(x,y)$ and a local transformation $T_L(x,y)$:

$$(x',y') = T(x,y) = T_G(x,y) + T_L(x,y) , \tag{5.15}$$

where the parameters of T_L change with (x,y). The following example illustrates the use of such a transformation.

Example 5.14. Spatial Alignment in Cardiac MR Images [24, 25]. Magnetic imaging (MR) of the cardiovascular system is playing an increasingly important role is the diagnosis and treatment of cardiovascular diseases. During a cardiac cycle the heart changes size and shape. To perform spatial alignment of two MR images taken at different times during the cardiac cycle we require a transformation $T(x,y)$ which includes both a global transformation and a local transformation. The purpose of the local transformation $T_L(x,y)$ is to correct for changes in the size and shape of the heart.

The thin-plate spline is widely used to model non-rigid spatial transformations. This is illustrated in the following example.

Example 5.15. Fingerprint Warping Using a Thin-plate spline [29]. The performance of a fingerprint matching system is affected by the nonlinear deformations introduced in the fingerprint during image acquistion. This nonlinear deformation is represented using a thin-plate spline function (TPS) [39]:

$$x' = a_1 + a_2 x + a_3 y + \sum_{i=1}^{N} \alpha_i r_i^2 \ln r_i^2 ,$$

$$y' = a_4 + a_5 x + a_6 y + \sum_{i=1}^{N} \beta_i r_i^2 \ln r_i^2 ,$$

where $(x_i, y_i), i \in \{1, 2, \ldots, N\}$, is a set of known anchor points and $r_i^2 = (x - x_i)^2 + (y - y_i)^2 + d^2$. Apart from the parameter d, the TPS model has six parameters, $a_1, a_2, \ldots, a_6$, corresponding to the global (rigid) transformation and $2N$ parameters $(\alpha_i, \beta_i), i \in \{1, 2, \ldots, N\}$, corresponding to the local (non-rigid) deformation, and which satisfy the following constraints:

$$\sum_{i=1}^{N} \alpha_i = 0 = \sum_{i=1}^{N} \beta_i ,$$

$$\sum_{i=1}^{N} x_i \alpha_i = 0 = \sum_{i=1}^{N} x_i \beta_i ,$$

$$\sum_{i=1}^{N} y_i \alpha_i = 0 = \sum_{i=1}^{N} y_i \beta_i .$$

5.8 Uncertainty Estimation *

In this section we consider the accuracy of the spatial alignment process. In many multi-sensor data fusion applications we use the registration accuracy to determine whether, and to what extent, we may rely on the registration results. Kybic [18] suggests using bootstrap resampling (see Sect. 4.6) for this purpose. The algorithm is as follows:

Example 5.16. Bootstrap Resampling for Image Registration Uncertainty [18]. Given two $M \times N$ input images A and B we create K bootstrapped sets of pixels $A^{(k)}_{\text{boot}}, B^{(k)}_{\text{boot}}, k \in \{1, 2, \ldots, K\}$. For each k, we perform registration on $(A^{(k)}_{\text{boot}}, B^{(k)}_{\text{boot}})$. Suppose $\hat{\theta}^{(k)}$ denotes the corresponding estimated transformation parameters. Then the uncertainty in θ is given by the covariance matrix Σ_{boot}:

$$\Sigma_{\text{boot}} = \frac{1}{K} \sum_{k=1}^{K} \left(\hat{\theta}_{\text{boot}} - \bar{\theta}_{\text{boot}}\right) \left(\hat{\theta}_{\text{boot}} - \bar{\theta}_{\text{boot}}\right)^T ,$$

where

$$\bar{\theta}_{\mathrm{boot}} = \frac{1}{K}\sum_{k=1}^{K} \hat{\theta}^{(k)}_{\mathrm{boot}} \ .$$

5.9 Image Fusion

Image fusion is concerned with combining information from multiple *spatially registered* images. In Ex. 5.11 we described demosaicing which we may regard as the fusion of spatially registered and subsampled red, green and blue images. Two more applications which use image fusion are fusion of PET and MRI images and shape averaging. The reader will find many additional examples in the author's accompanying textbook [23].

5.9.1 Fusion of PET and MRI Images

The fusion of PET and MRI images is concerned with image fusion between complementary sensors in which the output is a decision surface or image. In the decision image each pixel gray-level represents a given decision. In forming the decision surface we reduce the effects of spatial misregistration by using a low resolution polar map or common representational format (see Ex. 4.9).

Example 5.17. Multi-Modal Cardiac Imaging [2]. Diseases of the heart are characterized by a reduced flow of blood to the heart muscles. This reduced flow can be studied using different imaging sensors or modalities, each of which gives a specific view of the phenomena. For example, the reduced flow of blood causes a deterioration of the myocardial perfusion which can be analyzed with positron emission tomography (PET) or with magnetic resonance imaging (MRI). Other changes which result from the reduced flow are metabolic changes in the heart tissues which can be highlighted using a PET sensor and the reduced capacity of the heart to eject blood into the body which can be monitored using an ultrasonic (US) sensor.

To assess the myocardial viability of the patient and to determine the proper therapeutic action it is important to fuse together images of all of the above phenomenom. In this example we consider the fusion of PET and MRI images [2]. The images are first converted to low-resolution polar maps which are then spatially registered. Parameters are extracted from each sector in the co-aligned PET and MRI polar maps which are then fused together using a single fusion cell F (Fig. 5.4). The fusion in F is performed using the fuzzy logic truth table shown in Fig. 5.5. The input to the table are two parameters: Inoptropic Reserve (IR) and 18-fluorodeoxyglucose (FDG) uptake and the output of the table is the myocardial viability (V). All three parameters are normalized by converting them to *linguistic values* using fuzzy membership functions similar to that shown in Fig. 5.5.

The output is a decision image in which each pixel gray-level corresponds to a linguistic myocardial viability value.

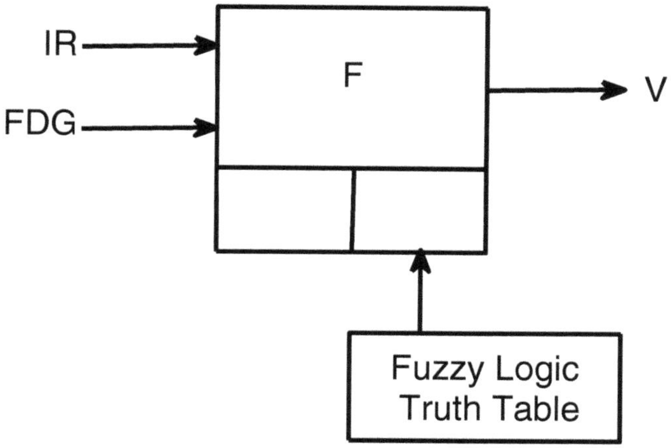

Fig. 5.4 Shows the fusion of the MRI and PET images using a single fusion cell F. Input to F is the Inoptroic Reserve, IR, which is derived from the MRI image and the 18-fluorodeoxyglucose, FDG, which is derived from the PET image. The output from F is the myocardial viability V. The fusion is performed using a fuzzy logic truth table which is supplied as external knowledge.

5.9.2 *Shape Averaging*

Averaging of multiple segmentations of the same image is an effective method to obtain segments that are more accurate than any of the individual segmentations. The idea is that each individual segmentation makes different errors, so that errors made by any one of the segmentation algorithms are corrected by the others. Suppose I_m denotes the mth segmented image. Each segmented image has L labels, $l \in \{1,2,\ldots,L\}$. Let $d_{m,l}(\mathbf{y})$ be the signed Euclidean distance of the pixel $\mathbf{y}$ in image I_m with respect to the label l, where $d_{m,l}(\mathbf{y})$ is the distance of $\mathbf{y}$ from the nearest pixel with label $\mathbf{y}$ less the distance of $\mathbf{y}$ from the nearest pixel with label which is

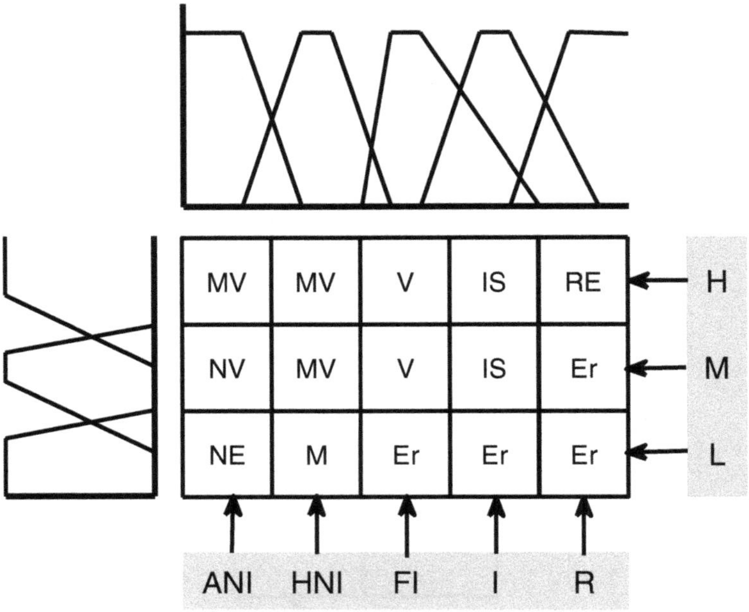

Fig. 5.5 Shows the fuzzy logic truth table used to fuse the inoptropic reserve (*IR*) and the 18-fluorodeoxyglucose (FDG) parameters. Inputs to the table are *IR* (horizontal axis) which is derived from the MRI image and FDG (vertical axis) which is derived from the PET image. The squares in the table contain the linguistic values of output variable V: Necrosis (NE), Maimed (*M*), Metabolic Viable (*MV*), Viable (*V*), Ischemic (*IS*), Remote (*RE*) and Error (Er), where the error Er is used to express an impossible combination of input values. The graph at the top of the figure shows the five membership functions associated with the linguistic values (*ANI*, *HNI*, FI, I, *R*) of the *IR* parameter. The graph at the side of the figure shows the three membership functions associated with the linguistic values (H, M, L) of the FDG parameter.

different from l. Based on the $d_{m,l}(\mathbf{y})$ image we let $D_l(\mathbf{y})$ be the mean distance of $\mathbf{y}$ from label l:

$$D_l = \frac{1}{M}\sum_{m=1}^{M} d_{m,l}(\mathbf{y}) \,. \tag{5.16}$$

The shape averaged segmented image, $S(\mathbf{y})$, is found by minimizing the mean distance over all label l:

$$S(\mathbf{y}) = \arg \min_{l\in\{1,2,\ldots,L\}} D_l(\mathbf{y}) \,. \tag{5.17}$$

Fig. 5.6. shows the behaviour of the shape average algorithm.

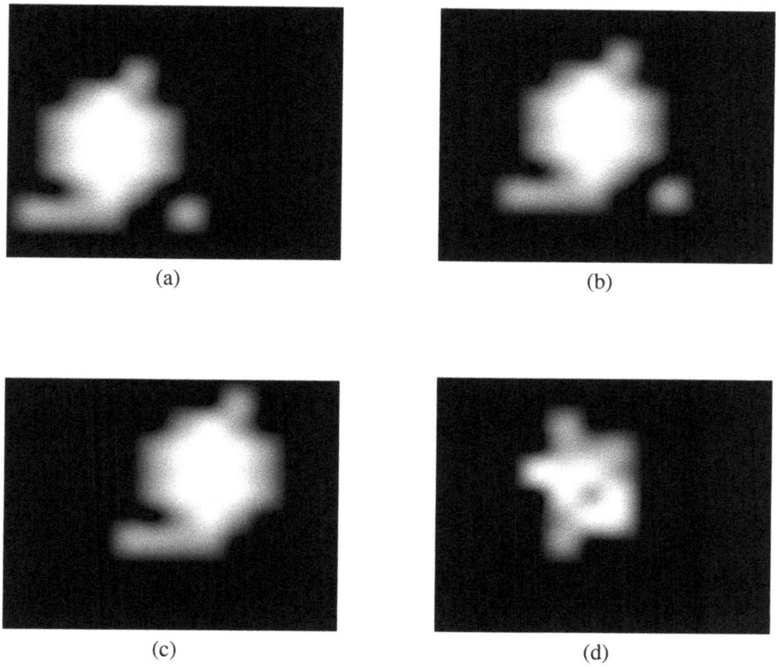

Fig. 5.6 Shows the shape average algorithm. **(a), (b)** and **(c)** show, respectively, three binary segmentations. The segmentations are identical except being having different displacements. **(d)** Shows the shape average segmentation $S(\mathbf{y})$. Observe how the shape-based averaging helps preserve the contiguous nature of the input images.

5.10 Mosaic Image

Thus far we have considered the problem of registering a pair of images. In some applications we are interested in building a single panoramic or "mosaic" image I^* from multiple images $I_m, m \in \{1,2,\ldots,M\}$. To do this we need to find functions T_m which transform each input image I_m onto the image I^*.

Building a mosaic image from a sequence of partial views is a powerful means of obtaining a broader view of a scene than is available with a single view. Research on automated mosaic construction is ongoing with a wide range of different applications.

Example 5.18. Mosaic Fingerprint Image [15]. Fingerprint-based verification systems have gained immense popularity due to the high level of uniqueness attributed to fingerprints and the availability of compact solid-state fingerprint sensors. However, the solid-state sensors sense only a limited portion of the fingerprint pattern and this may limit the accuracy of the user verification. To deal with this problem we construct a mosaic fingerprint image from multiple fingerprint impressions.

At first sight we may assume that the ability to register a pair of images is sufficient to solve the problem of forming a mosaic of the entire scene from multiple partial views. Theoretically, if one image can be established as an "anchor image" I_0 on which to base the mosaic image I^*, then the transformation of each remaining images onto this anchor may be estimated using pairwise registration. The mosaic image I^* is then formed by stitching the transformed images $T_m(I_m)$(Sect. 4.2). Unfortunately, in practice, this approach may not work for the following reasons:

1. Some images may not overlap with the anchor image at all. This makes a direct computation of the transformation impossible. In other cases, images may have insufficient overlap with the anchor image to compute a stable transformation. The straightforward solution is to compose transformations using an "intermediate" image. This is problematic, however, since repeated application of the transformation will often magnify the registration error.
2. The transformations T_m may be mutually inconsistent. This may happen even if all the image-to-anchor transformations have been accurately estimated. The reason for this is as follows: Although each image may individually register accurately with the anchor image and the non-anchor images may register accurately with each other, this does not ensure that the transformations onto the anchor image are *mutually consistent*.

One approach to solving this problem is to constrain the transformations so that they are all mutually consistent.

Example 5.19. Transformation Constraints in a Mosaic Image [10]. Given a sequence of N images $I_1, I_2, \ldots, I_N$, we estimate $N(N-1)$ pairwise transformations

$$(x',y') = T_{ij}(x,y) ,$$

where (x,y) and (x',y') denote, respectively, the coordinates of corresponding points in I_i and I_j. The T_{ij} must satisfy the following relationships:

$$T_{ik} = T_{ij} \circ T_{jk} ,$$
$$T_{ij} = T_{ji}^{-1} ,$$

where $\circ$ denotes the composition operator.

For an affine transformation, the transformation T_{ij} can be written in matrix form as $\begin{pmatrix} x' \\ y' \end{pmatrix} = A_{ij} \begin{pmatrix} x \\ y \end{pmatrix} + B_{ij}$. In this case, the above relationships become

$$A_{ik} = A_{ij} A_{jk} \; ,$$
$$B_{ik} = A_{ij} B_{jk} + B_{ij} \; .$$

5.11 Software

MATLAB IMAGE PROCESSING TOOLBOX. The mathworks matlab image processing toolbox. The toolbox contains various m-files for performing image registration, resampling and interpolation.

KDE, KDE2. Cross-entropy kernel density estimation. Author: Z. I. Botev [3].

MILCA. A matlab toolbox for performing independent component analysis. Authors: Sergey Astakhov, Peter Grassberger, Alexander Kraskov, harald Stogbauer. It contains an m-file for calculating $MI(u,v)$ using a binless method.

MUTIN. A matlab m-file for calculating the mutual information $MI(u,v)$ using the method of variable partitioning [8]. Author: Petr Tichavsky.

5.12 Further Reading

The subject of image registration has been intensely investigated for many years. A modern review of image warping is [12]. Specific references on the use of mutual information for image registration are [6, 21, 26]. The calculation of mutual information has been considered by many authors including [7, 8]. Such bounds are useful in evaluating different image registration techniques and determining parameter regions where accurate registration is possible. A modern reference on non-separable interpolation is [28].

Problems

5.1. Define mutual information and explain the different approaches to measuring it.

5.2. What are the main drawbacks to using mutual information for image registration? Explain how neighborhood and regional MI help alleviate these drawbacks.

5.3. Resampling is an important step in spatial alignment. Explain what resampling is and why it is necessary.

5.4. Explain why mutual information is useful for aligning sensors of different modality. Suggest alternative approaches which may be used instead of MI.

5.5. Describe the multi-sensor data fusion of PET and MRI using fuzzy logic.

5.6. Develop the optical flow equations for gray-scale images and color images.

5.7. What is the SIFT operator. How do we use the SIFT operator to perform image registration.

References

1. Andrews, R.J., Lovell, B.C.: Color optical flow. In: Proc. Workshop Digital Image Comput., pp. 135–139 (2003)
2. Behloul, F., Lelieveldt, B.P.E., Boudraa, A., Janier, M., Revel, D., Reiber, J.H.C.: Neuro-fuzzy systems for computer-aided myocardial viability assessment. IEEE Trans. Med. Imag. 20, 1302–1313 (2001)
3. Botev, Z.I., Kroese, D.P.: The generalized cross-entropy method with applications to probability density estimation (2006)
4. Can, A., Stewart, C.V., Roysam, B., Tanenbaum, H.L.: A feature-based technique for joint, linear estimation of high-order image-to-mosaic transformations: mosaicing the curved human retina. IEEE Trans. Patt. Anal. Mach. Intell. 24, 412–419 (2002)
5. Can, A., Stewart, C.V., Roysam, B., Tanenbaum, H.L.: A feature-based, robust, hierarchical algorithm for registering pairs of images of the curved human retina. IEEE Trans. Patt. Anal. Mach. Intell. 24, 347–364 (2002)
6. Chen, H.-M., Arora, M.K., Varshney, P.K.: Mutual information based image registration for remote sensing data. Int. J. Remote Sens. 24, 3701–3706 (2003)
7. Darbellay, G.A.: An estimator for the mutual information based on a criterion for independence. Comp. Stat. Data Anal. 32, 1–17 (1999)
8. Darbellay, G.A., Vajda, I.: Estimation of the information by an adaptive partitioning of the observation space. IEEE Trans. Inform. Theory 45, 1315–1321 (1999)
9. Elgammal, A., Duraiswami, R., Harwood, D., Davis, L.S.: Background and foreground modeling using nonparametric kernel density estimation for visual surveillance. Proc. IEEE 90, 1151–1163 (2002)
10. Farsiu, S.: A fast and robust framework for image fusion and enhancement. PhD thesis, University of California (2005)
11. Fleet, D.J., Weiss, Y.: Optical flow estimation. In: Paragios, et al. (eds.) Handbook of Mathematical models in Computer Vision. Springer, Heidelberg (2006)
12. Glasbey, C.A., Mardia, K.V.: A review of image warping methods. App. Stat. 25, 155–171 (1998)
13. Govindu, V.M.: Revisiting the Brightness Constraint: Probabilistic Formulation and Algorithms. In: Leonardis, A., Bischof, H., Pinz, A. (eds.) ECCV 2006. LNCS, vol. 3953, pp. 177–188. Springer, Heidelberg (2006)
14. Hseih, C.-K., Lai, S.-H., Chen, Y.-C.: An optical flow-based approach to robust face recognition under expression variations. IEEE Trans. Im. Proc. 19, 233–240 (2010)
15. Jain, A., Ross, A.: Fingerprint mosaicking. In: Proc. IEEE Int. Conf. Acquis. Speech Sig. Proc., ICASSP (2002)
16. Jones, M.C., Marron, J.S., Sheather, S.J.: A brief survey of bandwidth selection for density estimation. Am. Stat. Assoc. 91, 401–407 (1996)
17. Knuth, K.H.: Optimal data-based binning for histograms. arXiv:physics/0605197v1 (2006)

18. Kybic, J.: Bootstrap resampling form image registration uncertainty estimation without ground truth. IEEE Trans. Im. Proc. 19, 64–73 (2010)
19. Lehmann, T.M., Gonner, C., Spitzer, K.: Survey: Interpolation methods in medical image processing. IEEE Trans. Med. Imag. 18, 1049–1075 (1999)
20. Lowe, D.G.: Distinctive image features from scale invariant key-points. Int. J. Comp. Vis. 60, 91–100 (2004)
21. Maes, F., Vandermeulen, D., Suetens, P.: Medical image registration using mutual information. Proc. IEEE 91, 1699–1722 (2003)
22. Menon, D., Andriani, S., Calvagno, G.: Demosaicing with directional filtering and a posteriori decision. IEEE Trans. Imag. Process. 16, 132–141 (2007)
23. Mitchell, H.B.: Image Fusion: Theories, Techniques and Applications. Springer, Heidelberg (2010)
24. Perperidis, D.: Spatio-temporal registration and modeling of the heart using of cardiovascular MR imaging. PhD thesis, University of London (2005)
25. Perperidis, D., Mohiaddin, R., Rueckert, D.: Spatio-temporal free-form registration of cardiac MR image sequences. Med. Imag. Anal. 9, 441–456 (2005)
26. Pluim, J.P.W., Maintz, J.B.A., Viergever, M.A.: Mutual information based registration of medical images: a survey. IEEE Trans. Med. Imag. 22, 986–1004 (2003)
27. Ramachandran, M., Zhou, S.K., Jhalani, D., Chellappa, R.: A method for converting a similing face to a neutral face with applications to face recognition. In: Proc. ICASSP, pp. 18–23 (2005)
28. Reichenbach, S.E., Geng, F.: Two-dimensional cubic convolution. IEEE Trans. Image. Process. 12, 857–865 (2003)
29. Ross, A., Dass, S., Jain, A.: A deformable model for fingerprint matching. Patt. Recogn. 38, 95–103 (2003)
30. Russakoff, D.B., Tomasi, C., Rohlfing, T., Jr., C.R.M.: Image Similarity Using Mutual Information of Regions. In: Pajdla, T., Matas, J(G.) (eds.) ECCV 2004. LNCS, vol. 3023, pp. 596–607. Springer, Heidelberg (2004)
31. Scott, D.W.: On optimal and data-based histograms. Biometrika 66, 605–610
32. Scott, D.W.: Multivariate Density Estimation. Wiley (1992)
33. Senaratne, R., Jap, B., Lal, S., Halgamuge, S., Fischer, P.: Comparing two video-based techniques for driver fatigue detection: classification versus optical flow approach. Mach. Vis. App. 22, 597–618 (2011)
34. Silverman, B.: Density estimation for statistical data analysis. Chapman-Hall (1986)
35. Steuer, R., Kurths, J., Daub, C.O., Weise, J., Selbig, J.: The mutual information: detecting and evaluating dependencies between variables. Bioinform. 18 (Supplement 2), S231–S240 (2002)
36. Thevenaz, P., Blu, T., Unser, M.: Interpolation Revisited. IEEE Trans. Med. Imag. 19, 739–758 (2000)
37. Thevenaz, P., Unser, M.: A pyramid approach to sub-pixel image fusion based on mutual information. In: Proc. IEEE Int. Conf. Image Proc., Switzerland, vol. 1, pp. 265–268 (1996)
38. Tomazevie, D., Likar, B., Pernus, F.: Multi-feature mutual information. In: Proc. SPIE, vol. 5370, pp. 143–154 (2004)
39. Zagorchev, L., Goshtasby, A.: A comparative study of transformation functions for nonrigid image registration. IEEE Trans. Imag. Proc. 15, 529–538 (2006)
40. Zitova, B., Flusser, J.: Image registration methods: a survey. Imag. Vis. Comp. 21, 977–1000 (2003)
41. Zuliani, M.: Computational methods for automatic image registration. PhD thesis, University of California (2006)

Chapter 6
Temporal Alignment

6.1 Introduction

The subject of this chapter is *temporal alignment*, or registration, which we define as the transformation $T(t)$ which maps local sensor observation times t to a common time axis t'. Temporal alignment is one of the basic processes required for creating a common representational format. It often plays a critical role in applications involving in many multi-sensor data fusion applications. This is especially true for applications operating in real-time (see Sect. 2.4.2).

The following example illustrates the use of temporal registration in medical imaging.

Example 6.1. Cardiac MR Image Sequences [14]. Cardiovascular diseases are a very important cause of death in the developed world. Their early diagnosis and treatment is crucial in order to reduce mortality and to improve patient's quality of life. The recent advances in the development of cardiac imaging modalities have led to an increased need for cardiac registration, whose aim is to bring K sequences of cardiac images acquired from the same subject, or from the same subject at different times, into the same spatial-temporal coordinate system. To do this we must perform temporal registration which corrects for any *temporal misalignment* caused by different acquisition parameters, different lengths of cardiac cycles and different motion patterns of the heart.

Let $O = \langle E, \mathbf{u}, t, \mathbf{y}, \Delta\mathbf{y} \rangle$ denote a given sensor observation. Then, strictly speaking the transformation $T(t)$ is a function of the position $\mathbf{u}$ and the time t. However, it is common practice to assume (at least as a first approximation) that $T(t)$ is independent of $\mathbf{u}$ [13].

The following example illustrates an application in which perform temporal alignment on sequences of input images. In this application we reduce each input image to a single point in space which reduces the errors involved in assuming $T(t)$ is independent of **u**.

Example 6.2. Visualization of Dysphagia (Disorders Associated with Swallowing) [19]. We consider an application in which we form a high resolution video sequence of a patient swallowing. In this application, a patient repeatedly swallows. For a given swallow S, a magnetic resonance imaging (MRI) camera records the swallow in a sequence of MRI images $I_i, i \in \{1,2,\ldots,N\}$, taken at times t_i, where $t_i < t_{i+1}$. For each sequence (swallow) we segment the bolus and track its path in the space-time volume of the swallow. We do this by extracting the centroid, $(\bar{x},\bar{y})$, of the segmented region from each MRI image and creating a path of the centroid motion for each swallow (see Fig. 6.1).

After performing temporal registration, we form a high resolution video of the patient swallowing by combining the multiple swallows. The temporal registration is performed by matching the spatial-temporal paths of the bolus in multiple swallows.

To keep our discussion focused we shall limit ourselves to the temporal alignment of a time series, **P**, to a common time-axis t', where $\mathbf{P} = \{P_1,P_2,\ldots,P_M\}$ consists of M sensor observations, $P_i = \langle E_i,*,t_i,y_i,*\rangle, i \in \{1,2,\ldots,M\}$, made with the same sensor and ordered in time, i. e. $t_i < t_{i+1}$. Furthermore, we shall assume (as is common in many multi-sensor data fusion systems) that we do not have an independent time-axis t'. Instead, we shall define the time-axis t' by a second time series **Q**, where **Q** consists of N sensor observations, $Q_j = \langle E_j,*,t'_j,y'_j,*\rangle, j \in \{1,2,\ldots,N\}$. *Note.* Sometimes we normalize the time series $\mathbf{P} = \{P_1,P_2,\ldots,P_M\}$ such that $\sum_{i=1}^{M} y_i = 1$. In this case the time series is, in fact, a *probability* distribution.

In many multi-sensor data fusion applications temporal alignment is the primary fusion algorithm. In Table 6.1 we list some of these applications together with the classification of the type of fusion algorithm involved.

Table 6.1 Applications in which Temporal Alignment is the Primary Fusion Algorithm

Class	Application
DaI-DaO	Ex. 6.1 Cardiac MR Image Sequences.
DaI-FeO	Ex. 6.2 Visualization of the Dysphagia (Disorders Associated with Swallowing).
	Ex. 6.4 Speech recognition in an embedded speech recognition system.
	Ex. 6.5 Word spotting in historical manuscripts.
	Ex. 6.6 Generating a digital elevation map using DTW.
	Ex. 6.7 Identifying a face profile using the DDTW algorithm.
	Ex. 6.8 Signature verification using continuous dynamic time warping.
	Ex. 6.9 Compressed Video Sequence.

The designations DaI-DaO, DaI-FeO refer, respectively, to the Dasarathy input/output classifications: "Data Input-Data Output" and "Data Input-Feature Output" (Sect. 1.3.3).

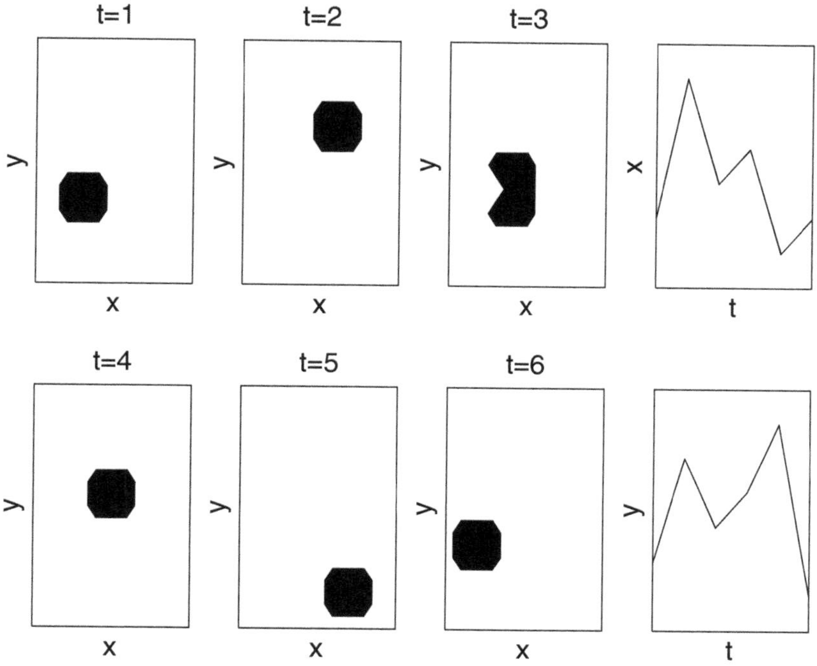

Fig. 6.1 Shows six two-dimensional images $I_i, i \in \{1,2,\ldots,6\}$, taken at times $t = \{1,2,\ldots,6\}$. In each image we segment the bolus (which is shown as a dark irregularly shaped area). To the right of the images I_i we show the x and y position of centroid of the bolus as a function of the time t.

6.2 Dynamic Time Warping

Dynamic Time Warping (DTW) [16, 17] is a general technique for performing temporal alignment between two time series, **P** and **Q**. DTW was originally developed for use in speech recognition. However, recently, it has been used in many different areas, including data mining, pattern and shape recognition and biometric identification.

DTW finds the *optimal* alignment between two time series **P** and **Q** in the sense that it minimizes the sum of the local distances $d(i,j)$ between the aligned observation pairs (P_i, Q_j). In many applications we define $d(i,j)$ as the square of the Euclidean distance:

$$d(i,j) = (P_i - Q_j)^2 \,. \tag{6.1}$$

However, in fact any appropriate local distance measure may be used instead.

The procedure is called time warping because it warps the time axes, t and t', in such a way that corresponding sensor observations appear at the same location on a common time axis. In general, the time axes, t and t', are aligned in a non-linear way.

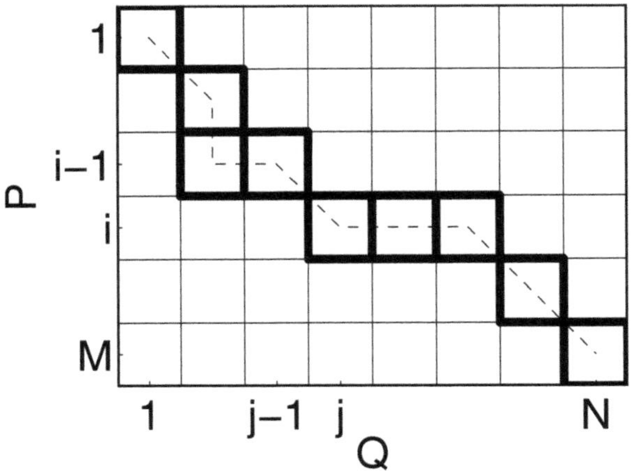

Fig. 6.2 Shows the Dynamic Time Warping (DTW) algorithm. We start at $w_1 = (1,1)$ where $D(1,1)$ is equal to $d(1,1)$. We then proceed top-to-bottom and left-to-right updating the cumulative distance $D(i,j)$ using the cumulative distances of the neighbouring cells. The optimal matching cost, $DTW(\mathbf{P},\mathbf{Q})$, between the time series **P** and the second time series **Q** is $D(M,N)$. The optimal warping path W_{DTW} is represented by the dashed line and the thick-bordered cells which can be recovered by tracing back the sequence of optimal neighbours. $\mathbf{W}_{\text{DTW}}$ provides information about the optimal correspondence (alignment) between the two time series **P** and **Q**.

In the DTW algorithm we define a given alignment of the time series, **P** and **Q**, by means of a "warping path" **W** (Fig. 6.2). This is a set of matrix elements $w_k = (i_k, j_k), k \in \{1,2,\ldots,K\}$, where w_k defines a mapping between the observations P_{i_k} and Q_{j_k}. Formally,

$$\mathbf{W} = \{w_1, w_2, \ldots, w_K\} , \tag{6.2}$$

where $\max(M,N) \leq K \leq M+N-1$.

In general, we are only interested in warping paths (i.. e. alignments) which satisfy the following constraints:

Boundary Conditions. $w_1 = (1,1)$ and $w_K = (M,N)$. This requires the warping path to start by matching the first observation of **P** with the first observation of **Q** and end by matching the last observation of **P** with the last observation of **Q**.

Continuity. Given $w_k = (a,b)$, then $w_{k-1} = (a',b')$, where $a - a' \leq 1$ and $b - b' \leq 1$. This restricts the allowable steps in the warping path to adjacent matrix elements (including diagonally adjacent elements).

Monoticity. Given $w_k = (a,b)$ then $w_{k-1} = (a',b')$, where $a - a' \geq 0$ and $b - b' \geq 0$. This forces the warping path sequence to increase monotonically in time.

There are exponentially many warping paths that satisfy the above conditions, however, we are interested only in the path which minimizes the sum of the local distances $d(i_k, j_k) \equiv d(w_k)$. This is the optimal ("DTW") warping path and is defined as

$$\mathbf{W}_{\mathrm{DTW}} \equiv \mathbf{W}_{\mathrm{DTW}}(\mathbf{P},\mathbf{Q}) = \arg\min_{\mathbf{W}} \sum_{k=1}^{K} d(w_k) \ . \tag{6.3}$$

6.3 Dynamic Programming

A brute force method for finding the optimal warping path requires the evaluation of an exponential number of warping paths. Fortunately dynamic programming can be used to find the optimal warping path, $\mathbf{W}_{\mathrm{DTW}}$, and its cost, $DTW(\mathbf{P},\mathbf{Q})$, in $O(MN)$ time using the recursion:

$$D(i_k, j_k) = d(i_k, j_k) + \min(D(i_k, j_k-1), D(i_k-1, j_k), D(i_k-1, j_k-1)) \ , \tag{6.4}$$

where $D(i_k, j_k)$ is the cost of the optimal warping path from $(1,1)$ to (i_k, j_k) and $D(1,1) = d(1,1)$.

The DTW algorithm starts by building a $M \times N$ table $D(m,n), m \in \{1,2,\ldots,M\}, n \in \{1,2,\ldots,N\}$, column by column. We start with $n = 1$ and compute $D(1,1)$ using (6.4) and the boundary conditions

$$D(0,0) = 0 \ , \tag{6.5}$$

$$D(0, j_k) = \infty = D(i_k, 0) \ . \tag{6.6}$$

We then proceed from $m = 2$ to $m = M$ and compute $D(2,1), D(3,1), \ldots, D(M,1)$. We then increase n by one and proceed from $m = 1$ to $m = M$ computing $D(1,2)$, $D(2,2), \ldots, D(M,2)$. We continue the process until the last column $n = N$ and row $m = M$ are reached. By definition $D(M,N)$ is equal to $DTW(\mathbf{P},\mathbf{Q})$, the cost of the optimal alignment of the sequences $\mathbf{P}$ and $\mathbf{Q}$. The optimal warping path, $\mathbf{W}_{\mathrm{DTW}}$, is obtained by tracing the recursion backwards from $D(M,N)$.

The following example illustrates the construction of the cumulative cost matrix $D(m,n)$ used in Fig. 6.2.

Example 6.3. Dynamic Programing. Given two time series $\mathbf{P} = (1.7, 4.1, 3.1, 2.2, 3.4, 1.6)^T$ and $\mathbf{Q} = (2.4, 3.5, 2.7, 3.2, 4.5, 4.7, 4.8, 5.9)^T$, we create a matrix of squared Euclidean distances $d(i,j) = (P_i - Q_j)^T$:

0.49	3.24	1.00	2.25	7.84	9.00	9.61	17.64
2.89	0.36	1.96	0.81	0.16	0.36	0.49	3.24
0.49	0.16	0.16	0.01	1.96	2.56	2.89	7.84
0.04	1.69	0.25	1.00	5.29	6.25	6.76	13.69
1.00	0.01	0.49	0.04	1.21	1.69	1.96	6.25
0.64	3.61	1.21	2.56	8.41	9.61	10.24	18.49

The corresponding cumulative matrix D is

0.49	3.73	4.73	6.98	14.82	23.82	33.43	51.07
3.39	0.85	2.81	3.62	3.73	4.14	4.63	7.87
3.87	1.01	1.01	1.02	2.98	5.54	7.03	12.47
3.91	2.70	1.26	2.01	6.31	9.23	12.30	20.72
4.91	2.71	1.75	1.30	2.51	4.20	6.16	12.41
5.50	6.32	2.96	3.86	9.71	12.12	14.44	24.65

The cost of the optimal alignment of **P** and **Q** is $DTW(\mathbf{P},\mathbf{Q}) = 24.65$. By tracing back (see Fig. 6.2) we find the corresponding optimal warp path:

$$\mathbf{W}_{\mathrm{DTW}} = ((1,1),(2,2),(3,3),(4,3),(5,4),(5,5),(5,6),(5,7),(6,8))^T .$$

The original use of the DTW algorithm was in speech recognition. Although more advanced speech-recognition techniques are now available, the DTW is still widely used in small-scale embedded-speech recognition systems, such as those embedded in cellular telephones, which require minimal hardware.

Example 6.4. Speech-Recognition in an Embedded-Speech Recognition System [1]. A DTW speech-recognition system works as follows. Let $\omega^{(l)}$ and $\mathbf{Q}^{(l)}, l \in \{1,2,\ldots,L\}$, denote, respectively, a set of L reference words and their corresponding time series. Given an unknown input time series **P** we use the DTW algorithm to optimally match **P** to each time series $\mathbf{Q}^{(l)}, l \in \{1,2,\ldots,L\}$, in turn. If $DTW(\mathbf{P},\mathbf{Q}^{(l)})$ denotes the cost of matching **P** with $\mathbf{Q}^{(l)}$, then we assign **P** to the word $\Omega = \omega_{\mathrm{OPT}}$, where

$$\omega_{\mathrm{OPT}} = \arg\min_l DTW(\mathbf{P},\mathbf{Q}^{(l)}) .$$

The next example illustrates the use of the DTW to match images.

Example 6.5. Word-Spotting in Historical Manuscripts [18]. Easy access to handwritten historical manuscripts requires an index similar to that found in the back of this book. The current approach - manual transcription followed by index generation from the transcript - is extremely expensive and time-consuming.

For collections of handwritten manuscripts written by a single author the images of multiple instances of the same word are likely to look similar. For such collections, the *word spotting* algorithm provides an alternative approach to index generation. This works as follows. Each page in the collection of manuscripts is segmented into words. Then different instances of a given word are clustered together using an image matching algorithm. By annotating the clusters we create an index which links words to the locations where they occur.

A critical issue in word spotting is matching the word images. We may use the DTW algorithm to match the word images as follows [18]: We divide each word image into columns of equal width Δ. In each column we extract four features, $(a_i, b_i, c_i, d_i)^T$, where i denotes the corresponding index of the column. In this manner we convert each word image into a multivariate series of feature values in which the horizontal image axis functions as a time axis. Once the conversion is complete we match any two word images **P** and **Q** by matching their multivariate feature values using the DTW algorithm, where local distance between the ith multivariate feature in P and the jth multivariate feature in Q is the squared Euclidean distance

$$d(i,j) = (a_i^{(P)} - a_j^{(Q)})^2 + (b_i^{(P)} - b_j^{(Q)})^2 + (c_i^{(P)} - c_j^{(Q)})^2 + (d_i^{(P)} - d_j^{(Q)})^2 .$$

The following example illustrates an application of DTW to multi-sensor image fusion.

Example 6.6. Generating a Digital Elevation Map (DEM) using DTW [8]. Let A and B denote two epipolar images taken from slightly different viewpoints. Then each row i in A corresponds to the same row i in B. If $\mathbf{r} = \big(a(i,1), a(i,2), \ldots, a(i,M)\big)^T$ and $\mathbf{s} = \big(b(i,1), b(i,2), \ldots, b(i,M)\big)^T$ denote, respectively, the pixel gray-values in the ith row in A and B, then we find the optimal warping path W_{DTW} between $\mathbf{r}$ and $\mathbf{s}$ using the DTW algorithm where the distance matrix $D(m,n)$ is defined as

$$D(m,n) = |r(m) - s(n)| .$$

The path D_{DTW} defines the *disparity* between pixels in $\mathbf{r}$ and in $\mathbf{s}$ from which the DEM may be calculated.

6.3.1 Derivative Dynamic Time Warping

Although the DTW has been successfully used in many domains, it can produce pathological results. The reason is the algorithm may try to explain a difference between **P** and **Q** by unnaturally warping the two time axes t and t'. This can lead to un-intuitive alignments. For example, a single point on one times series may map onto a large subsection of another time series (Fig. 6.3). One way of dealing with

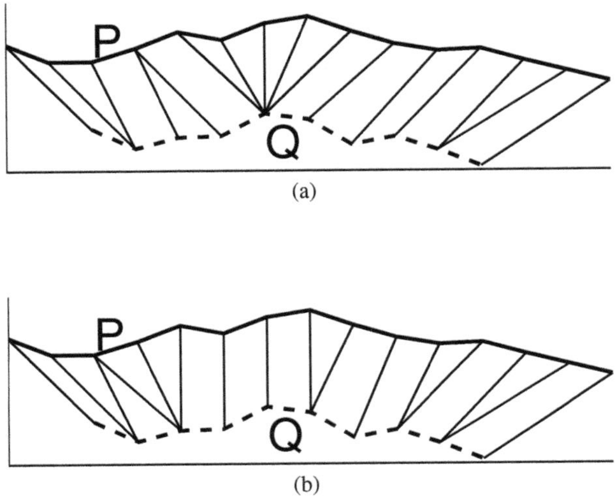

Fig. 6.3 **(a)** Shows the matching of similar time series **P** and **Q** using the traditional DTW algorithm with an Euclidean local cost function. **(b)** Shows the matching of the same curves using a DDTW algorithm.

these singularities is to place an additional constraint on **W**. Two constraints which are sometimes used for this purpose are:

Windowing. Locally we constrain **W** so it does not deviate too far from the diagonal path (as defined by the end points w_1 and w_K). The distance that the path is allowed to deviate is the window length R.

Slope Constraint. Locally we constrain **W** so it is not too steep and not too shallow. Mathematically we do this by replacing (6.4) by $D(i_k, j_k) = d(i_k, j_k) + \min(\alpha D(i_k, j_k - 1), \alpha D(i_k - 1, j_k), D(i_k - 1, j_k - 1))$, where α is positive real number.

An alternative approach is to use a different local cost function. In the derivative dynamic time warping (DDTW) algorithm [6] we use a local distance $d(i, j)$, which is defined as the square of the difference between the slopes of the curves **P** and **Q** at the times t_i and t'_j. Mathematically, the local distance is

$$d(i,j) = \left(\left.\frac{dP}{dt}\right|_i - \left.\frac{dQ}{dt'}\right|_j \right)^2 , \tag{6.7}$$

where $dP/dt|_{t_i}$ and $dQ/dt'|_{t'_j}$ are, respectively, the local slopes of **P** and **Q** at the times t_i and t'_j.

Example 6.7. Identifying a Face Profile Using the DDTW Algorithm [3]. Most current face profile recognition algorithms rely on the detection of fiducial points. Unfortunately features such as a concave nose, protrubing lips or a flat chin make the detection of such points difficult and unreliable. Also the number and position of the fiducial points vary when the expression changes even for the same person. As an alternative we may perform face profile recognition by matching an unknown face profile **P** to a data base of known face profiles $\mathbf{Q}^{(l)}, l \in \{1,2,\ldots,L\}$, using the DDTW algorithm.

6.3.2 Continuous Dynamic Time Warping

The continuous DTW [11, 12] is a continuous counter-part of the dynamic time warping algorithm in which an observation P_i is allowed to match a point which lies between two observations Q_j and Q_{j-1} and an observation Q_j is allowed to match a point which lies between two observations P_i and P_{i-1}. In order to generate the intermediate matching points we must assume a particular interpolation model. In [11, 12] the authors used a linear interpolation model. An alternative to the CDTW algorithm is to simply increase the number of points in **P** and **Q** by (linear) interpolation.

Example 6.8. Signature Verification Using Continuous Dynamic Time Warping [11, 12]. An area of increasing importance is the use of biometric techniques for personal verification in which we try to ascertain whether the unknown person is who he claims to be. Signature verification is one such technique. In [11, 12] the authors demonstrated the use of the continuous DTW in on-line signature verification. *Note:* In biometrics we distinguish between *identification* and *verification* tasks. In identification, the system identifies the unknown person by computing the similarities between the unknown person and all the known persons in the database D. In verification, the system verifies the unknown person by computing the similarity between the unknown person and the person he claims to be.

6.4 One-Sided DTW Algorithm

An important variant of the DTW algorithm occurs when we are only interested in warping one of the time series, say **Q**, onto the second time series, **P**. In this case, we use a warping path $\mathbf{V} = (v_1, v_2, \ldots, v_N)^T$, where $v_j = i$ specifies the mapping $Q_j \rightarrow P_i$. Two applications of this model are now considered.

6.4.1 Video Compression

In video compression we have a original time series **P** which consists of a sequence of M images $P_i, i \in \{1,2,\ldots,M\}$, taken at times t_i. This is compressed by forming a new time series **Q** which consists of N images $Q_j, j \in \{1,2,\ldots,N\}$, at times t'_j. The images Q_j are created by selectively

1. "Dropping" the input images P_i.
2. "Repeating" the input images P_i.
3. Interpolating consecutive input images P_{i-1} and P_i.

A simple model which describes all of the above effects is (6.8):

$$Q_j = \alpha_j P_i + (1-\alpha_j)P_{i-1} \;, \tag{6.8}$$

where $\alpha_j \in [0,1]$ is a given interpolation weight. The index i changes with the compressed image Q_j and its acquisition time t'_j. To show this explicitly we rewrite (6.8) using a warping path $\mathbf{V} = (v_1, v_2, \ldots, v_N)^T$, where $v_j = i$ denotes a mapping between the compressed image Q_j and a pair of original images P_{v_j} and P_{v_j-1}:

$$Q_j = \alpha_j P_{v_j} + (1-\alpha_j)P_{v_j-1} \;. \tag{6.9}$$

Example 6.9. Compressed Video Sequence [4]. Fig. 6.4 shows an original video sequence, $\mathbf{P} = \{P_1, P_2, \ldots, P_8\}$, which we compress by "frame drop", "frame repeat" and "frame interpolation". The corresponding compressed sequence is $\mathbf{Q} = \{Q_1, Q_2, \ldots, Q_4\}$, where Q_1 and Q_2 correspond, respectively, to the original images P_2 and P_4; Q_3 is a "repeat" of Q_2 (i. e. $Q_3 = Q_2$); Q_4 is a weighted average of the images P_7 and P_8; and we ignore, or "drop", the images P_1, P_3, P_5 and P_6 . *Note:* Images P_1, P_3, P_5, P_6 are "drop" images since no compressed images Q_j are mapped to P_1, P_3, P_5, P_6.

The mappings v_j and weights α_j are: $v_j = \{2,4,4,8\}$ and $\alpha_j = \{1.0, 1.0, 1.0, 0.3\}, j \in \{1,2,3,4\}$.

In general we are only interested in warping paths, **V**, which satisfy the following constraints:

Boundary Conditions. $v_1 = 2$ and $v_N = M$. This requires the warping path to start by matching the first compressed image, Q_1, with P_1 and P_2 and to end by matching the last compressed image, Q_N, with P_{M-1} and P_M.

Monoticity. Given $v_j = a$, then $v_{j-1} = a'$, where $a - a' \geq 0$. This forces the warping path to increase monotonically in time.

Note: Because of frame "dropping" we do *not* use a continuity constraint as is common in DTW.

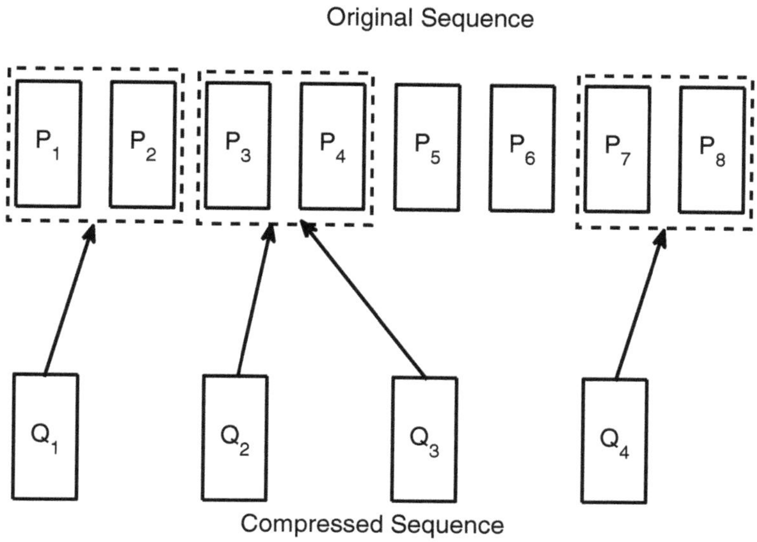

Fig. 6.4 Shows the use of the temporal model to represent a compressed video sequence.

There are exponentially many warping paths, $\mathbf{V}$, which satisfy the boundary and monoticity constraints, however, we are only interested in the path which minimizes the sum of the square errors between $\mathbf{Q}$ and $\mathbf{P}$. This jointly defines the optimal warping path, $\mathbf{V}_{OPT}$, and the optimal interpolation weights, α_{OPT}:

$$(\mathbf{V}_{OPT}, \alpha_{OPT}) = \arg\min\left(\sum_{j=1}^{N} |Q_j - (\alpha_j P_{v_j} + (1-\alpha_j)P_{v_j-1})|^2\right) . \tag{6.10}$$

Graphically, we may represent $\mathbf{V}_{OPT}$ as a monotonically non-decreasing path in a two-dimensional, (i, j), space. For example, the solution of Ex. 6.9 is the monotonically non-decreasing path shown in Fig. 6.5.

Apart from the boundary conditions and monoticity, we may include additional constraints which reflect *a priori* information concerning the application and what constitutes an optimal solution. For example, in video compression, frame repeat and frame drop are used infrequently and they are seldom repeated more than once. In [4] the authors included this information as an additional constraint C.

Example 6.10. Frame Repeat Constraint C [4]. In video compression, frame repeat is used infrequently and is seldom repeated more than once. This means that the most likely reason for a large number of similar frames is that the scenes are relatively static and not that frame repeat has been used multiple times. An appropriate constraint, or cost function, for this case is shown in Fig. 6.6. This cost function enforces:

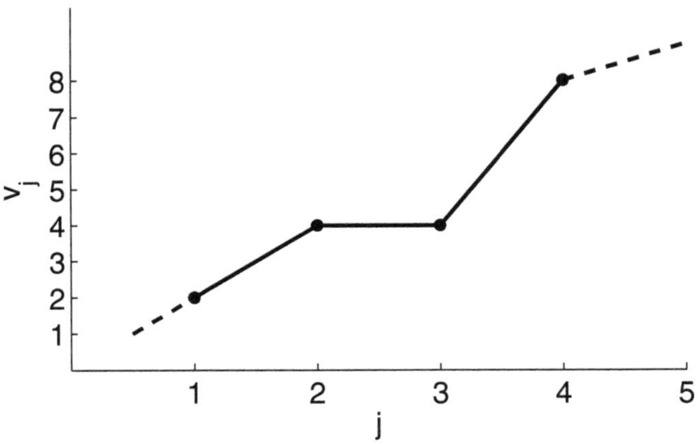

Fig. 6.5 Shows the monotonically non-decreasing path $\mathbf{V}_{\mathrm{OPT}}$ which represents the optimal matching indices v_j in Ex. 6.9.

1. The causal constraint by setting $C(\Delta v)$ to the maximal distortion when $\Delta v < 0$, i. e. when $v_j < v_{j-1}$.
2. A smooth transition among static scenes by setting $C(\Delta v) = 0$ when $\Delta v = 1$, i. e. when $v_j = v_{j-1} + 1$.
3. Penalizes frame repeat by setting $C(\Delta\alpha)$ equal to a positive cost when $\Delta v = 0$, i. e. when $v_j = v_{j-1}$.

As in the case of $\mathbf{W}_{\mathrm{OPT}}$, we may find $\mathbf{V}_{\mathrm{OPT}}$ using *dynamic programming*. By definition, the solution is a monotonically non-decreasing path $\mathbf{V}_{\mathrm{OPT}}$ that has the minimal sum of square errors. Let $D(n)$ denote the accumulated square error over an initial segment of $\mathbf{V}_{\mathrm{OPT}}$ from $j = 1$ to $j = n$, then

$$
\begin{aligned}
D(n) &= \min_{\substack{\alpha_1,\alpha_2,\ldots,\alpha_n \\ v_1 \le \ldots \le v_n}} \sum_{j=1}^{n} |Q_j - (\alpha_j P_{v_j} + (1-\alpha_j) P_{v_j-1})|^2 , \\
&= \min_{v_{n-1} \le v_n} \left\{ \min_{\substack{\alpha_1,\alpha_1,\ldots,\alpha_{n-1}, \\ v_1 \le \ldots \le v_{n-1}}} \sum_{j=1}^{n-1} |Q_j - (\alpha_j P_{v_j} + (1-\alpha_j) P_{v_j-1})|^2 \right. \\
& \left. + \min_{\alpha_n} |Q_n - (\alpha_n P_{v_n} + (1-\alpha_n) P_{v_n-1})|^2 \right\} , \\
&= D(n-1) + \min_{v_{n-1} \le V_n} \left\{ \min_{\alpha_n} |Q_n - (\alpha_n P_{v_n} + (1-\alpha_n) P_{v_n-1})|^2 \right\} . \qquad (6.11)
\end{aligned}
$$

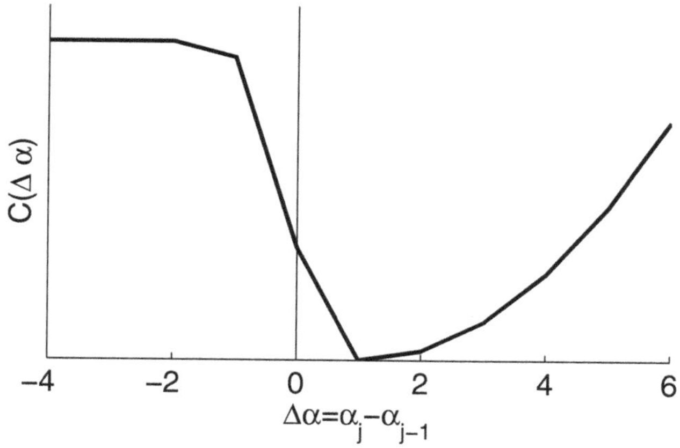

Fig. 6.6 Shows a contextual cost function $C(\Delta\alpha)$, where $\Delta\alpha = \alpha_j - \alpha_{j-1}$. This cost function enforces causal constraint and encourages smooth transition when frames have little motion or are static and penalizes the decision of frame repeat.

Inspection of (6.11) shows that $\mathbf{V}_{\mathrm{OPT}}$ and $\boldsymbol{\alpha}_{\mathrm{OPT}}$ may be efficiently found using dynamic programming as follows:

1. Find the optimum weight, α_j, for each compressed image $Q_j, j \in \{1,2,\ldots,N\}$:

$$\alpha_j = \arg\min_{\alpha_j} |Q_j - (\alpha_j P_{v_j} + (1-\alpha_j) P_{v_j-1})|^2 \,. \tag{6.12}$$

2. Recursively compute $D(j), j \in \{1,2,\ldots,N\}$, using the optimum weight α_j, in (6.11).
3. After completing the calculation of $D(N)$, trace back to find the optimum warping path, $\mathbf{V}_{\mathrm{OPT}}$, and the optimum weight vector, $\boldsymbol{\alpha}_{\mathrm{OPT}}$.

6.4.2 Video Denoising

We consider a an efficient method for video denoising which is based on the the one-sided DTW algorithm [2]. Although we can apply static image denosing methods to the case of image sequences (intraframe methods) we can do much better by including temporal information (interframe methods). This temporal information is crucial since our perception is very sensitive to temporal distortions like edge displacement: the disregard of temporal information may lead to inconsistencies in the result.

The input to the denoising algorithm is a video sequence of $M \times N$ images, or frames, $I^t, t \in \{1,2,\ldots\}$. We partition each frame I^t into a disjoint set of horizontal lines $\mathscr{L}_i^t = \{I^t(i,1), I^t(i,2), \ldots, I^t(i,N)\}$, where each line is of length N pixels. For each line $\mathscr{L}_i^t$ we consider the family of lines which are close to $\mathscr{L}_i^t$ in the same frame and in the neighouring frames. We warp each of these lines so they match with $\mathscr{L}_i^t$. Let $\phi(\mathscr{L}_m^s)$ denote the warped version of the line $\mathscr{L}_m^s$ onto the line $\mathscr{L}_i^t$. We then obtain a denoised version of $\mathscr{L}_i^t$ by performing an average (mean, weighted mean or median) of the lines $\phi(\mathscr{L}_m^s)$.

6.5 Multiple Time Series

Until now we have limited ourselves to the temporal alignment of two time series P and Q. We now consider how we may perform the simultaneous temporal alignment of N time series $P_1, P_2, \ldots, P_N$. Specifically, our aim is to create an average time series Q such that the sum of the squared DTW distances from Q to $P_1, P_2, \ldots, P_N$ is minimal. One algorithm [15]for doing this is the DBA algorithm.

Example 6.11. Dynamic Barycenter Averaging (DBA) [15]. Let $P_1, P_2, \ldots, P_N$ denote a set of N one-dimensional time series, where $P_n(i)$ denotes the ith observation in P_n. Let Q be an estimate of the average time series, where $Q(j)$ denotes the jth observation in Q. We compute the DTW between each time series $P_n, n \in \{1, 2, \ldots, N\}$ and Q. Each DTW gives the associations between $P_n(i)$ and $Q(j)$. Then we update $Q(j)$ as the average of those observations $P_n(i)$ which are associated with $Q(j)$. The process is repeated until convergence.

6.6 Software

DTW. A matlab m-file for performing dynamic time warping. Author: Timothy Felty.

6.7 Further Reading

Temporal alignment has been intensely investigated over the years. In recent years it has been increasingly used in various multi-sensor data fusion applications which involve medical imaging [7, 9], speech recognition [1, 16], image sequence compression or watermark tracking in video and digital cinema [4, 5, 10]. The DTW has also been used in applications which do not explicitly involve temporal alignment but require curve matching in one form or another. Examples of these applications include: word spotting, identifying facial profiles and signature verification.

Problems

6.1. Describe the process of dynamic time warping. Explain the role the boundary conditions and the constraints of continuity and monoticity play in the algorithm.

6.2. Describe the derivative DTW algorithm. Suggest when DDTW should be used instead of DTW.

6.3. Describe the continuous DTW algorithm.

6.4. Show how we may use the framework of DTW to explain video compression.

References

1. Abdulla, W.H., Chow, D., Sin, G.: Cross-words reference template for DTW-based speech recognition systems. In: Proc. IEEE Tech. Conf., TENCON 2003 (2003)
2. Bertalmio, M., Casselles, V., Pardo, A.: Movie denoising by average of warped lines. IEEE Trans. Imag. Proc. 16, 2333–2347 (2007)
3. Bhabu, B., Zhou, X.: Face recognition from face profile using dynamic time warping. In: Proc. 17th Int. Conf. Patt. Recogn. (ICPR 2004), vol. 4, pp. 499–502 (2004)
4. Cheng, H.: Temporal registration of video sequences. In: Proc. ICASSP 2003, Hong Kong, China (2003)
5. Cheng, H.: A review of video registration for watermark detection in digital cimema applications. In: Proc. ISCAS (2004)
6. Keogh, E.J., Pazzani, M.J.: Derivative dynamic time warping. In: 1st SIAM Int. Conf. Data Mining (2001)
7. Klein, G.J.: Four-dimensional processing of deformable Cardiac PET data. Med. Image. Anal. 6, 29–46 (2002)
8. Krauss, T., Reinartz, Lehner, M., Schroeder, M., Stilla, U.: DEM generation from very high resolution stereo satellite data in urban areas using dynamic programming. In: Proc. ISPRS (2005)
9. Ledesma-Carbayo, M.J., Kybic, J., Desco, M., Santos, A., Suhling, M., Hunziker, P., Unser, M.: Spatio-temporal nonrigid registration for ultrasound cardiac motion estimation. IEEE Trans. Med. Imag. 24, 1113–1126 (2005)
10. Lubin, J., Bloom, J.A., Cheng, H.: Robust, content-dependent, high-fidelity watermark for tracking in digital cinema. In: Proc. SPIE, vol. 5020 (2003)
11. Munich, M.E.: Visual input for pen-based computers. PhD thesis, California Institute of Technology, Pasadena, California, USA (2000)
12. Munich, M.E., Perona, P.: Continuous dynamic time warping for translation-invariant curve alignment with application to signature verification. In: Proc. 7th Int. Conf. Comp. Vis. (ICCV 1999), Corfu, Greece (1999)
13. Perperidis, D.: Spatio-temporal registration and modeling of the heart using cardiovascular MR imaging. PhD thesis, University of London (2005)
14. Perperidis, D., Mohiaddin, R., Rueckert, D.: Spatio-temporal free-form registration of cardiac MR image sequences. Med. Imag. Anal. 9, 441–456 (2005)
15. Petitjean, F., Ketterlin, A., Gancarski, P.: A global averaging method for dynamic time warping with applications to clustering. Patt. Recogn. 44, 678–693 (2011)
16. Rabiner, L., Juang, B.: Fundamentals of Speech Processing. Prentice-Hall (1993)
17. Ratanahatano, C.A., Keogh, E.: Three myths about dynamic time warping. In: SIAM 2005 Data Mining Conf., CA, USA (2005)
18. Rath, T.M., Manmatha, R.: Features for word spotting in historical manuscripts. In: Proc. 7th Int. Conf. Document Anal. Recogn. (ICDAR), Edinburgh, Scotland, vol. 1, pp. 218–222 (2003)
19. Singh, M., Thompson, R., Basu, A., Rieger, J., Mandal, M.: Image based temporal registration of MRI data for medical visualization. In: IEEE Int. Conf. Image. Proc., Atlanta, Georgia (2006)

Chapter 7
Semantic Alignment

7.1 Introduction

The subject of this chapter is semantic alignment. This is the conversion of multiple input data or measurements which do not refer to the same object, or phenomena, to a common object or phenomena. The reason for performing semantic alignment is that different inputs can only be fused together if the inputs refer to the same object or phenomena. In general, if the observations have been made by sensors of the same type, then the observations should refer to the same object or phenomena. In this case, no semantic alignment is required, although radiometric normalization may be required.

Example 7.1. Multiple Edge Maps. Consider an input image on which we separately apply a Sobel edge detector and a Canny edge detector. The two detectors work on different principles but both measure the presence, or absence, of an edge in the input image. The corresponding edge images, F_{sobel} and F_{canny}, are said to be "semantically equivalent" and no semantic alignment is required. However, in general, F_{sobel} and F_{canny} do not use the same radiometric scale, and radiometric normalization may be required before they can be fused together.

However, when the sensors are of different types, the observations may refer to different phenomena. If this is true, then it is not possible to perform data fusion unless the input data are semantically aligned to a common object or phenomena. The following example illustrates the semantic alignment of several inputs, each of which refers to a different phenomena.

Example 7.2. Multi-Feature Infra-Red Target Detection in an Input Image [20]. We consider the detection of a small target in an infra-red input image I. At each pixel (m,n) in I we test for the presence of a target by extracting the following local features:

1. $F^{(1)}$: Maximum Gray Level.
2. $F^{(2)}$: Contrast Mean Difference.
3. $F^{(3)}$: Average Gradient.
4. $F^{(4)}$: Gray-level Variation.
5. $F^{(5)}$: Entropy.

The features $F^{(k)}, k \in \{1,2,\ldots,5\}$, clearly do not measure the same phenomena. However, according to the theory of target detection in an infra-red image, they are all causally linked to the presence of a target. In this case, we may semantically align the features by converting $F^{(k)}(x,y), k \in \{1,2,\ldots,K\}$, into evidence that a target is present at (x,y).

In some cases, the inputs may refer to the same phenomena but each sensor uses a different set of names or symbols. In this case, we must semantically align the inputs so that for each phenomena all the sensors will use the same names or symbols. This is commonly known as "solving the label correspondence problem".

Example 7.3. Label correspondence. We consider an input image $\mathbf{X} = (x_1,x_2,\ldots,x_N)^T$, which we segment using two different algorithms. The first algorithm, $\mathfrak{A}$, partitions the pixels $\{x_1,x_2,\ldots,x_N\}$ into K clusters which are labeled $\{\alpha_1,\alpha_2,\ldots,\alpha_K\}$. The corresponding segmented image is $\mathbf{A} = (a_1,a_2,\ldots,a_N)^T$, where $a_i = k$ if $\mathfrak{A}$ assigns the pixel x_i to the kth cluster α_k. Similarly, the second algorithm, $\mathfrak{B}$, partitions the pixels $\{x_1,x_2,\ldots,x_N\}$ into L clusters which are labeled $\{\beta_1,\beta_2,\ldots,\beta_L\}$. The corresponding segmented image is $\mathbf{B} = (b_1,b_2,\ldots,b_N)^T$, where $b_i = l$ if $\mathfrak{B}$ assigns x_i to the lth cluster β_l. Then, one way of semantically aligning $\mathbf{A}$ and $\mathbf{B}$ is to find which labels α_k in $\mathbf{A}$ are associated with which labels β_l in $\mathbf{B}$ and vice versa. A convenient way of defining the associations is through an assignment matrix λ, where

$$\lambda(k,l) = \begin{cases} 1 \text{ if labels } \alpha_k \text{ and } \beta_l \text{ are associated with each other} , \\ 0 \text{ otherwise.} \end{cases}$$

In many multi-sensor data fusion applications semantic alignment is the primary fusion algorithm. In Table 7.1 we list some of these applications together with the classification of the fusion algorithm involved. We now consider the concept of an assignment matrix in detail and the essential role it plays in semantic alignment.

Table 7.1 Applications in which Semantic Alignment is the Primary Fusion Algorithm

Class		Application
FeI-FeO	Ex. 5.10	Point-based image registration.
	Ex. 7.5	Handwritten character recognition.
	Ex. 7.6	Closed contour matching using the shape context.
	Ex. 11.15	Target tracking initialization using the Hough transform.
	Ex. 12.8	Tracking pedestrians across non-overlapping cameras.
DeI-DeO	Ex. 7.9	Image segmentation via the method of cluster ensembles.
	Ex. 7.10	Color image segmentation.

The designations FeI-FeO and DeI-DeO refer, respectively, to Dasarathy input/output classifications: "Feature Input-Feature Output" and "Decision Input-Decision Output" (Sect. 1.3.3).

7.2 Assignment Matrix

If $C(k,l)$ denotes the cost of matching a label α_k with a label β_l, then the *optimal* assignment matrix $\widetilde{\lambda}$ is defined as the assignment matrix with the minimum overall cost:

$$\widetilde{\lambda} = \arg\min \sum_{k,l} C(k,l)\lambda(k,l) , \tag{7.1}$$

where $C(k,l)$ is the cost of associating label α_k with label β_l. In applications, in which each label represents a different physical object, it is common practice to use a distance measure for $C(k,l)$ and to solve (7.1) subject to the following one-to-one constraints:

$$\sum_{k=1}^{K} \lambda(k,l) \leq 1 , \tag{7.2}$$

$$\sum_{l=1}^{L} \lambda(k,l) \leq 1 . \tag{7.3}$$

Fast algorithms for finding the optimal assignment matrix $\widetilde{\lambda}$ are available. Among them is the Hungarian algorithm [12] which is widely used in many applications. Traditionally these algorithms are used when $K = L$ and we require that each label $\alpha_k, k \in \{1,2,\ldots,K\}$, is matched to a corresponding label $\beta_l, l \in \{1,2,\ldots,L\}$, and vice versa. The algorithms may, however, be used when $K \neq L$, or when we wish to make the assignment robust against outliers by finding the best $M, M \leq \min(K,L)$, associations[1]. In this case we use an enlarged cost matrix C_e:

$$C_e(k,l) = \begin{cases} C(k,l) & \text{if } k \in \{1,2,\ldots,K\}, l \in \{1,2,\ldots,L\} , \\ P & \text{otherwise} , \end{cases}$$

[1] This case is used as a guard against outliers.

where P is the cost, or penalty, of not associating a label α_k with any label β_l or of not associating a label β_l with any label α_k.

In some applications it may not be possible to define a penalty P. In this case, we may use the following algorithm to find the sub-optimal label permutation.

Example 7.4. Sub-Optimal Assignment Algorithm. Given a cost matrix $C(k,l), k \in \{1,2,\ldots,K\}, l \in \{1,2,\ldots,L\}$, we find the best M one-to-one associations, where $M \leq \min(K,L)$, using the following pseudo code:

1. Find the association pair (k_1,l_1) with the smallest cost.
2. Find the association pair (k_2,l_2) with the second smallest cost, where $k_2 \neq k_1$ and $l_2 \neq l_1$.
3. Find the association pair (k_3,l_3) with the third smallest cost, where $k_3 \neq \{k_1,k_2\}$ and $l_3 \neq \{l_1,l_2\}$.
4. Continue this process until we have M pairs: $(k_1,l_1),(k_2,k_2),\ldots,(k_M,l_M)$.

In many applications, solving the assignment problem is the fusion algorithm itself.

Example 7.5. Handwritten Character Recognition [1]. Shape matching is a powerful method for automatic handwritten character recognition. In this technique we match the contour of an unknown character with the contour of a known character. Let A and B denote the two contours. We sample the contour A at K points $a_k, k \in \{1,2,\ldots,K\}$. We regard each point a_k as a "physical" object with its own unique label α_k. Similarly, B is sampled at L points $b_l, l \in \{1,2,\ldots,L\}$ each of which has its own label β_l. For each pair of labels (α_k,β_l), i. e. points (a_k,b_l), we let

$$C(k,l) = d(S_k,S_l) ,$$

where S_k and S_l are, respectively, the shape contexts (see Ex. 4.13) of the points a_k and b_l and $d(S_k,S_l)$ is the χ^2 distance metric. If $\widetilde{\lambda}$ denotes the optimal one-to-one assignment matrix, then we may use

$$\widetilde{C} = \sum_{(k,l)} C(k,l)\widetilde{\lambda}(k,l) ,$$

as a similarity measure: the smaller $\widetilde{C}$ the more likely A and B are the same character.

In some applications we may impose additional constraints on (7.1). For example, in the above example, we may require the order of the points to be preserved [15].

Example 7.6. Closed Contour Matching using the Shape Context [15]. Suppose the point a_k and b_l in Ex. 7.5 are labeled in increasing order (see Fig. 7.1). Then, we match the two contours A and B assuming:

1. No point a_k may be paired with more than one point b_l and no point b_l may be paired with more than one point a_k. This is the one-to-one constraint (Eqns. 7.2 and 7.3).
2. If $b_l = \pi(a_k)$ and $b_{l'} = \pi(a_{k'})$, then $k' > k$ if $l' > l$, where $b_l = \pi(a_k)$ denotes that the point a_k is matched with b_l and $\pi(a_k) = 0$ denotes that the point a_k is not matched to any of the points b_l. This is an increasing-order constraint.

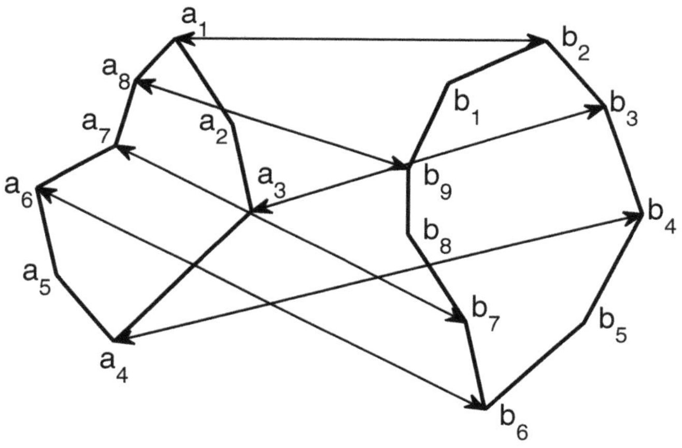

Fig. 7.1 Illustrates the matching of two closed contours $\mathbf{A} = (a_1, a_2, \ldots, a_8)^T$ and $\mathbf{B} = (b_1, b_2, \ldots, b_9)^T$. The optimum matching is $\pi(1,2,3,4,5,6,7,8) = (2,0,3,4,0,6,7,9)$, i.e. point $a_1 \rightarrow b_2$, $a_3 \rightarrow b_3, \ldots, a_8 \rightarrow b_9$, while points a_2 and a_5 are not associated with any points on $\mathbf{B}$ and points b_1, b_5 and b_8 are not associated with any points on $\mathbf{A}$.

Sometimes we may solve (7.1) subject to only one of the constraints (7.2) or (7.3). This may occur when the sensors have different resolutions: For example, if sensor A has a low resolution, then a point a_m may be associated with several points $\{b_{n'}, b_{n''}, \ldots\}$. Similarly, if sensor B has a low resolution, then a point b_n may be associated with several points $\{a_{m'}, a_{m''}, \ldots\}$.

In Exs. 7.5 and 7.6, the labels α_k and β_l represented physical objects. We now turn our attention to the case when the labels represent symbolic class names, and not specific physical objects.

7.3 Clustering Algorithms

Clustering algorithms constitute the most popular method for partitioning a set of objects, $\{O_1, O_2, \ldots, O_N\}$, into K classes. Of the different clustering algorithms which are available, the K-means algorithm [9] is the workhorse. This algorithm is described in the next example.

Example 7.7. K-means cluster algorithm [9]. The K-means cluster algorithm works in an iterative fashion as follows. Let $\{O_1, O_2, \ldots, O_N\}$ denote a set of N objects. Each object is characterized by L features. We may thus represent each object O_n as a point $O_n = (a_n, b_n, \ldots, d_n)^T$ in L-dimensional feature space. The cluster algorithm partitions the objects into K clusters, in which each cluster is represented as a point, $P_k = (A_k, B_k, \ldots, D_k)^T, k \in \{1, 2, \ldots, K\}$, in L-dimensional feature space. Let $dist(O_n, P_k)$ denote the distance between $P_k = (A_k, B_k, \ldots, D_k)^T$ and $O_n = (a_n, b_n, \ldots, d_n)^T$. If $P_k^{(r)}$ denotes the kth cluster at the rth iteration, then at the next iteration $P_k^{(r+1)} = (A_k^{(r+1)}, B_k^{(r+1)}, \ldots, D_k^{(r+1)})^T$, where

$$A_k^{(r+1)} = A_k^{(r)} + \frac{\sum_{n=1}^{N} \lambda(n,k) a_n}{\sum_{n=1}^{N} \lambda(n,k)} ,$$

$$B_k^{(r+1)} = B_k^{(r)} + \frac{\sum_{n=1}^{N} \lambda(n,k) b_n}{\sum_{n=1}^{N} \lambda(n,k)} ,$$

$$\vdots$$

$$D_k^{(r+1)} = D_k^{(r)} + \frac{\sum_{n=1}^{N} \lambda(n,k) d_n}{\sum_{n=1}^{N} \lambda(n,k)} ,$$

and

$$\lambda(n,k) = \begin{cases} 1 \text{ if } P_k^{(r)} \text{ is the cluster representative point which is closest to } O_n , \\ 0 \text{ otherwise} . \end{cases}$$

In recent years spectral clustering algorithms have become increasingly popular. These techniques are known to perform well in cases where classical clustering methods such as the K-means fail. The idea behind spectral clustering is to map the observations $O_i, i \in \{1, 2, \ldots, N\}$, into a K-dimensional feature space in which the observations are easily partitioned into K groups. The K-dimensional feature space is formed by extracting the eigenvectors of a normalized affinity matrix. The following example illustrates the NJW spectral clustering algorithm including the construction of the affinity matrix $\mathscr{A}$.

Example 7.8. NJW spectral clustering algorithm [14, 22]. Let K be the desired number of clusters. Then, given a set of observations, $\{O_1, O_2, \ldots, O_N\}$, we proceed as follows:

(1). Form an $N \times N$ affinity matrix $\mathscr{S}$:

$$\mathscr{A}(i,j) = \exp\left(-\frac{dist^2(O_i,O_j)}{\sigma^2}\right),$$

where $dist(O_i,O_j)$ is the distance between O_i and O_j and σ is a global scale parameter defined for the observations $O_i, i \in \{1,2,\ldots,N\}$. A convenient formula [2] for σ is

$$\sigma = \frac{1}{N}\sum_{i=1}^{N} dist(O_i,O_{(1)}),$$

where $O_{(M)}$ denotes the Mth nearest neighbour of O_i. In place of a global scale parameter, Zelnik-Manor and Perona [22], recommend using a local scale σ_i, where

$$\sigma_i = dist(O_i,O_{(M)}).$$

In this case,

$$\mathscr{A}(i,j) = \exp\left(-\frac{dist^2(O_i,O_j)}{\sigma_i\sigma_j}\right).$$

(2). Set the diagonal elements $\mathscr{A}(i,i), i \in \{1,2,\ldots,N\}$, to zero:

$$\mathscr{A}(i,i) = 0.$$

(3). Construct a normalized $N \times N$ affinity matrix $\mathscr{N}$:

$$\mathscr{N}(i,j) = D^{-1/2}\mathscr{A}D^{-1/2},$$

where D is a $N \times N$ diagonal degree matrix:

$$D(i,j) = \begin{cases} \Sigma_{h=1}^{N}\mathscr{A}(i,h) & \text{if } j = i, \\ 0 & \text{otherwise}. \end{cases}$$

Thus

$$\mathscr{N}(i,j) = \begin{cases} \mathscr{A}(i,i)/D(i,i) & \text{if } j = i, \\ \mathscr{A}(i,j)/\sqrt{D(i,i) \times D(j,j)} & \text{if } j \neq i. \end{cases}$$

(4). Calculate the eigenvector solutions of $\mathscr{N}\mathbf{u} = \lambda\mathbf{u}$. If $\mathbf{u}_k$ is the eigenvector of $\mathscr{N}$ with the kth largest eigenvalue, then $U = (\mathbf{u}_1, \mathbf{u}_2, \ldots, \mathbf{u}_K)$. For numerical stability, Verma and Meila [18] recommend defining $\mathbf{u}_k$ as the solution of the generalized eigenvector equation $\mathscr{A}\mathbf{u} = \lambda D\mathbf{u}$ with the kth largest eigenvalue.

(5). Renormalize the rows of U to have unit length. This yields a $N \times K$ matrix V, where

$$V(i,j) = U(i,j) \Big/ \sqrt{\sum_{h=1}^{K} U^2(i,h)} \, .$$

(6). Treat each row of V as a K-dimensional point. Cluster the N rows with the K-means algorithm.

(7). Assign the object O_i to the kth cluster if, and only if, the corresponding row i of the matrix V was assigned to the kth cluster.

Spectral clustering is expensive both regarding computational cost and storage requirements: For N observations, the complexity of the eigenvector calculation is $O(N^3)$ and the the size of the affinity matrix is $N \times N$. In order to reduce the cost and storage we may perform spectral clustering using the Nystrom approximation [4].

The main drawback to the K-means and the NJW clustering algorithms is the need to define the number of partitions K. One effective method for finding K is described in [22].

In the next section we show how we may perform semantic alignment on cluster ensembles.

7.4 Cluster Ensembles

Cluster ensembles is a subject of increasing importance and involves the semantic alignment, or label correspondence, of symbolic labels.

Example 7.9. Image Segmentation via the Method of Cluster Ensembles [10, 7]. Image segmentation is a critical task in image fusion. In order to deal with the great variability of features encountered in different images, specific segmentation methods have been designed for different types of images. In [7] we apply several different segmentation algorithms (or the same segmentation algorithm with different parameters) to the same input image. We then combine the multiple segmented images into a single consensus segmented image. In general the consensus segmented image is found to have better overall quality than any of the input segmented images.

Formally, we may define the method of cluster ensembles as follows. Let $O = \{O_1, O_2, \ldots, O_N\}$ denote a set of N objects. We partition the objects using an ensemble of M cluster algorithms $\{\mathfrak{A}, \mathfrak{B}, \ldots, \mathfrak{D}\}$. The algorithm $\mathfrak{A}$ partitions the N objects into clusters: $\{\alpha_1, \alpha_2, \ldots\}$. The corresponding identity vector is $\mathbf{A} = (a_1, a_2, \ldots, a_N)^T$, where $a_n = \alpha_k$ if O_n is assigned to the kth cluster $\alpha_k, k \in \{1, 2, \ldots\}$. Similarly, algorithms $\mathfrak{B}$ and $\mathfrak{D}$ partition the objects into clusters $\{\beta_1, \beta_2, \ldots\}$ and $\{\delta_1, \delta_2, \ldots\}$. The corresponding identity vectors are $\mathbf{B} = (b_1, b_2, \ldots)^T$ and $\mathbf{D} = (d_1, d_2, \ldots)^T$, where $b_n = \beta_l$ if $\mathfrak{B}$ assigns O_n to cluster β_l and $d_n = \delta_h$ if $\mathfrak{D}$ assigns O_n to δ_h. *Note*: Each cluster algorithm may generate a different number of clusters. Finally, we combine the identity vectors $\{\mathbf{A}, \mathbf{B}, \ldots, \mathbf{D}\}$ into a single "consensus" vector $\widetilde{\mathbf{Y}} = (\widetilde{y}_1, \widetilde{y}_2, \ldots, \widetilde{y}_N)^T$.

In order to find the vector $\widetilde{\mathbf{Y}}$ we need to transform the individual vectors $\{\mathbf{A}, \mathbf{B}, \ldots, \mathbf{D}\}$, into a common representational format. One way to do this is to designate one of the cluster algorithms as a "reference" algorithm $\mathfrak{R}$. If $\{\phi_1, \phi_2, \ldots\}$ denotes the clusters associated with $\mathfrak{R}$, then we calculate the optimum assignment matrices $\{\lambda_A(k,r), \lambda_B(k,r), \ldots, \lambda_D(k,r)\}$ for the remaining algorithms $\{\mathfrak{A}, \mathfrak{B}, \ldots, \mathfrak{D}\}$, where

$$\lambda_A(k,r) = \begin{cases} 1 \text{ if } \alpha_k \text{ corresponds to } \phi_r \ , \\ 0 \text{ otherwise} \ , \end{cases}$$

$$\lambda_B(k,r) = \begin{cases} 1 \text{ if } \beta_k \text{ corresponds to } \phi_r \ , \\ 0 \text{ otherwise} \ , \end{cases}$$

$$\vdots$$

$$\lambda_D(k,r) = \begin{cases} 1 \text{ if } \delta_k \text{ corresponds to } \phi_r \ , \\ 0 \text{ otherwise} \ . \end{cases}$$

A convenient cost function which may be used to calculate $\{\lambda_A(k,r), \lambda_B(k,r), \ldots, \lambda_D(k,r)\}$, is to maximize the overlap of $\{\mathbf{A}, \mathbf{B}, \ldots, \mathbf{D}\}$ with $\mathbf{R} = (r_1, r_2, \ldots, r_N)^T$. Alternatively, we may side-step the need to calculate the pairwise assignment matrices $\{\lambda_A(k,r), \lambda_B(k,r), \ldots, \lambda_D(k,r)\}$ (and the need to define an appropriate cost function) as follows:

Concatenation. For each object $O_n, n \in \{1, 2, \ldots, N\}$, we form a feature vector $\mathbf{f}_n$ by concatenating the identities $a_n, b_n, \ldots, d_n$:

$$\mathbf{f}_n = (a_n, b_n, \ldots, d_n)^T \ . \tag{7.4}$$

We may then cluster the vectors $\mathbf{f}_n, n \in \{1, 2 \ldots, N\}$, to give a consensus identity vector $\widetilde{\mathbf{Y}}$. See Ex. 7.10 for an example of this technique.

Co-Association Matrix. For each cluster algorithm $\{\mathfrak{A}, \mathfrak{B}, \ldots, \mathfrak{D}\}$ we rewrite the identity vectors $\{\mathbf{A}, \mathbf{B}, \ldots, \mathbf{D}\}$ as a set of $N \times N$ co-association matrices $\{\mathscr{A}, \mathscr{B}, \ldots, \mathscr{D}\}$. The co-association matrices are special similarity matrices which all use the same binary language. We then fuse together the matrices $\{\mathscr{A}, \mathscr{B}, \ldots, \mathscr{D}\}$ and generate the fused co-association matrix $\widetilde{\mathscr{Y}}$. Finally, we perform spectral analysis on $\widetilde{\mathscr{Y}}$ to generate the corresponding consensus identity vector $\widetilde{\mathbf{Y}}$. This is described in more detail in Sect. 7.4.1.

Example 7.10. Color Image Segmentation [13]. Given a color *RGB* input image we transform it into M different color spaces. In each color space we segment the pixels into $K_m, m \in \{1,2,\ldots,M\}$, clusters using a given cluster algorithm. As a result, each pixel (i,j) is represented by a M-dimensional feature vector $\mathbf{f}(i,j)$:

$$\mathbf{f}(i,j) = \left(y^{(1)}(i,j), y^{(2)}(i,j), \ldots, y^{(M)}(i,j)\right) ,$$

where $y^{(m)}(i,j) = k_m$ if in the mth color space, the cluster algorithm assigns the pixel (i,j) to the k_mth cluster. The M-dimensional feature map $\mathbf{f}(i,j)$ is then segmented using a K-means cluster algorithm (see Ex. 7.7). Using this technique, Mignotte [13] obtained state-of-the-art image segmentation performance

7.4.1 Co-association Matrix

Co-association matrices are special similarity measures which enable us to fuse together multiple identity vectors $\{\mathbf{A},\mathbf{B},\ldots,\mathbf{D}\}$ without calculating an optimum assignment matrix. In this method, we convert each identity vector into a corresponding *co-association* matrix [5, 6], as follows:

$$\mathscr{A}(i,j) = \begin{cases} 1 \text{ if } a_i = a_j , \\ 0 \text{ otherwise} . \end{cases}$$

$$\mathscr{B}(i,j) = \begin{cases} 1 \text{ if } b_i = b_j , \\ 0 \text{ otherwise} . \end{cases}$$

$$\vdots$$

$$\mathscr{D}(i,j) = \begin{cases} 1 \text{ if } d_i = d_j , \\ 0 \text{ otherwise} . \end{cases}$$

The co-association matrices all use the same binary language and may therefore be fused together without further processing. This is illustrated in the following example in which we form a mean co-association matrix $\overline{\mathscr{Y}}$ from a set of (binary) co-association matrices $\{\mathscr{A},\mathscr{B},\ldots,\mathscr{D}\}$.

Example 7.11. Mean co-association matrix [21]. Let $\mathbf{X} = (x_1, x_2, \ldots, x_7)^T$ denote an input vector. We partition $\mathbf{X}$ using three different cluster algorithms $\{\mathfrak{A}, \mathfrak{B}, \mathfrak{C}\}$. The resulting identity vectors are:

$$\mathbf{A} = (\,\alpha_1\ \alpha_1\ \alpha_2\ \alpha_2\ \alpha_2\ \alpha_3\ \alpha_3\,)^T$$
$$\mathbf{B} = (\,\beta_3\ \beta_3\ \beta_2\ \beta_2\ \beta_3\ \beta_3\ \beta_1\,)^T\ ,$$
$$\mathbf{C} = (\,\gamma_2\ \gamma_3\ \gamma_2\ \gamma_2\ \gamma_1\ \gamma_1\ \gamma_1\,)^T\ ,$$

and the corresponding co-association matrices are:

$$\mathscr{A} = \begin{pmatrix} 1\,1\,0\,0\,0\,0\,0 \\ 1\,1\,0\,0\,0\,0\,0 \\ 0\,0\,1\,1\,1\,0\,0 \\ 0\,0\,1\,1\,1\,0\,0 \\ 0\,0\,1\,1\,1\,0\,0 \\ 0\,0\,0\,0\,0\,1\,1 \\ 0\,0\,0\,0\,0\,1\,1 \end{pmatrix}\ , \quad \mathscr{B} = \begin{pmatrix} 1\,1\,0\,0\,1\,1\,0 \\ 1\,1\,0\,0\,1\,1\,0 \\ 0\,0\,1\,1\,0\,0\,0 \\ 0\,0\,1\,1\,0\,0\,0 \\ 1\,1\,0\,0\,1\,1\,0 \\ 1\,1\,0\,0\,1\,1\,0 \\ 0\,0\,0\,0\,0\,0\,1 \end{pmatrix}\ ,$$

$$\mathscr{C} = \begin{pmatrix} 1\,0\,1\,1\,0\,0\,0 \\ 0\,1\,0\,0\,0\,0\,0 \\ 1\,0\,1\,1\,0\,0\,0 \\ 1\,0\,1\,1\,0\,0\,0 \\ 0\,0\,0\,0\,1\,1\,1 \\ 0\,0\,0\,0\,1\,1\,1 \\ 0\,0\,0\,0\,1\,1\,1 \end{pmatrix}\ .$$

The co-association matrices all use the same binary language and may therefore be fused together by summing them together. The fused mean co-association matrix is

$$\overline{\mathscr{Y}} = \begin{pmatrix} 1 & \frac{2}{3} & \frac{1}{3} & \frac{1}{3} & \frac{1}{3} & \frac{1}{3} & 0 \\ \frac{2}{3} & 1 & 0 & 0 & \frac{1}{3} & \frac{1}{3} & 0 \\ \frac{1}{3} & 0 & 1 & 1 & \frac{1}{3} & 0 & 0 \\ \frac{1}{3} & 0 & 1 & 1 & \frac{1}{3} & 0 & 0 \\ \frac{1}{3} & \frac{1}{3} & \frac{1}{3} & \frac{1}{3} & 1 & \frac{2}{3} & \frac{1}{3} \\ \frac{1}{3} & \frac{1}{3} & 0 & 0 & \frac{2}{3} & 1 & \frac{2}{3} \\ 0 & 0 & 0 & 0 & \frac{1}{3} & \frac{2}{3} & 1 \end{pmatrix}\ ,$$

which may be regarded as a "fuzzy" co-association, or affinity, matrix whose elements vary continuously between 0 and 1. Physically, the closer $\overline{\mathscr{Y}}(i,j)$ is to 1, the more the original decision maps put the elements x_i and x_j into the same cluster and the stronger the bond between the elements x_i and x_j, and vice versa.

Example 7.12. Bagged co-association matrix [3]. Tumor classification in DNA microarray experiments requires both the partition of the tumor samples into clusters and the confidence of the cluster assignments. For this purpose Dudoit and Fridlyand [3] recommend a bagging procedure. Given N DNA observations $\mathbf{X} = \{x_1, x_2, \ldots, x_N\}$ we create B bootstrap sets of observations: $\mathbf{X}^{(b)}, b \in \{1, 2, \ldots, B\}$, by sampling $\mathbf{X}$ with replacement (see Sect. 4.6). Let $\mathbf{A}^{(b)} = (a_1^{(b)}, a_2^{(b)}, \ldots, a_N^{(b)})^T$ denote the identity vector obtained as a result of clustering $\mathbf{X}^{(b)}$, where

$$a_i^{(b)} = \begin{cases} k \text{ if } x_i \in \mathbf{X}^{(b)} \text{ and } x_i \text{ has been placed in the } k\text{th partition} , \\ 0 \text{ if } x_i \notin \mathbf{X}^{(b)} . \end{cases}$$

If $\mathscr{A}^{(b)}$ is the corresponding co-association matrix, then

$$\mathscr{A}^{(b)}(i, j) = \begin{cases} 1 \text{ if } a_i^{(b)} = a_j^{(b)} \text{ and } a_i^{(b)} > 0 , \\ 0 \text{ otherwise} , \end{cases}$$

and the "bagged" co-association matrix is $\overline{\mathscr{Y}}$ where

$$\overline{\mathscr{Y}}(i, j) = (\mathscr{A}^{(1)}(i, j) + \mathscr{A}^{(2)}(i, j) + \ldots, \mathscr{A}^{(B)}(i, j)) / M(i, j) ,$$

where $M(i, j)$ is the number of bootstrap samples for which both $x_i \in \mathbf{X}^{(b)}$ and $x_j \in \mathbf{X}^{(b)}$.

By viewing the fused co-association matrix $\widetilde{\mathscr{Y}}$ as a special similarity matrix $\mathscr{S}$, we may generate the corresponding fused identity vector, $\tilde{\mathbf{y}}$, by using the NJW algorithm [14] (see Ex. 7.8).

7.5 Software

CLUSTERPACK. A matlab toolbox for cluster ensemble algorithms. Authors: A. Strehl, J. Ghosh [16].

HUNGARIAN ALGORITHM FOR LINEAR ASSIGNMENT PROBLEM. A matlab routine for solving the linear assignment problem. Author: Yi Cao.

CLUSTER ENSEMBLE TOOLBOX. Matlab toolbox for cluster analysis and cluster ensemble. Author: Vincent de Sapio.

COPAP. Matlab m-files for order-preserved contour matching. Authors: C. Scott and R. Nowak.

SPECTRAL CLUSTERING ALGORITHMS. A set of m-files for spectral clustering. Author: Asad Ali.

SPECTRAL CLUSTERING TOOLBOX. Matlab toolbox for spectral clustering. Authors: Deepak Verma and Marina Meila.

SELF-TUNING SPECTRAL CLUSTERING. A set of m-files for self-tuning spectral clustering [22]. The spectral clustering includes a method for automatically estimating the number of clusters K. Author: Lihi Zelnik-Manor.

7.6 Further Reading

A comprehensive survey of spectral clustering is given in [19].

References

1. Belongie, S., Malik, J., Puzicha, J.: Shape matching and object recognition using shape contexts. IEEE Trans. Patt. Anal. Mach. Intell. 24, 509–522 (2002)
2. Brand, M., Huang, K.: A unifying theorem for spectral embedding and clustering. In: Ninth Int. Conf. Art Intell. Stat. (2002)
3. Dudoit, S., Fridlyand, J.: Bagging to improve the accuracy of a clustering procedure. Bioinformatics 19, 1090–1099 (2003)
4. Fowlkes, C., Belongie, S., Chung, F., Malik, J.: Spectral grouping using the Nystrom method. IEEE Trans. Patt. Anal. Mach. Intell. 26, 214–225 (2004)
5. Fred, A.L.N.: Finding Consistent Clusters in Data Partitions. In: Kittler, J., Roli, F. (eds.) MCS 2001. LNCS, vol. 2096, pp. 309–318. Springer, Heidelberg (2001)
6. Fred, A.L.N., Jain, A.K.: Combining multiple clusterings using evidence accumulation. IEEE Trans. Patt. Anal. Mach. Intell. 27, 835–850 (2005)
7. Franek, L., Abdala, D.D., Vega-Pons, S., Jiang, X.: Image Segmentation Fusion Using General Ensemble Clustering Methods. In: Kimmel, R., Klette, R., Sugimoto, A. (eds.) ACCV 2010, Part IV. LNCS, vol. 6495, pp. 373–384. Springer, Heidelberg (2011)
8. Ghaemi, R., Sulaiman, M.N., Ibrahim, H., Mustapha, N.: A survey: clustering ensembles techniques. World Acad. Sci. Eng. Tech. 50, 636–645 (2009)
9. Jain, A.K.: Data clustering: 50 years beyond K-means. Patt. Recogn. Lett. 31, 651–666 (2010)
10. Jia, J., Liu, B., Jiao, L.: Soft spectral clustering ensemble applied to image segmentation. Front. Comp. Sci. China 5, 66–78 (2004)
11. Li, T., Ogihara, M., Ma, S.: On combining multiple clusterings: an overview and a new perspective. Appl. Intell. (2010)
12. Kuhn, H.W.: The Hungarian method for the assignment problem. Naval Research Logistics 52, 7–21 (2005)
13. Mignotte, M.: Segmentation by fusion of histogram-based K-means clusters in different color spaces. IEEE Trans. Im. Proc. 17, 780–787 (2008)
14. Ng, A., Jordan, M., Weiss, Y.: On spectral clustering: analysis and an algorithm. Adv. Neural Inform. Proc. Sys. 14, 849–856 (2001)
15. Scott, C., Nowak, R.: Robust contour matching via the order preserving assignment problem. IEEE Trans. Image Process. 15, 1831–1838 (2006)
16. Strehl, A., Ghosh, J.: Cluster ensembles - a knowledge reuse framework for combining multiple partitions. Mach. Learn. Res. 3, 583–617 (2002)
17. Vega-Pons, S., Ruiz-Shulcloper, J.: Int. J. Patt. Recogn. Art Intell. 25, 337–372 (2011)
18. A comparison of spectral clustering algorithms, Tech. Rept UW-CSE-03-05-01. Dept. Comp. Sci. Eng., Univ. Washington (2003)

19. von Luxburg, U.: A tutorial on spectral clustering. Stat. Comp. 17, 395–416 (2007)
20. Wang, Z., Gao, C., Tian, J., Lia, J., Chen, X.: Multi-feature distance map based feature detection of small infra-red targets with small contrast in image sequences. In: Proc. SPIE, vol. 5985 (2005)
21. Wang, X., Yang, C., You, J.: Spectral aggregation for clustering ensemble. In: Proc. Int. Conf. Patt. Recogn., pp. 1–4 (2008)
22. Zelnik-Manor, L., Perona, P.: Self-tuning spectral clustering. Adv. Neural. Inform. Proc. Sys. 17, 1601–1608 (2005)

Chapter 8
Radiometric Normalization

8.1 Introduction

The subject of this chapter is radiometric calibration, or normalization, which we define as the conversion of all sensor values to a common scale. This is the fourth and last function listed in Sect. 4.1 which is required for the formation of a common representational format. Although conceptually the radiometric normalization and semantic alignment are very different, we often use the same probabilistic transformation for both semantic alignment and radiometric normalization. The reader should be careful not to confuse the two terms.

In many multi-sensor data fusion applications radiometric normalization is the primary fusion algorithm. In Table 8.1 we list some of these applications together with the classification of the type of fusion algorithm involved.

Table 8.1 Applications in which Radiometric Normalization is the Primary Fusion Algorithm

Class		Application
DaI-DaO	Ex. 2.2	Approximate Lesion Localization in Demoscopy Images.
	Ex. 6.1	Cardiac MR Image Sequences.
	Ex. 8.16	Change Detection Thresholding using Mutual Information.
	Ex. 8.24	Remote Sensing.
DaI-FeO	Ex. 8.1	Similarity Measures for Virtual Screening.
	Ex. 8.25	Bhat-Nayar Ordinal Similarity Measures.
FeI-FeO	Ex. 8.10	Histogram Equalization for Robust Speech Recognition.
	Ex. 8.12	Midway Histogram Equalization.
	Ex. 8.22	Fusing Similarity Coefficients in a Virtual Screening Programme.
	Ex. 8.26	Converting SVM Scores in Probability Scores.
FeI-DeO	Ex. 8.4	χ^2-distance and Probability Binning.
	Ex. 8.13	Ridler-Calvard Image Histogram Thresholding.
	Ex. 8.15	Robust Threshold for Unimodal Histograms.

The designations DaI-DaO, DaI-FeO refer, respectively, to the Dasarathy input/output classifications: "Data Input-Data Output" and "Data Input-Feature Output" (Sect. 1.3.3).

8.2 Scales of Measurement

Physically, the scale of measurement of a variable describes how much information the values associated with the variable contains. Different mathematical operations on variables are possible, depending on the scale at which a variable is measured.

Four scales of measurement are usually recognized:

Nominal Scale. The variable values are names or labels. The only operations that can be meaningfully applied to the variable values are "equality" and "inequality".

Ordinal Scale. The variable values have all the features of a nominal scale and are numerical. The values represent the rank order (first, second, third, etc.) of the variables. Comparisons of "greater" and "less" can be made, in addition to "equality" and "inequality".

Interval Scale. The variable values have all the features of an ordinal scale and, additionally, are separated by the same interval. In this case, differences between arbitrary pairs of values can be meaningfully compared. Operations such as "addition" and "subtraction" are therefore meaningful. Additionally, negative values on the scale can be used.

Ratio Scale. The variable values have all the features of an interval scale and can also have meaningful ratios between arbitrary pairs of numbers. Operations such as "multiplication" and "division" are therefore meaningful. The zero value on a ratio scale is non-arbitrary. Most physical quantities, such as mass, length, or energy are measured on ratio scales.

In Table 8.2 we list the mathematical properties and types of statistics that can be applied to the different scales of measurement variables.

Table 8.2 Scales of Measurement

Name	Properties	Mathematical Operations	Descriptive Statistics	Inferential Statistics
Nominal	Identity	Count	Mode	Non-parametric, χ^2.
Ordinal	Identity, magnitude	Rank order	Mode, median, range statistics	Non-parametric, χ^2, Mann-Whiteney, Kruskal-Wallis, ANOVA.
Interval	Identity, magnitude, equal interval	Add, subtract	Mode, median, mean, range statistics, variance	Non-parametric, parametric, t-test, ANOVA.
Ratio	Identity, magnitude, equal interval, true zero	Add, subtract, multiply, divide	Mode, median, mean, range statistics, variance	Non-parametric, parametric, t-test, ANOVA.

8.3 Degree-of-Similarity Scales

Degree-of-similarity is a popular interval scale which is widely used for radiometric normalization. In fact most of the data fusion applications listed in Table 8.1 use a degree-of-similarity scale. The following example illustrates the use of different degree-of-similarity scales in virtual screening.

Example 8.1. Similarity Measures for Virtual Screening. [19, 44] "Virtual screening" involves the use of a computational scoring scheme to rank molecules in decreasing order of probability of activity. One way of carrying out virtual screening is to use similarity searching. This involves taking a molecule (normally called the target compound) with the required activity and then searching a database to find molecules that are structurally most similar to it.

A molecular similarity measure has two principal components: (1) the structural representation used to characterize the molecules and (2) the similarity coefficient used to compute the degree of resemblance between pairs of such representations. Most current systems for similarity based virtual screening use molecular "fingerprints" which are binary vectors encoding the presence, or absence, of substructural fragments within the molecule. Table 8.3 lists twenty-two similarity coefficients for the comparison of pairs of molecular fingerprints.

Although all twenty-two similarity coefficients are calculated using the same binary vectors their meanings are not all the same. The coefficients labeled "*D*" are *distance coefficients*, in which the greater the degree of similarity between the two molecules the smaller the value of the coefficient, the coefficients labeled "*S*" are similarity measures in which the greater the degree of similarity between the molecules the larger the value of the coefficient.

Example 8.2. Positive Matching Index (PMI) [37]. The PMI is a recent similarity measure devised for applications which require the matching of two input lists. The PMI may also be used for virtual screening. Using the same notation as that used in Table 8.3, the PMI is defined as follows:

$$S = \frac{a}{|b-c|} \ln\left(\frac{a+max(b,c)}{a+min(b,c)}\right) .$$

In many applications the object of interest is described by a vector of M feature values. In this case, we may compare two objects by measuring the distance between their corresponding feature vectors. Let $\mathbf{A} = (a_1, a_2, \ldots, a_M)^T$ and

Table 8.3 Similarity Coefficients for a Virtual Screening Programme

Name	Type	Formula
Jaccard/Tanimoto	S	$a/(a+b+c)$.
Dice	S	$2a/(2a+b+c)$.
Russell/Rao	S	a/n.
Sokal/Sneath$_1$	S	$a/(a+2b+2c)$.
Sokal/Sneath$_2$	S	$2(a+d)/(a+d+n)$.
Sokal/Sneath$_3$	S	$(a+d)/(b+c)$.
Kulczynski$_1$	S	$a/(b+c)$.
Kulczynski$_2$	S	$a(2a+b+c)/(2(a+b)(a+c))$.
Simple Match	S	$(a+d)/n$.
Hamann	S	$(a+d-b-c)/n$.
Rogers/Tanimoto	S	$(a+d)/(b+c+n)$.
Baroni-Urbani/Buser	S	$(\sqrt{ad}+a)/(\sqrt{ad}+a+b+c)$.
Ochiai/Cosine	S	$a/\sqrt{(a+b)(a+c)}$.
Forbes	S	$na/((a+b)(a+c))$.
Fossum	S	$n(a-0.5)^2/((a+b)(a+c))$.
Simpson	S	$a/\min(a+b,a+c)$.
Mean Manhattan	D	$(b+c)/n$.
Pearson	S	$(ad-bc)/\sqrt{(a+b)(a+c)(b+d)(c+d)}$.
Yule	S	$(ad-bc)/(ad+bc)$.
Hamming	D	$b+c$.
Vari	D	$(b+c)/(4n)$.
Bray Curtis	D	$(b+c)/(2a+b+c)$.
McConnaughey	S	$(a^2-bc)/((a+b)(a+c))$.
Stiles	S	$\log_{10} n+2\log_{10}(\lvert ad-bc\rvert-n/2)-\log_{10}((a+b)(a+c)\times (b+d)(c+d))$.
Dennis	S	$(ad-bc)/\sqrt{n(a+b)(a+c)}$.

$n=a+b+c+d$ is the total number of attributes in two molecules A and B, where a is the number of attributes which are present in both molecules A and B, b is the number of attributes which are not present in molecule A but which are present in molecule B, c is the number of attributes which are present in molecule A and not present in molecule B, d is the number of attributes which are not present in A and are not present in B.

$\mathbf{B}=(b_1,b_2,\ldots,b_M)^T$ denote the feature vectors for the two objects. Then, in table 8.4 we list several distance measures which we may use to compare the feature vectors.

In some applications the feature vector is in fact a histogram of feature values. In this case, it is important to choose an appropriate set of histogram bins. The following method is a simple and effective method for optimally choosing the number of histogram bins in a *regular* histogram (see also Table 5.2).

Example 8.3. Optimum number of bins in a regular histogram [7]. Let $x_n, n \in \{1,2,\ldots,N\}$, denotes a set of N measurements taken from an unknown density f on $[0,1]$. Then Birge and Rozenholc [7] find the optimum number of

Table 8.4 Feature Vector Distance Functions

Name	Formula
Euclidean distance	$\sqrt{\sum_{m=1}^{M} \lvert a_m - b_m \rvert^2}$.
City-block distance	$\sum_{m=1}^{M} \lvert a_m - b_m \rvert$.
Chebyshev distance	$\max_m(\lvert a_m - b_m \rvert)$.
Minkowski distance	$\left[\sum_{m=1}^{M} \lvert a_m - b_m \rvert^p\right]^{1/p}$.
Canberra distance	$\sum_{m=1}^{M} (\lvert a_m - b_m \rvert / (\lvert a_m \rvert + \lvert b_m \rvert))$.
Bray Curtis distance	$\sum_{m=1}^{M} \lvert a_m - b_m \rvert / \sum_{m=1}^{M} (a_m + b_m)$.

histogram bins, K^*, by optimizing a penalized likelihood: Given K histogram bins, let $H_k, k \in \{1,2,\ldots,K\}$, denote the number of measurements which fall in the kth bin. Then, the optimum number of histogram bins is

$$K^* = \max_K (L_K - P_K) ,$$

where

$$L_K = \sum_{k=1}^{K} H_k \log_2 \left(\frac{K \times H_k}{N} \right) ,$$

$$P_K = D - 1 + (\log_2 K)^{2.5} .$$

If the bins are allowed to vary in size, then a popular rule-of-thumb is to choose bins which have at least 5 or 10 observations.

Given two histograms F and G we may measure their similarity using the χ^2-distance.

Example 8.4. χ^2-distance and Probability Binning [34]. Suppose f represents M observations $x_1, x_2, \ldots, x_M$ and g represents N observations $y_1, y_2, \ldots, y_N$. We divide the observations f and g into K non-overlapping bins $B_k = [b'_k, b''_k), k \in \{1,2,\ldots,K\}$, where $b'_{k+1} = b''_k$. If m_k and n_k represent, respectively, the number of observations which fall in the kth bin, then the corresponding histograms are

$$F = (m_1, m_2, \ldots, m_K)^T ,$$

$$G = (n_1, n_2, \ldots, n_K)^T ,$$

and the χ^2 distance used for comparing F and G is

$$\chi^2 = \sum_{k=1}^{K} \frac{|m_k - n_k|^2}{m_k + n_k} .$$

In probability binning [34] we use a modified χ^2 distance as follows. We use variable width bins B_k such that each bin contains the same number of observations m_k. Then we convert the histograms F and G into probability distributions by dividing, respectively, through by M and N. Let

$$\mathscr{F} = (\widetilde{m}_1, \widetilde{m}_2, \ldots, \widetilde{m}_K)^T ,$$
$$\mathscr{G} = (\widetilde{n}_1, \widetilde{n}_2, \ldots, \widetilde{n}_K)^T ,$$

where $\widetilde{m}_k = m_k/M$ and $\widetilde{n}_k = n_k/N$. Then the modified χ^2 measure used for comparing $\mathscr{F}$ and $\mathscr{G}$ is

$$\widetilde{\chi}^2 = \sum_{k=1}^{K} \frac{|\widetilde{m}_k - \widetilde{n}_k|^2}{\widetilde{m}_k + \widetilde{n}_k} .$$

The main advantage of using $\widetilde{\chi}^2$ is that we can define a metric for $\widetilde{\chi}^2$:

$$T(\widetilde{\chi}^2) = \max\left(0, \frac{\widetilde{\chi}^2 - \mu}{\sigma}\right) ,$$

where $T(\widetilde{\chi}^2)$ represents the difference between $\mathscr{F}$ and $\mathscr{G}$ as the number of standard deviations above μ, where $\mu = K/\min(M,N)$ is the minimum statistical significant value of $\widetilde{\chi}^2$ (i. e. the minimum value for which a confident decision of histogram difference can be made) and $\sigma = \sqrt{K}/\min(M,N)$ is an appropriate standard deviation for $\widetilde{\chi}^2$. *Note:* Baggerly [1] suggests the following is a more accurate metric:

$$T(\widetilde{\chi}^2) = \frac{2\widetilde{\chi}^2\left(\frac{MN}{M+N}\right) - (K-1)}{\sqrt{2(K-1)}} .$$

Additional histogram similarity measures are listed in Table 8.5 [10]. The histogram similarity measures may be divided into two classes: bin-to-bin distances and cross-bin distances. The bin-to-bin distances compare corresponding bins in the two histograms. They thus ignore all correlations between neighbouring bins. As a consequence, these measures may be sensitive to changes in the positions and sizes of the histogram bins. In contrast, the cross-bin distances overcome these shortcomings, by comparing both corresponding bins and non-corresponding bins in the two histograms. Among the different cross-bin measures, the earth mover's (EMD) is recognized as a powerful and robust distance measure.

The following example explains the working of the EMD.

Table 8.5 Signature and Histogram Similarity Measures

Name	Formula
L_p distance	$\left(\sum_{k=1}^{K} \lvert m_k - n_k \rvert^p\right)^{1/p}$.
χ^2 Distance	$\chi^2 = \sum_{k=1}^{K} \lvert m_k - n_k \rvert^2 / (m_k + n_k)$.
Kullback-Leibler (KL) Distance	$\sum_{k=1}^{K} \widetilde{m}_k \log(\widetilde{m}_k / \widetilde{n}_k)$, where $\widetilde{m}_k = m_k/M$, $\widetilde{n}_k = n_k/N$, $M = \sum_k m_k$ and $N = \sum_k n_k$.
Jeffrey divergence	$\sum_{k=1}^{K} \left(m_k \log(m_k/n_k) + n_k \log(n_k/m_k)\right)$.
Hausdorff Distance	$\max\left(h(f,g), h(g,f)\right)$, where $h(f,g) = \max_k\left(\min_l\left(\lvert m_k - n_l \rvert\right)\right)$.
Partial HD	$\max(h_p(f,g), h_p(g,f))$ where $h_p(f,g)$ is the p-th largest value of $\min_{k,l} \lvert m_k - n_l \rvert$.
Earth Mover's Distance	See Ex. 8.5

Example 8.5. Earth Mover's Distance (EMD) [36]. EMD defines the distance between two histograms $F = \{(m_1,u_1),(m_2,u_2),\ldots,(m_K,u_K)\}$ and $G = \{(n_1,v_1),(n_2,v_2),\ldots,(n_L,v_L)\}$ as the minimum "work" required to transform F into G, where m_k is the number of elements in F which fall in the kth bin whose position is u_k and n_l is the number of elements in G which fall in the lth bin whose position is v_l.

EMD treats the elements in F as "supplies" located at u_k and treats the elements in G as "demands" located at v_l. The work expended in moving one element from u_k to v_l is defined as the ground distance d_{kl} between u_k and v_l. The EMD is then given by

$$EMD = \min_{F=\{f_{kl}\}} \left(\sum_{k=1}^{K} \sum_{l=1}^{L} f_{kl} d_{kl} \Big/ \sum_{k=1}^{K} \sum_{l=1}^{L} f_{kl} \right) ,$$

subject to

$$f_{kl} \geq 0 ,$$

$$\sum_{l=1}^{L} f_{kl} \leq m_k ,$$

$$\sum_{k=1}^{L} f_{kl} \leq n_l ,$$

$$\sum_{k=1}^{K} \sum_{l=1}^{L} f_{kl} = \min\left(\sum_{k=1}^{K} m_k, \sum_{l=1}^{L} n_l \right) ,$$

where the flow f_{kl} is number of elements moved from the kth supply at u_k to the lth demand at v_l.

The main disadvantage of the EMD is its high computational complexity: $O(N^3 \log_2(N))$. For this reason, there is continuing interest in the development of fast cross-bin similarity measures. A recent example is the variable bin size distance (VSBD).

Example 8.6. Variable Bin Size Distance (VSBD) [24]. The VSBD is calculated iteratively as follows. First we calculate a bin-to-bin distance d_1 for the two histograms F and G using a refined histogram bin. Then the intersection $I = |F \wedge G|$ of the two histograms is substracted from each histogram:

$$F' = F - I ,$$
$$G' = G - I .$$

The residual histograms F' and G' are recalculated using a coarser bin size and the corresponding bin-to-bin distance d_2 is calculated for the new coarser histograms. The entire process is repeated for several iterations. At each iteration k we obtain a bin-to-bin distance d_k. The VSBD is then given by the sum of the individual bin-to-bin distances d_k:

$$d_{\mathrm{VSBD}} = \sum_k d_k .$$

Although the histogram similarity measures can also be used to measure the similarity between two pdf's, statistical measures based on mutual information or the Kullback-Leibler (KL) distance, d_{KL}, are more common. Given two probability distributions f and g, the Kullback-Leibler (KL) distance is defined as:

$$d_{KL}(f|g) = \int f(x) \log\big(f(x)/g(y)\big)\,dx , \tag{8.1}$$

(see Ex. 9.10).

In some applications, we use a "hybrid" representation, which requires a special purpose degree-of-similarity scale. The following examples illustrate two hybrid representations and their similarity scales.

Example 8.7. A Joint Hue and Saturation Degree-of-Similarity Scale [25]. For illumination-tolerant image processing we often use only the hue and saturation images. Given a RGB color image, the hue and saturation images are defined as follows:

$$H = \tan^{-1}\left(\frac{\sqrt{3}(G-B)}{2R-G-B}\right) ,$$
$$S = 1 - \frac{3}{R+G+B}\min(R,G,B)$$

The joint Hue-Saturation degree-of-similarity (distance) measure between two pixels with hue and saturation values (H_1,S_1) and (H_2,S_2) is defined as follows:

$$d((H_1,S_1),(H_2,S_2)) = \frac{\sqrt{(H_1-H_2)^2+(S_1-S_2)^2}}{\sqrt{H_1^2+S_1^2}+\sqrt{H_2^2+S_2^2}} .$$

Example 8.8. A Spatiogram Similarity Measure [17, 30]. The spatiogram [6, 31] is a special histogram in which we augment the histogram, $H = (H_1,H_2,\ldots,H_K)^T$, with spatial means and covariances. Given an input image I, let H_k be the number of pixels which fall in the kth histogram bin and let $\boldsymbol{\mu}_k$ and $\boldsymbol{\Sigma}_k$ be the mean location and covariance matrix of these pixels. Then, formally, we define the spatiogram as:

$$S = (S_1,S_2,\ldots,S_K)^T ,$$

where $S_k = (H_k,\boldsymbol{\mu}_k,\boldsymbol{\Sigma}_k)$.

To compute the degree-of-similarity between two spatiograms, $F = \{m_k,\boldsymbol{\mu}_k^{(F)},\boldsymbol{\Sigma}_k^{(F)}\}$ and $G = \{n_k,\boldsymbol{\mu}_k^{(G)},\boldsymbol{\Sigma}_k^{(G)}\}$, we need a measure which combines histogram bin similarity and spatial information similarity:

$$\rho(F,G) = \sum_{k=1}^{K} \psi_k\phi_k ,$$

where ψ_k and ϕ_k are, respectively, the histogram similarity of the kth bin and the spatial information similarity of the kth bin. ψ_k may be computed from m_k and n_k using any of the histogram bin-to-bin, or cross-bin, similarity measures and ϕ_k is computed from $(\boldsymbol{\mu}_k^{(F)},\boldsymbol{\Sigma}_k^{(F)})$ and $(\boldsymbol{\mu}_k^{(G)},\boldsymbol{\Sigma}_k^{(G)})$.

A formula which has been proposed for ϕ_k is:

O'Conaire *et al.* **[30]**:

$$\phi_k = \left(\det(\boldsymbol{\Sigma}_k^{(F)}\boldsymbol{\Sigma}_k^{(G)})\right)^{1/4} \Big/ \sqrt{\det(\widetilde{\boldsymbol{\Sigma}}_k)}\,\Big) \exp\left(-\frac{1}{2}\widetilde{\boldsymbol{\mu}}_k^T(2\widetilde{\boldsymbol{\Sigma}}_k)^{-1}\widetilde{\boldsymbol{\mu}}_k\right) ,$$

where $\widetilde{\boldsymbol{\mu}}_k = (\boldsymbol{\mu}_k^F - \boldsymbol{\mu}_k^G)$ and $\widetilde{\boldsymbol{\Sigma}}_k = \boldsymbol{\Sigma}_k^{(F)} + \boldsymbol{\Sigma}_k^{(G)}$.

8.4 Radiometric Normalization

The process of converting the values of different pieces of information to a common scale is known as radiometric normalization. Suppose we use a set of logical sensors, or sources-of-information, $S_i, i \in \{1,2,\ldots,N\}$, to make N measurements x_i on an object O. Then, we transform the x_i into a common representational format by changing the statistical parameters of the x_i.[1] Mathematically we write the transformation of an input measurement x to a normalized value y as

$$y = f(x|\alpha,\beta,\gamma,\ldots,\delta) , \tag{8.2}$$

where f is a parametric *transformation function*, α and β are the scale and location parameters and γ, δ, etc. are the higher order statistical parameters.

In most cases the parameters $\alpha,\beta,\ldots,\delta$ are learnt from measurements made on a set of *training samples* $D = \{x_1,x_2,\ldots,x_N\}$. In this case the procedure is referred to as *fixed sensor value normalization*. In a few cases, the transformation is estimated from the current measurement values. This is known as *adaptive sensor value normalization* and its main advantage is its ability to adapt to variations in the input data.

Example 8.9. Whitening: An adaptive normalization procedure. In whitening we transform a set of input measurements $x_i, i \in \{1,2,\ldots,N\}$, into normalized values y_i which are distributed according to a Gaussian distribution $\mathcal{N}(\bar{x},\sigma^2)$, where

$$\bar{x} = \frac{1}{N}\sum_{i=1}^{N} x_i ,$$

$$\sigma^2 = \frac{1}{N-1}\sum_{i=1}^{N}(x_i - \bar{x})^2 .$$

For a good normalization scheme, the parameters $\alpha,\beta,\ldots,\delta$ must be *robust* and *efficient*, where *robustness* refers to insensitivity in the presence of outliers[2] and *efficiency* refers to the proximity of the obtained estimate to the optimal estimate when the distribution of the data is known.

The following example illustrates the normalization of the input data x_i in which we match the pdf of the x_i to a reference pdf.

---

[1] The most important statistical parameters are the scale and location parameters.

[2] See Chapt. 11 for a full discussion concerning robust statistics and outliers.

Example 8.10. Histogram Equalization for Robust Speech Recognition [41]. The performance of automatic speech recognition (ASR) systems degrades significantly if there is an acoustic mismatch between the training and test data. One method for compensating for the acoustic mismatch is histogram equalization (HEQ). *Note:* Another widely used method of reducing the mismatch is dynamic time warping (see Chapt. 6).

The goal of HEQ is to transform the speech features in such a way that the acoustic environment does not affect its probability distribution. This is achieved by transforming the distribution of each test feature into the corresponding training set, or reference, distribution. Suppose x denotes a given test feature whose probability density function (pdf) and cumulative distribution function (cdf) are, respectively, $p_1(x)$ and $c_1(x)$, where $c_1(x) = \int_{-\infty}^{x} p_1(x)dx$. We seek a function $y = F(x)$ which maps $p_1(x)$ into the corresponding training pdf, $p_2(y)$. This is obtained by equating $c_1(x)$ and $c_2(y)$, where $c_2(y) = \int_{-\infty}^{y} p_2(y)dy$:

$$c_1(x) = c_2(y) = c_2(F(x)) ,$$

or

$$y = F(x) = c_2^{-1}(c_1(x)) ,$$

where c_2^{-1} denotes the inverse of c_2 and is is defined as follows: If $y = c_2(x)$, then $c_2^{-1}(y) = x$.

The following example describes a special histogram matching algorithm which uses a principal component analysis (PCA) face recognition algorithm (see Sect. 4.5.1).

Example 8.11. "Perfect" histogram matching for improved face recognition [39]. In "perfect" histogram matching [39] the histogram of all input images are transformed into the same Gaussian distribution:

$$a \exp\left[-\frac{1}{2}\left(\frac{k-b}{c}\right)^2\right] .$$

For an $M \times N$ 8-bit input image, Ref. [39] recommends: $b = 127.5$, $c = 100$ and

$$a \sum_{k=0}^{255} \exp\left[-\frac{1}{2}\left(\frac{k-b}{c}\right)^2\right] = M \times N .$$

Perfect histogram matching is found to yield superior recognition results when applied as a pre-processing module prior to conventional PCA face recognition.

A critical issue in HEQ is the reliable estimation of the cdf's $c_1(x)$ and $c_2(y)$. In speech recognition applications, the cdf $c_2(y)$ can be estimated quite reliably by computing cumulative histograms with a large amount of training data. However, when short utterances are used as test data, the length of each utterance may be insufficient for a reliable estimation. In this case, we may use an order-statistic estimate for the test cdf, $c_1(x)$: Given a test sequence consisting of K examples $\{x_1, x_2, \ldots, x_K\}$ of a given feature, then the order-statistic estimate of the cdf $c_1(x)$ is:

$$\hat{c}_1(x_k) \approx \frac{r_k - \frac{1}{2}}{K} ,$$

where r_k denotes the rank of x_k and $r_k = j$ if x_k is the jth largest input data.

The following example illustrates the normalization of two input vectors $\mathbf{x}$ and $\mathbf{y}$ by simultaneously matching their pdf's.

Example 8.12. Midway Histogram Equalization [14]. Midway histogram equalization is defined as any method which warps two input histograms $p_1(x)$ and $p_2(y)$ to a common "intermediate" histogram $p_3(z)$, such that $p_3(z)$ retains as much as possible of the shapes of $p_1(x)$ and $p_2(y)$. Mathematically, midway image equalization may be implemented as follows: Given the two cumulative probability distributions $c_1(x)$ and $c_2(y)$ we define the inverse cumulative distribution function of the intermediate distribution $p_3(z)$ by:

$$c_3^{-1}(z) = \frac{c_1^{-1}(z) + c_2^{-1}(z)}{2}$$

We now warp $p_1(x)$ by matching it to $p_3(z)$. Let $c_3(z) = \int_0^z p(z)dz$ denote the cumulative distribution of $p_3(z)$, then the warped distribution is $p_1'(x')$, where

$$p_1'(x') = p_1(x) ,$$
$$x' = c_3^{-1}(c_1(x)) .$$

Similarly we warp $p_2(y)$ by matching it to $p_3(z)$. The corresponding warped distribution is $p_2'(y')$, where

$$p_2'(y') = p_2(y) ,$$
$$y' = c_3^{-1}(c_2(y)) .$$

8.5 Binarization

Binarization is probably the simplest radiometric normalization technique. It is robust and is widely used when different sensor measurements $x_n, n \in \{1, 2, \ldots, N\}$, are made on the same object: Given an object O we threshold each measurement x_n using a local threshold t_n, or a global threshold t_G. The corresponding normalized values are y_n, where

$$y_n = \begin{cases} 1 \text{ if } x_n \geq t_n \ , \\ 0 \text{ otherwise} \ . \end{cases} \quad (8.3)$$

The thresholds $t_n, i \in \{1,2,\ldots,N\}$, may be established *a priori* on physical grounds or may be learnt on a training set D ("supervised" learning) or learnt on the current set of measurements ("unsupervised" learning). The following example illustrates unsupervised learning of global threshold t_G for an input image.

Example 8.13. Ridler-Calvard Image Histogram Thresholding [33, 40]. Let I denote an $M \times N$ input image, where $g(m,n)$ denotes the gray-level of pixel (m,n). Then the pixels in the image are divided into two groups: foregound pixels and background pixels by using a global gray-level threshold T. The threshold is found iteratively by analyzing the image histogram $H = (H_1, H_2, \ldots, H_K)^T$, where H_k is the number of pixel whose gray-levels fall in the k histogram bin, i. e. in the interval $[b'_k, b''_k)$.

The Ridler-Calvard thresholding may be implemented using the following matlab code:

Example 8.14. Matlab Code for Ridler-Calvard Thresholding.

Input

$H(k), k \in \{1,2,\ldots,K\}$ is the histogram defined with K bins

$b(k) = (b'_k + b''_k)/2, k \in \{1,2,\ldots,K\}$, is average gray-level of kth histogram bin

niter is the number of iterations

Output

Threshold t

Code

```
t = sum(b.*H)./sum(H);
for iter = 1 : niter
    A = sum(b(1 : t).*H(1 : t));        na = sum(H(1 : t));
    B = sum(b(t+1 : K).*H(t+1 : K));    nb = sum(H(t+1 : K));
    if (na == 0)
        B = B/nb;    A = B;
    elseif (nb == 0)
        A = A/na;    B = A;
    else
```

```
        A = A/na;      B = B/nb;
    end
    t = (A+B)/2;
end
```

Some common global thresholding techniques for bimodal histograms are listed in Table 8.6. These techniques assume the image histogram consists of two dominant modes which respectively represent the object and background.

Table 8.6 Thresholding Techniques

Name	Description
Otsu [40]	$T = \arg\min_t \left(\omega_{low}(t)s^2_{low}(t) + \omega_{high}(t)s^2_{high}(t)\right)$, where $\omega_{low}(t)$ and $s_{low}(t)$ are the number and standard deviation of pixels whose gray-levels are less than, or equal to, t and $\omega_{high}(t)$ and $s_{high}(t)$ are the number and standard deviation of pixels whose gray-levels are greater than t.
Range-constrained Otsu (RCoOtsu) [47]	RCoOtsu is used when $\lvert s^2_{low}(T) - s^2_{high}(T)\rvert >> 0$: Calculate initial threshold T_{init} by applying Otsu algorithm to input image. If $s^2_{high}(T_{init}) >> s^2_{low}(T_{init})$, calculate RCoOtsu threshold T by applying Otsu algorithm to pixels whose gray-levels lie below T_{init} and if $s^2_{high}(T_{init}) << s^2_{low}(T_{init})$, calculate RCoOtsu threshold T by applying Otsu algorithm to pixels whose gray-levels lie above T_{init}.
Weighted Otsu [20]	$T = \arg\min_t \left[p(t)\left(\omega_{low}(t)s^2_{low}(t) + \omega_{high}(t)s^2_{high}(t)\right)\right]$, where $p(k)$ is the fraction of pixels whose gray-levels fall in the kth histogram bin.
Robust-Otsu [48]	See Ex. 11.5.
Kittler and Illingworth [40]	$T = \arg\min_t \left[\omega_{low}(t)\log\left(s_{low}(t)/\omega_{low}(t)\right) + \omega_{high}(t)\log\left(s_{high}(t)/\omega_{high}(t)\right)\right]$.
Robust-KI [48]	See Ex. 11.5.
Kapur, Sahoo and Wong [40]	$T = \arg\max_t \left[-\Sigma_{k=0}^{t}\left((h(k)/\omega_{low}(t))\log(h(k)/\omega_{low}(t))\right) - \left(\Sigma_{k=t+1}^{K}(h(k)/\omega_{high}(t))\log(h(k)/\omega_{high}(t))\right)\right]$, where $h(k)$ is the number of pixels in the kth histogram bin.

Sometimes, the modes overlap such that the histogram is unimodal in shape. In this case, the dominant pixel population produces the main peak of the histogram while the secondary population contributes to the tail of the histogram. The threshold T is assumed to be located somewhere after main peak and before the elongated part of the histogram tail. Special global thresholding techniques are for this case [2, 13, 35].

Example 8.15. Robust Threshold for Unimodal Histograms [13]. After the main peak, the descending slope of the histogram $H_k, k \in \{1,2,\ldots,K\}$, is decomposed into two parts: a steeply descending slope immediately after the peak and a slightly descending slope in the flat tail. We determine the two lines which best describe the two descending slopes (Fig. 8.1). The gray-level that minimizes the error between the descending slope of the histogram and the two lines is chosen as the optimum threshold T. If M is the location of the histogram peak, then each bin, $t \in \{M, M+1, \ldots, K\}$, is successively considered as the threshold t. For each t, the two lines are computed using the least mean square error. If $m_1(t), c_1(t)$ and $m_2(t), c_2(t)$ denote, respectively, the slope and intercept of the two lines, then the linear estimate of the kth bin is:

$$h_k = \begin{cases} m_1(t)g_k + c_1(t) & \text{if } k \in \{M, M+1, \ldots, t\}\ , \\ m_2(t)g_k + c_2(t) & \text{if } k \in \{t+1, t+2, \ldots, K\}\ . \end{cases}$$

For each t, the sum of the square errors between h_k and H_k is:

$$\varepsilon(t) = \sum_{k=M}^{K} (H_k - h_k(t))^2\ ,$$

and the optimum threshold is

$$T = \arg\min_t (\varepsilon(t))\ .$$

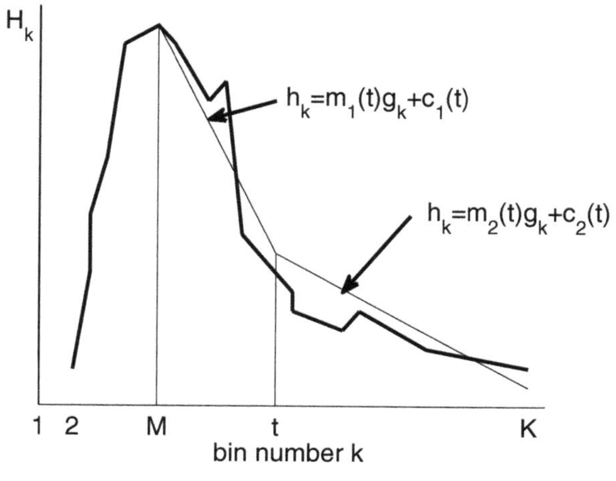

Fig. 8.1 Shows a unimodal histogram $H = (H_1, H_2, \ldots, H_K)^T$ and the calculation of the optimum threshold T.

Example 8.16. Change Detection Thresholding using Mutual Information [29]. The most common methodology to carry out an unsupervised change detection in remotely sensed imagery is to compare two co-registered multi-temporal remote sensing images $I_1(m,n)$ and $I_2(m,n)$ taken at two different dates over the same geographical area [8]. The result of the comparison is a binary image $y(m,n)$, where $C(m,n) = 1$ if the pixel (m,n) is classified as "changed", otherwise $C(m,n) = 0$. One way of generating C is to threshold the difference image $D = |I_1 - I_2|$ using one of the formulas listed in Table 8.6. A recent alternative is to separately threshold I_1 and I_2 using thresholds T_1 and T_2. If B_1 and B_2 are the corresponding thresholded images:

$$B_1(m,n) = \begin{cases} 1 \text{ if } I_1(m,n) > T_1 \,, \\ 0 \text{ otherwise} \,, \end{cases}$$

$$B_2(m,n) = \begin{cases} 1 \text{ if } I_2(m,n) > T_2 \,, \\ 0 \text{ otherwise} \,, \end{cases}$$

then T_1 and T_2 are chosen such that the mutual information $MI(B_1,B_2)$ between B_1 and B_2 is maximized [29]. Mathematically,

$$MI(B_1,B_2) = \sum_{u=0}^{1} \sum_{v=0}^{1} p_{12}(u,v) \log\left(\frac{p_{12}(u,v)}{p_1(u)p_2(v)}\right) ,$$

where $p_{12}(u,v)$ is the joint probability that $B_1(m,n) = u$ and $B_2(m,n) = v$, $p_1(u)$ is the *a posteriori* probability that $B_1(m,n) = u$, $p_2(v)$ is the *a posteriori* probability that $B_2(m,n) = v$ and $u, v \in \{0,1\}$.

The following example illustrates the use of a "soft" binarization procedure in multi-sensor data fusion.

Example 8.17. "Soft" Binarization. In many multi-sensor data fusion applications it is not possible to establish the correct threshold t_i exactly. To reduce the impact of choosing an incorrect threshold, we may replace (8.3) with a trimmed min-max transfer function (see Table 8.9):

$$y_i = \begin{cases} 1 & \text{if } x_i > t_i + \delta \,, \\ 0 & \text{if } x_i < t_i - \delta \,, \\ \frac{1}{2}(x_i - t_i + \delta)/\delta & \text{otherwise} \,. \end{cases}$$

or with a sigmoid-like transfer function (see Fig. 8.2).

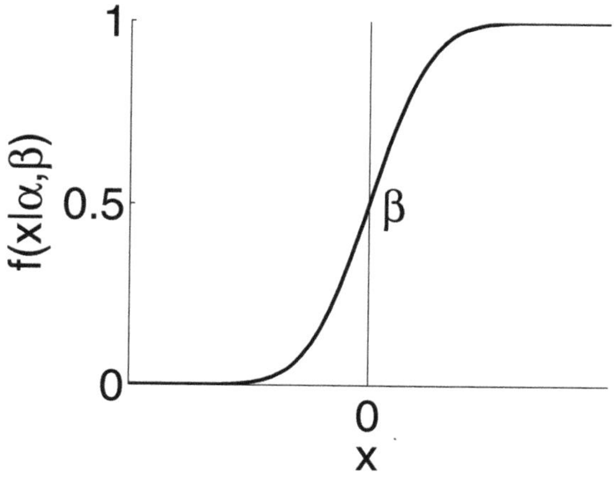

Fig. 8.2 Shows a Logistic sigmoid transfer function. See Table 8.7 for definition of this and other similar sigmoid-like transfer functions.

Table 8.7 Psychometric Transfer Functions $f(x|\alpha,\beta)$ and Slopes $df(x|\alpha,\beta)/dx$

Name	$f(x\|\alpha,\beta)$	$df(x\|\alpha,\beta)/dx$.
Logistic	$1/(1+\exp(\alpha(\beta-x)))$	$\alpha f(x\|\alpha,\beta)\times(1-f(x\|\alpha,\beta))$.
Probit	$(\alpha/\sqrt{2\pi})\int_{-\infty}^{x}\exp-\frac{1}{2}(\alpha(t-\beta))^2dt$	$(\alpha/\sqrt{2\pi})\exp-(\alpha(x-\beta))^2/2)$.
Gumbel	$1-\exp(-\exp(\alpha\ln(x)-\alpha\ln(\beta)))$	$(\alpha/x)(1-f(x\|\alpha,\beta))\exp(\alpha\ln(x)-\alpha\ln(\beta))$.
Weibull	$1-\exp(-(x/\beta)^{\alpha})$	$\alpha x^{\alpha-1}(1-f(x\|\alpha,\beta))/\beta^{\alpha}$.
Quick	$1-2^{-(x/\beta)^{\alpha}}$	$\ln(2)\alpha x^{\alpha-1}(1-f(x\|\alpha,\beta))/\beta^{\alpha}$.

Example 8.18. Psychometric Transfer Functions [15]. Psychophysics explores the connection between physical stimuli and subjective human responses. The psychometric transfer function, $f(x|\alpha,\beta)$, relates the intensity of the stimulus, x, to an observers's response, y. It is a parametric function with two parameters: a scale parameter α and a location parameter β. The psychometric function is commonly chosen to have a sigmoid-like shape. Five sigmoid-like psychometric transfer functions are listed, together with their slopes $df(x|\alpha,\beta)/dx$ in Table 8.7

Binarization is often used as a preliminary step where it functions as a *detection* step. For example, in calculating the similarity coefficients in Ex. 8.1 we assumed that all the features had undergone a preliminary binarization, or detection, process.

Although very effective, problems may arise if, for a given object O, we do not have a complete set of N measurement values. The reason is that when we perform binarization we do not differentiate between (1) $y_i = 0$ because $x_i \leq t_i$ and (2) $y_i = 0$ because, for one reason or another, no measurement value x_i is available. For example, the sensor S, which is responsible for making the measurement x_i, may have malfunctioned when the object O was being processed. To handle such situations we use a *censored* binary scale which is defined as follows:

$$y_i = \begin{cases} 1 \text{ if } x_i \geq t_i \ , \\ 0 \text{ if } x_i < t_i \ , \\ * \text{ if } x_i \text{ is missing} \ , \end{cases} \tag{8.4}$$

where $*$ is a special symbol which is handled differently from 0 and 1 by the data fusion algorithm. Censored binary scales are widely used when the information concerning the x_i are *subjective* in nature. The following example illustrates the use of a censored binary scale when the x_i are subjective replies to questions contained in a questionnaire.

Example 8.19. Automatic Disease Classification: Syndromic Surveillance [18, 21]. The concept of *syndromic surveillance* was developed to ensure early detection of an epidemic in a given population. Syndrome surveillance works by monitoring the temporal and spatial trends of a syndrome, or a group of related symptoms or diseases, in the local population. For example we may use a censored questionnaire, similar to that shown in Table 8.8, to monitor cases of acute gastrointestinal syndrome which were admitted to the local hospital emergency departments.

Table 8.8 Questionnaire Used for Detecting Acute Gastrointestinal Syndrome

Question	Medical Diagnosis	Possible Answer
Diarrhea present?	Yes, No	[1,0].
Diarrhea duration?	Acute, Chronic, Unknown	[1,0,*].
Diarrhea etiology?	Infectious, Non-infectious, Unknown	[1,0,*].
Stool culture taken?	Yes, No	[1,0].
Vomiting present?	Yes, No	[1,0].
Vomiting duration?	Acute, Chronic, Unknown	[1,0,*].
Vomiting etiology?	Infectious, Non-infectious, Unknown	[1,0,*].
Acute GI disorder?	Yes, No	[1,0].

8.6 Parametric Normalization Functions

The most common normalization technique is to use a *parametric* normalization function $y = f(x|\alpha,\beta,\ldots,\delta)$, whose parameters $\alpha,\beta,\ldots,\delta$, are learnt from a training set D of labelled objects $O_i, i \in \{1,2,\ldots,N\}$. One example of a parametric normalization function is the family of pyschometric transfer functions listed in Table 8.7. Additional parametric transfer functions are listed in Table 8.9[3].

Table 8.9 Parametric Normalization Functions

Function	Formula
Min-Max	$y = (x-a)/(b-a)$, where $a = \min_i(x_i)$, $b = \max_i(x_i)$. In trimmed min-max we replace, respectively, a and b by the lth smallest and largest x_i values. The resulting normalized value is $y = \min(\max(0,(a-x)/(b-a)),1)$. *Note*: Only min-max values y_i retains the same distribution as the input x_i values.
Z-Transform	$y = (x-\mu)/\sigma$, where $\mu = \sum_k x_k/K$, $\sigma^2 = \sum_k (x_k-\mu)^2/(K-1)$. In robust Z-transform we replace μ by median$\{x_i\}$ and Q by the interquartile distance $(x_{(3N/4)} - x_{(N/4)})/2$, where $x_{(l)}$ is the lth largest value in $\{x_i\}$.
Robust Tanh	$y = \frac{1}{2}\tanh(\alpha(x-\mu_H)/\sigma_H + 1)$, where α determines the spread of the normalized scores and μ_H and σ_H are robust Hampel mean and standard deviation estimates of the $x_i, i \in \{1,2,\ldots,N\}$ [22].

8.7 Fuzzy Normalization Functions*

Fuzzy logic provides us with a very powerful normalization technique. In this technique, an input value x is converted into a *vector* of M components:

$$\mathbf{y} = \left(y^{(1)}, y^{(2)}, \ldots, y^{(M)}\right)^T , \tag{8.5}$$

where $y^{(m)} = \mu_m(x)$ represents the degree to which x "belongs" to the mth membership function $\mu_m(x)$. The membership functions $\mu_m(x), m \in \{1,2,\ldots,M\}$, are continuous functions of x for which $0 \leq \mu_m(x) \leq 1$. *Note*: In fuzzy logic there is no requirement that the components $y^{(m)}$ should sum to one. In fact, $\sum_m y^{(m)}$ can take any value between 0 and M. The following example illustrates the normalization of a given parameter using fuzzy logic.

Example 8.20. Normalization of the Inoptropic Reserve Using Fuzzy Logic [3]. In Ex. 5.17, we considered the fusion of the Inoptropic Reserve (IR) and

[3] Only the trimmed min-max, robust z-transform and robust tanh normalization functions are robust against outliers (see Chapt. 11). The remaining functions are not robust against outliers and should not be used if outliers are likely to be present.

the 18-fluorodeoxyglucose (FDG). Before fusion both variables were normalized by using fuzzy membership functions each of which was interpreted as a *linguistic variable*. Fig. 8.3 shows the five membership functions used to normalize IR together with their linguistic interpretation.

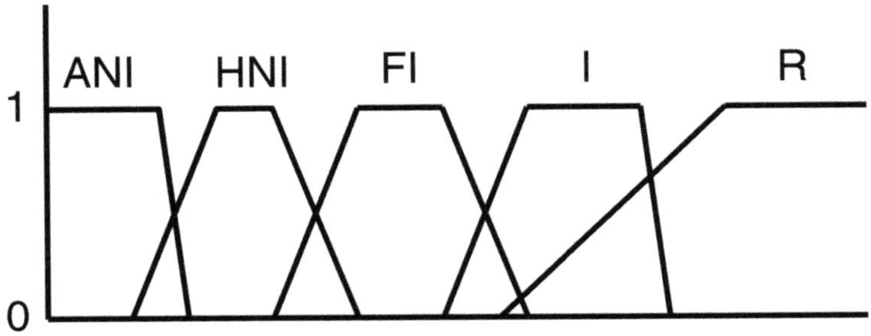

Fig. 8.3 Shows the five membership functions $\mu_i(x)$ used to normalize the Inoptropic Reserve (IR). Each membership function is given a name, ANI, HNI, FI, I, R, corresponding to its linguistic, or medical, interpretation, "Akinetic and No Improvement", "Hypokinetic with No Improvement", "Function Improvement", "Ischemic", "Remote".

8.8 Ranking

Ranking is a robust normalization technique which has enjoyed a growing popularity in recent years. Ranking is an example of *adaptive score normalization*, i. e. the normalization is based on the current input measurements and it does not require the use of a training set D. Its main advantages are: it is simple and fast to implement, linear, robust against outliers and can be applied to all kinds of input information without the need to make additional assumptions regarding their distributions etc[4].

Mathematically, rank normalization is defined as follows. Suppose the input data consists of N objects $O_n, n \in \{1,2,\ldots,N\}$. On each object O_n we make a measurement x_n. Then the corresponding rank is r_n, where

$$r_n = m \text{ if } x_n \text{ is } m\text{th smallest input value} . \qquad (8.6)$$

Note: If there are ties among the x_n, then the ranks r_n are chosen so that all x_n with the same value have the same rank. Matlab code for calculating the ranks r_n is:

4 Robust techniques are discussed at length in Chapt. 11.

Example 8.21. Matlab Code for Ranks r_n.

Input

Input vector of N values $x(n), n \in \{1,2,\ldots,N\}$

Output

Rank vector whose nth element is $rnk(n), n \in \{1,2,\ldots,N\}$

Code

$N = \text{length}(x);$

$[junk, invr] = \text{sort}(x, \text{'descend'});$ $[junk, r] = \text{sort}(invr, \text{'descend'});$

$xflip = x(N : -1 : 1);$

$[junk, invrflip] = \text{sort}(x, \text{'descend'});$ $[junk, R] = \text{sort}(invrflip, \text{'descend'});$

$R = R(N : -1 : 1);$ $rnk = (r + R)/2;$

The following example illustrates the use of ranking in an application involving the fusion of different similarity measures.

Example 8.22. Fusing Similarity Coefficients in a Virtual Screening Program [16]. In similarity-based virtual screening (Ex. 8.1) we match a bio-active target structure A against a database of molecules $B_n, n \in \{1,2,\ldots,N\}$, using any one of several different similarity operators $S_k, k \in \{1,2,\ldots,K\}$. *A priori* we cannot know which operator will be best for a given target structure or for a given type of activity. Let $x_k(A,B_n)$ denote the corresponding similarity value as measured by the kth operator S_k. Then, by fusing together the $x_k(A,B_n), k \in \{1,2,\ldots,K\}$, we may produce a more effective and robust activity measure [16]. Since in general, the $x_k(A,B_n)$ are incommensurate we must normalize them before fusion. This we do by ranking $x_k(A,\mathbf{B}) = \{x_k(A,B_1), x_k(A,B_2),\ldots,x_k(A,B_N)\}$. Let $r_k(A,B_n)$ denotes the rank of $x_k(A,B_n)$, where $r_k(A,B_n) = m$ if $x_k(A,B_n)$ is the mth smallest value in $\{x_k(A,\mathbf{B})$ value, then the fused similarity value, $\tilde{r}(A,B_n)$, is defined as the average of the $r_k(A,B_n), k \in \{1,2,\ldots,K\}$:

$$\tilde{r}(A,B_n) = \frac{1}{K}\sum_{k=1}^{K} r_k(A,B_n) .$$

The following example illustrates the use of ranking to fuse together PCA and LDA subspaces for biometric face recognition.

Example 8.23. Biometric Face Recognition by Rank Fusion of PCA and LDA subspaces [27]. As in Ex. 4.20 we are required to identify a test face $\mathbf{y}$ by matching it against a set of face images $\mathbf{Y}_n \in \{1,2,\ldots,N\}$. We project $\mathbf{y}$ and $\mathbf{Y}_n, n \in \{1,2,\ldots,N\}$, on K-dimensional PCA and LDA subspaces. Let d_n and D_n denote, respectively, the Euclidean distance between $\mathbf{y}$ and $\mathbf{Y}_n$ projected onto the PCA and LDA subspaces. The Euclidean distances are converted to ranks:

$$\begin{aligned} r_n &= k \text{ if } d_n \text{ is } k\text{th smallest PCA distance} , \\ R_n &= k \text{ if } D_n \text{ is } k\text{th smallest LDA distance} , \end{aligned}$$

which are then fused together:

$$F_n = r_n + R_n ,$$

The test face is classified as belonging to the n^*th training face, where

$$n^* = \arg\min (F_1, F_2, \ldots, F_N) .$$

Example 8.24. Remote Sensing [28]. Remotely sensed data are increasingly used for mapping and monitoring the physical environment. One of the advantages of monitoring with remotely sensed data is that temporal sequences can accurately indicate environmental changes, assuming that the input data is radiometrically consistent for all scenes. Factors contributing to the potential inconsistency in measured radiance include changes in surface condition, illumination geometry, sensor calibration, observation geometry and atmospheric condition. By using a radiometric normalization technique, we may however, correct for data inconsistencies resulting from many different effects. Image normalization is carried out in one step by converting image values to ordinal ranks. Ordinal ranking allows us to assign each pixel a new value based on its reflectance value, relative to all other pixels. When image pairs are converted to ordinal ranks the global characteristics of the distributions of pixel values are matched.

Pixel ranking does not require atmospheric details, sensor information, or selection of subjective pseudo-invariant features, and therefore allows images to be simply and efficiently normalized and processed for changes with minimal *a priori* knowledge. In general, for small pictures, pixel ranking is an effective image normalization technique. It is, however, less effective on very large digital images because in this case we obtain many tied ranks.

In many multi-sensor image fusion applications we require a measure of association, or similarity, between two images or two image patches [5, 38]. It is desirable for such a measure to be robust in the presence of outliers while being invariant under reasonable image transformations. Let $A = \{a_1, a_2, \ldots, a_N\}$ and $B = \{b_1, b_2, \ldots, b_N\}$ denote two input images, or two input image patches, written as one-dimensional vectors. Let $\mathbf{r} = \{r_1, r_2, \ldots, r_N\}$ and $\mathbf{R} = \{R_1, R_2, \ldots, R_N\}$ denote the corresponding rank images written as one-dimensional vectors. Then in Table 8.10 we list some ordinal measures for the distance, or correlation, between $\mathbf{r}$ and $\mathbf{R}$ (see also Table 11.6). The following code illustrates the calculation of the Bhat-Nayar ordinal similarity measure.

Table 8.10 Ordinal Measures for Image Correspondence

Name	Formula
Kendall τ	$1 - \left(\sum_i \lvert r_i - R_i \rvert\right) / \left(6N(N-1)\right)$.
Spearman ρ	$\left(\sum_i \sum_j \mathrm{sgn}(a_i - a_j)\mathrm{sgn}(b_i - b_j)\right) / \left(N(N-1)\right)$.
Bhat-Nayar κ	(1) $\kappa' = 1 - 2\max_i(d_i)/[N/2]$, where $d_i = i - \sum_{j=1}^{i} \delta_{ij}$, $\delta_{ij} = 1$ if $s_j \leq i$, otherwise 0. $\mathbf{s}$ is a composition permutation vector and is defined by $s_i = r^B_k, k = \tilde{r}_i$, where $\tilde{r}_i$ is the inverse of $\mathbf{R}$. (2) $\kappa'' = 1 - 2d_{N/2}/[N/2]$.
Scherer *et. al.* λ	$\lambda = 1 - (\max_i(d_i)/[N/2] + \sum_i d_i/[N^2/4]$.

Example 8.25. Matlab Code for the Bhat-Nayar Ordinal Similarity Measure [5, 38].

Input

Two input images with pixel gray-levels $a(n)$ and $b(n), n \in \{1, 2, \ldots, N\}$.

Output

lambdaBN: Bhat-Nayar distance function

lambdaSWP: Scherer-Werth-Pinz distance function

Code

```
[junk, invrA] = sort(a);    [junk, rA] = sort(invrA) ;
[junk, invrB] = sort(b);    [junk, rB] = sort(invrB) ;
s = rB(invrA);   S = ones(N,1) * s(:)' ;
G = (1:N)' * ones(1,N);   d = (1:N)' - sum(tril(S < G),2) ;
lambdaBN = 1 - 2 * max(d)/floor(N/2) ;
lambdaSWP = 1 - (max(d)/floor(N/2) + sum(d)/floor(N*N/4)) ;
```

8.9 Conversion to Probabilities

Until now we have only considered normalized techniques in which a sensor measurement x, and its corresponding normalized value y, have the same physical properties. We now consider a new normalization technique in which this is no longer true. In particular we give y an explicit physical meaning as an *a posteriori* probability[5].

We start with the simplest case in which all objects O belong to either a class $C = c_1$ or to a class $C = c_2$. Suppose x is the measured value of O. Then our aim is to find two functions $y = f_1(x)$ and $y = f_2(x)$ such that $f_k(x)$ closely approximates the *a posteriori* probability density function $p(C = c_k|x, I)$:

$$f_1(x) \approx p(C = c_1|x, I) , \tag{8.7}$$

$$f_2(x) \approx p(C = c_2|x, I) . \tag{8.8}$$

Since, by definition,

$$p(C = c_1|x, I) = 1 - p(C = c_2|x, I) , \tag{8.9}$$

we need only calculate one function $f_2(x)$. We now consider several different approaches to finding $f_2(x)$. For notational convenience we shall, from now on, write $f(x)$ in place of $f_2(x)$. Methods for calculating $f(x)$ are listed in Table 8.11. The following examples illustrate, respectively, the methods of Platt calibration, histogram binning and isotonic regression.

Table 8.11 Conversion to Probabilities

Method	Description
Platt calibration	Model $f(x)$ using a sigmoid function.
Histogram binning	Approximate $f(x)$ using a discrete histogram.
Kernel density estimation	Approximate $f(x)$ using a kernel density function, i. e. a continuous bin histogram.
Isotonic Regression	Model $f(x)$ using an isotonic function, i. e. a function of unknown shape which is non-decreasing.

Example 8.26. Converting SVM Scores into Probability Estimates [32, 51]. In Platt calibration we use a parametric function, $f(x|\alpha, \beta, \ldots, \delta)$, with a given shape. We learn the parameters $\alpha, \beta, \ldots, \delta$, using a database D of labeled objects $O_i, i \in \{1, 2, \ldots, N\}$. Given a measured value x derived from a support vector machine (SVM) we convert it into a estimate of $p(C = c_2|x, I)$ by assuming a sigmoid function for $f(x|\alpha, \beta)$:

$$f(x|\alpha, \beta) = \frac{1}{1 + \exp(\alpha(\beta - x))} ,$$

[5] Theoretically, this is the preferred normalization technique since, if successful, we have at our disposal all of the methods of Bayesian analysis (Chapt. 9). However, in practice, conversion to *a posteriori* probabilities is often found to perform badly and for this reason the methods considered previously are often used instead.

where α and β are two parameters which we adjust for maximum likelihood as follows. Suppose the object O_i has a value x_i and a class label y_i, where

$$y_i = \begin{cases} 0 \text{ if } O_i \text{ belongs to class } c_1 , \\ 1 \text{ if } O_i \text{ belongs to class } c_2 . \end{cases}$$

Then we find the parameters α and β by minimizing the negative log likelihood of the data set D:

$$(\alpha,\beta) = \arg\min \left[-\sum_{i=1}^{N} \Big(y_i \ln f(y_i|\alpha,\beta) + (1-y_i) \ln\big(1 - f(x_i|\alpha,\beta)\big) \Big) \right] .$$

In performing the above optimization procedure we must be careful not to *overfit*. This may happen if the number of objects O_i which belong to $C = c_1$ or to $C = c_2$ is too low. In this case, we mitigate the effects of overfitting by using modified labels, $\widetilde{y}_i$, in the above minimization procedure, where

$$\widetilde{y}_i = \begin{cases} 1/(N_1+1) & \text{if } O_i \text{ belongs to class } C = c_1 , \\ (N_2+1)/(N_2+2) & \text{if } O_i \text{ belongs to class } C = c_2 . \end{cases}$$

where $N_k, k \in \{1,2\}$, is the number of objects which belong to the class $C = c_k$.

Example 8.27. Histogram Binning. In *histogram binning* we do not know the shape of the function $f(x)$ and we cannot, therefore, use Platt calibration. Instead we use a non-parametric method such as frequency histogram or *binning*: Suppose the objects $O_i, i \in \{1,2,\ldots,N\}$, in D are arranged in order of increasing value, i. e. $x_1 \leq x_2 \leq \ldots \leq x_N$. We divide the x_i into M non-overlapping subsets, or *bins*, $B_m = [b'_m, b''_m), i \in \{1,2,\ldots,M\}$, where $b'_{m+1} = b''_m$. Then, we define $f(x) = p(C = c_2|x,I), b'_m \leq x \leq b''_m$, as

$$f(x) = \frac{\widetilde{n}_m^{(2)}}{\widetilde{n}_m^{(1)} + \widetilde{n}_m^{(2)}} ,$$

where $\widetilde{n}_m^{(k)} = n_m^{(k)} / \sum_{m=1}^{M} n_m^{(k)}$ and $n_m^{(k)}$ is the number of objects $O_i, i \in \{1,2,\ldots,N\}$, which simultaneously belong to the class $C = c_k$ and fall in the mth bin B_m. *Note*: To avoid problems of overfitting we often use the following equation:

$$f(x) = \frac{(\widetilde{n}_m^{(2)} + \delta)}{(\widetilde{n}_m^{(1)} + \widetilde{n}_m^{(2)} + 2\delta)} .$$

where $\delta = 1/N$.

Example 8.28. Isotonic Regression: Pair Adjacent Violation Algorithm [50]. In *isotonic regression* we do not know the shape of the function $y = f(x)$ except that it belongs to the class of *isotonic*, or non-decreasing, functions. Isotonic regression may therefore be regarded as an intermediary approach between sigmoid fitting and binning.

If we assume that the classifier ranks examples correctly, the function $f(x)$ from scores into probabilities is non-decreasing, and we can use isotonic regression to learn this mapping. A commonly used algorithm for computing the isotonic regression is the pair adjacent violation (PAV) algorithm. This algorithm finds the stepwise-constant isotonic function that best fits the data according to a mean-squared error criterion. Let $O_i, i \in \{1,2,\ldots,N\}$, be the training samples arranged such that the measurement values are in order of increasing value, i. e. $x_1 \leq x_2 \leq \ldots \leq x_N$. Let $f^*(x)$ be the sought after isotonic function. Let $f(x_i) = z_i$, then if f is already isotonic, we return $f^* = f$. Otherwise, there must be a subscript l such that $f(x_{l-1}) \leq f(x_l)$. The examples x_{l-1} and x_l are called *pair adjacent violators*, because they violate the isotonic assumption. The values of $f(x_{l-1})$ and $f(x_l)$ are then replaced by their average, so that the $l-1$th and lth examples now comply with the isotonic assumption. If this new set of $N-1$ values is isotonic, then $f^*(x_{l-1}) = f^*(x_l) = (f(x_{l-1}) + f(x_l))/2$ otherwise $f^*(x_l) = f(x_l)$. This process is repeated using the new values until an isotonic set of values is obtained. Fig. 8.4 illustrates the action of the PAV algorithm.

8.9.1 Multi-class Probability Estimates *

Thus far, we have considered two-class, or binary, probability estimates. We now apply the notion of calibration to *multi-class* probability estimates. Let $c_k, k \in \{1,2,\ldots,K\}$, denote K classes. Then our aim is to find a multi-dimensional function $f(x)$ such that $f(x)$ closely approximates the *a posteriori* probability density function $p(\{C = c_k\}|x,I)$:

$$f(x) \approx p(\{C = c_k\}|x,I) . \tag{8.10}$$

The function $f(x)$ represents a multidimensional mapping from one K-dimensional space to another K-dimensional space. In this case, it is not clear what function shape we should assume for the transformation function. For this reason, the direct calibration of the multi-class probabilities is not recommended. Instead we reduce the multi-class problem into $K(K-1)/2$ binary, or two-class, problems and then combine the binary probability estimates using a *voting* or other similar algorithm.

The following example illustrates the use of a simple voting algorithm to combine the binary probability estimates.

2.0	3.0	3.0	3.5	7.0	9.0	9.5	9.8	9.9	(a)
0	0	1	0	0	1	0	1	1	(b)
0	0	1	0	0	1	0	1	1	(c)
0	0	0.5	0.5	0	0.5	0.5	1	1	(d)
0	0	0.5	.25	.25	0.5	0.5	1	1	(e)
0.0	0.0	.33	.33	.33	0.5	0.5	1.0	1.0	(f)

Fig. 8.4 Shows the Pair Adjacent Violation (PAV) algorithm in action. **(a)** Shows the training samples O_i arranged in order of increasing value x_i. **(b)** Shows class label z_i value for each sample O_i. **(c)** Boxes with heavy margins show the adjacent samples which violate isotonicity. **(d)** and **(e)** Shows two iterations of the PAV algorithm in which we average the labels of the adjacent samples which violate isotonicity. **(f)** Shows the final isotonic function $f^*(y)$.

Example 8.29. Multi-Class Probability Estimates Using a Voting Algorithm [46, 49]. We consider a multi-class probability estimate involving K classes $c_k, k \in \{1,2,\ldots,K\}$. Let $f_{kl}(x|\alpha_{kl},\beta_{kl})$ denote a probabilistic calibration curve for classes $c_k, c_l, l > k$. The curves f_{kl} may be obtained directly from the classification algorithm or may be estimated using any one of the methods listed in Sect. 8.9. Then the *a posteriori* probabilities $p(c_k|x,I)$ for an input x are:

$$p(C = c_k|x) = \frac{1}{Z}\sum_{\substack{l=1\\ l\neq k}}^{K} f_{kl}(x|\alpha_{kl},\beta_{kl}) ,$$

where $f_{kl}(x|\alpha_{kl},\beta_{kl}) = 1 - f_{lk}(x|\alpha_{lk},\beta_{lk})$ and $Z = \sum_{m=1}^{K}\sum_{n=1}^{K} f_{mn}(y|\ \alpha_{mn},\beta_{mn})$ is a normalization factor which ensures the *a posteriori* probabilities sum to one.

Additional methods [46] for estimating the multi-class probabilities, $\mathbf{p} = (p_1, p_2, \ldots, p_K)^T$, are listed in Table 8.12.

Table 8.12 Methods for Estimating Multi-Class Probabilities **p**

Method	Description
PKPD	$p_k = -1/(K-2-\sum_{l,l\neq k} r_{kl}^{-1})$; where r_{kl} is the estimate that a given object O belongs to $C = c_k$ assuming it belongs to either $C = c_k$ or $C = c_l$.
Hastie-Tibshani	$\mathbf{p} = \arg\min_{\mathbf{p}} \sum_k [\sum_{l\neq k}(r_{kl}/K - p_k/2)]^2$.
WLW_1	$\mathbf{p} = \arg\min_{\mathbf{p}} \sum_k [\sum_{l\neq k}(r_{kl}p_l - r_{lk}p_k)]^2$.
WLW_2	$\mathbf{p} = \arg\min_{\mathbf{p}} \sum_k \sum_{l\neq k}(r_{kl}p_k - r_{lk}p_k)^2)$.

8.10 Software

LIBRA. A matlab toolbox for performing robust statistics. Authors: Sabine Verboven, Mia Hubert. The toolbox contains m-files on various robust normalization techniques.

LIBSVM. A matlab toolbox for performing SVM. Authors: Chih-Chung Chang, chih-Jen Lin. The toolbox contains an m-file for Platt calibration [23]. *Note:* This code is an improvement on the pseudo-code given in [32] which was found to give biased *a posteriori* probability estimates [26].

MATLAB STATISTICAL TOOLBOX. The mathworks matlab statistical toolbox. The toolbox contains m-files for performing various normalization procedures.

8.11 Further Reading

Similarity measures have been intensely investigated in many different fields. For the spatiogram, Gong *et al.* [17] have described a new similarity measure based on lie groups. In the chemical industry, Willett and co-workers [42, 43, 45] have published widely on the use of different similarity measures. For a comprehensive review of these similarity measures see [4, 11]. The advantages of using *calibrated* probability estimates are discussed at length in [12].

Problems

8.1. What is virtual screening? List some of the similarity coefficients which are used in virtual screening.

8.2. Define binarization and list some of its uses in multi-sensor data fusion.

8.3. Describe fuzzy normalization. What are the advantages/disadvantages of using a fuzzy normalization scheme.

8.4. Explain the advantages of using ranking as a normalization procedure.

8.5. Compare and contrast the following methods for converting measurements to probabilities: Platt calibration, binning, kernels, isotonic regression.

8.6. Explain Platt calibration. How may we reduce the impact of overfitting on Platt calibration.

8.7. Describe the process of isotonic regression.

References

1. Baggerly, K.A.: Probability binning and testing agreement between multivariate immunofluorescence histogram extending the chi-squared test. Cytometry 45, 141–150 (2001)
2. Baradez, M.-O., McGuckin, C.P., Forraz, N., Pettengell, R., Hope, A.: Robust and automated unimodal histogram thresholding and potential applications. Patt. Recogn. 37, 1131–1148 (2004)
3. Behloul, F., Lelieveldt, B.P.E., Boudraa, A., Janier, M., Revel, D., Reiber, J.H.C.: Neuro-fuzzy systems for computer-aided myocardial viability assessment. IEEE Trans. Med. Imag. 20, 1302–1313 (2001)
4. Bender, A., Glen, R.C.: Molecular similarity: a key technique in molecular informatics. Org. Biomol. Chem. 2, 3204–3218 (2004)
5. Bhat, D., Nayar, S.: Ordinal measures for image correspondence. IEEE Trans. Patt. Analy. Mach. Intell. 20, 415–423 (1998)
6. Birchfield, S.T., Rangarajan, S.: Spatial histograms for region-based tracking. ETRI Jrnl. 29, 697–699 (2007)
7. Birge, L., Rozenholc, Y.: How many bins should be put in a regular histogram. Euro. Series Appl. Indust. Math. Prob. Stat. 10, 24–45 (2006)
8. Bruzzone, L., Prieto, D.F.: Automatic analysis of the difference image for unsupervised change detection. IEEE Trans. Geosci. Rem. Sens. 38, 1171–1182 (2000)
9. Celebri, M.E., Iyatomi, H., Schaefer, G., Stoecker, W.V.: Approximate lesion localization in demoscopy images. Skin Res. Tech. 15, 314–322 (2009)
10. Cha, S.-H.: Taxonomy of nominal type histogram distance measures. In: Am. Conf. App. Math., pp. 325–330 (2008)
11. Choi, S.-S., Cha, S.-H., Tappert, C.C.: A survey of binary similarity and distance measures. Syst. Cyber Inform. 8, 43–48 (2010)
12. Cohen, I., Goldszmidt, M.: Properties and Benefits of Calibrated Classifiers. In: Boulicaut, J.-F., Esposito, F., Giannotti, F., Pedreschi, D. (eds.) PKDD 2004. LNCS (LNAI), vol. 3202, pp. 125–136. Springer, Heidelberg (2004)
13. Coudray, N., Buessler, J.-L., Urban, J.-P.: Robust threshold estimation for images with unimodal histograms. Patt. Recogn. Lett. 31, 1010–1019 (2010)
14. Delon, J.: Midway image equalization. Math. Imag. Vis. 21, 119–134 (2004)
15. Gilchrist, J.M., Jerwood, D., Ismaiel, H.S.: Comparing and unifying slope estimates across pyschometric function models. Perception and Psychophysics 67, 1289–1303 (2005)
16. Ginn, C.M.R., Turner, D.B., Willett, P., Ferguson, A.M., Heritage, T.W.: Similarity searching in files of three-dimensional chemical structures: evaluation of the EVA descriptor and combination of rankings using data fusion. Chem. Inform. Comp. Sci. 37, 23–37 (1997)
17. Gong, L., Wang, T., Liu, F., Chen, G.: A lie group based spatiogram similarity measure. In: Proc. Int. Conf. Multimedia Expo., ICME (2009)
18. Henning, K.J.: What is syndromic surveillance? In: Syndromic Surveillance: Reports from a National Conference (2003)
19. Holliday, J.D., Hu, C.-Y., Willett, P.: Grouping of coefficients for the calculation of intermolecular similarity and dissimilarity using 2D Fragment bit-strings. Combinatorial Chem. Comp. Sci. 42, 375–385 (2001)
20. Hongzhi, W., Ying, D.: An improved image segmentation algorithm based on Otsu method. In: Proc. SPIE, vol. 6625 (2008)

21. Ivanov, O., Wagner, M.M., Chapman, W.W., Olszewski, R.T.: Accuracy of three classifiers of acute gastrointestinal syndrome for syndromic surveillance. In: Proc. AMIA Symp., pp. 345–349 (2002)
22. Jain, A., Nandakumar, K., Ross, A.: Score normalization in multimodal biometric systems. Patt. Recogn. 38, 2270–2285 (2005)
23. Lin, H.-T., Lin, C.-J., Weng, R.C.: A note on Platt's probabilistic outputs for support vector machine. In: Tech Rept. Dept. Comp. Sci. Inform. Engng., National Taiwan University (2003)
24. Ma, Y., Gu, X., Wang, Y.: Histogram similarity measure using variable bin size distance. Comp. Vis. Image Under. 114, 981–989 (2010)
25. Madden, C., Cheng, E.D., Piccardi, M.: Tracking people across disjoint camera views by illumination-tolerant appearance representation. Mach. Vis. Appl. 18, 233–247
26. Milgram, J., Cheriet, M., Sabourin, R.: Estimating accurate multi-class probabilities with support vector machine. In: Int. J. Conf. Neural Network IJCNN, pp. 1906–1911 (2005)
27. Monwar, M.M., Gavrilova, M.L.: Multimodal biometric system using rank-level fusion approach. IEEE Trans. Sys. Man. Cybern. 39B, 867–878 (2009)
28. Nelson, T., Wilson, H.G., Boots, B., Wulder, M.A.: Use of ordinal conversion for radiometric normalization and change detection. Int. J. Rem. Sens. 26, 535–541 (2005)
29. O'Conaire, C., O'Connor, N., Cooke, E., Smeaton, A.: Detection thresholding using mutual information. In: Proc. 1st Int. Conf. Cmp. Vis. Theory App., vol. 2, pp. 408–415 (2006)
30. O'Conaire, C., O'Connor, N.E., Smeaton, A.: An improved spatiogram similarity measure for robust object localization. In: Int. Conf. Acc. Speech Sig. Proc. ICASSP (2007)
31. O'Conaire, C., O'Connor, N.E., Smeaton, A.: Thermo-visual feature fusion for object tracking using multiple spatiogram trackers. Mach. Vis. Appl. 19, 483–494 (2008)
32. Platt, J.: Probabilistic outputs for support vector machines and comparisons to regularized likelihood methods. In: Smola, A.J., Bartlett, P., Scholkopf, B., Schurmans, D. (eds.) Advances in Large Margin Classifiers, pp. 61–74. MIT Press (1999)
33. Ridler, T., Calvard, S.: Picture thresholding using an iterative selection method. IEEE Sys. Man Cyber. 8, 630–632 (1978)
34. Roederer, M., Treister, A., Moore, W., Herzenberg, L.A.: Probability binning comparison: a metric for quantitating univariate distribution differences. Cytometry 45, 37–46 (2001)
35. Rosin, P.L.: Unimodal thresholding. Patt. Recogn. 34, 2083–2096 (2001)
36. Rubner, Y., Tomasi, C., Guibas, L.J.: The earth movers distance as a metric for image retrieval. Int. J. Comp. Vis. 40, 99–121 (2000)
37. Dos Santosa, D.A., Deutsch, R.: The positive matching index: a new similarity measure with optimal characteristics. Patt. Recogn. Lett. 31, 1570–1576 (2010)
38. Scherer, S., Werth, P., Pinz, A.: The discriminatory power of ordinal measures - towards a new coefficient. In: Proc. IEEE Conf. Comp. Vis. Patt. Recogn. (1999)
39. Sevcenco, A.-M., Lu, W.-S.: Perfect histogram matching PCA for face recognition. Multidimen. Syst. Sig. Process. 21, 213–229 (2010)
40. Seggin, M., Sankur, B.: Survey over image thresholding techniques and quantitative performance evaluation. Elect. Imag. 13, 146–165 (2004)
41. Suh, Y., Ji, M., Kim, H.: Probabilistic class histogram equalization for robust speech recognition. IEEE Sig. Proc. Lett. 14, 287–290 (2007)
42. Willett, P.: Chemical similarity searching. Chem. Inform. Comp. Sci. 38, 983–996 (1998)
43. Willett, P.: Textual and chemical information processing: different domains but similar algorithms. Inform. Res. 5(2) (2000)
44. Willett, P.: Structural biology in drug metabolism and drug discovery. Biochem. Soc. Trans., Part 3 31, 603–606 (2003)
45. Wilton, D., Willet, P., Lawson, K., Mullier, G.: Comparison of ranking methods for virtual screening in lead-discovery programmes. Chem. Inform. Comp. Sci. 43, 469–474 (2003)
46. Wu, T.-F., Lin, C.-J., Weng, R.C.: Probability estimates for multiclass classification by pairwise coupling. Mach. Learn. Res. 5, 975–1005 (2004)
47. Xu, X., Xu, S., Jin, L., Song, E.: Characteristic analysis of Otsu threshold and its applications. Patt. Recogn. Lett. 32, 956–961 (2011)

48. Xue, J.-H., Titterington, M.: Median-based image thresholds. Preprint (2010)
49. Zadrozny, B.: Reducing multiclass to binary by coupling probability estimates. In: Neural Inf. Process. Sys. Conf., British Columbia, Vancouver (1999)
50. Zadrozny, B.: Policy mining: learning decision policies from fixed sets of data. PhD thesis, University of California, San Diego (2003)
51. Zadrozny, B., Elkan, C.: Transforming classifier scores into accurate multiclass probability estimates. In: SIGKDD 2002, Edmonton, Alberta, Canada (2002)

Chapter 9
Bayesian Inference

9.1 Introduction

In this chapter we give an overview of Bayesian statistics, and in particular, the methods of Bayesian inference as used in multi-sensor data fusion. The basic premise of Bayesian statistics is that all unknowns are treated as random variables and that the knowledge of these quantities is summarized via a probability distribution. The main advantages of using Bayesian statistics are

1. Bayesian statistics is the only known coherent system for quantifying objective and subjective uncertainties.
2. Bayesian statistics provides principled methods for the model estimation and comparison and the classification of new observations.
3. Bayesian statistics provides a natural way to combine different sensor observations.
4. Bayesian statistics provides principle methods for dealing with missing information.

In Table 9.1 we list some of the basic formulas which are used in Bayesian statistics. In a multi-sensor data fusion system, we use Bayesian statistics to represent the multiple sources of information. The corresponding probability distributions act as a powerful *common representational format*.

9.2 Bayesian Analysis

In Bayesian statistics we treat all quantities under consideration as random variables. This includes the observed data, $\mathbf{y}$ and any missing data, $\mathbf{z}$, unknown parameters, $\boldsymbol{\theta}$, and models, M. The full process of Bayesian analysis can be described as consisting of three stages:

Table 9.1 Basic Formulas Used in Bayesian Inference

Name	Formula
Probability Density Function (pdf)	$p(y\|I)dy = P(y \le Y \le y + dy\|I)$.
Normalization	$\int p(y\|I)dy = 1$.
Expectation of $f(Y)$	$E(f(Y)) = \int f(y)p(y\|I)dy$.
Expected Value	$E(Y) = \int yp(y\|I)dy$.
Moment of Order r	$M_r(Y) = \int y^r p(y\|I)dy$.
Variance	$\sigma^2 = \int (y - E(Y))^2 p(y\|I)dy$.
Product Rule	$p(x,y\|I) = p(x\|y,I)p(y\|I)$.
Independence	$p(x,y\|I) = p(x\|I)p(y\|I)$.
Marginalization	$p(x\|I) = \int p(x,y\|I)dy$.
Decomposition	$p(x\|I) = \int p(x\|y,I)p(y\|I)dy$.
Bayes' Rule	$p(x\|y,I) = p(y\|x,I)p(x\|I)/p(y\|I)$;
	$p(x\|y,z,I) = p(x\|I)p(y\|x,I)p(z\|x,y,I)/(p(y\|I)p(z\|I))$.
Likelihood	$L = p(y\|x,I)$.

In the table X and Y denote two random variables whose instantiations are, respectively, x and y. For variables with continuous values we do not normally distinguish between the random variable and its instantiation. Thus we write $p(y|I)$ and not $p(Y = y|I)$. In the case of variables with discrete values this may, however, cause confusion. For these variables we shall therefore continue to write $p(Y = y|I)$. In the table, I describes any background information we may have concerning the problem in hand. For definitions of additional terms, see e. g. the glossary of selected statistical terms in [28].

Probability Model. In this stage, we create a joint probability distribution that captures the relationship among all the variables under consideration.

A Posterior Distribution. In this stage, we summarize all of our information regarding the different quantities of interest in a set of *a posteriori* distributions. Typically the *a posteriori* distribution is a conditional probability density function (pdf) $p(\mathbf{y}|\boldsymbol{\theta},I)$.

Model Selection. In this stage, we evaluate the appropriateness of using a given model. We may then select the best model, or create a new model, or suggest improvements in one of the existing models.

The following example illustrates the role played by Bayesian analysis in the scientific process.

Example 9.1. Scientific Theories [16, 21]. Fig. 9.1 is adapted from [21] and illustrates that part of the scientific process in which data is collected and modeled. The two heavily-framed boxes denote processes which involve Bayesian inference. In the first box we infer what the model parameters might be given the model and the data. In the second box we infer what model is probably closer to the correct model given the data.

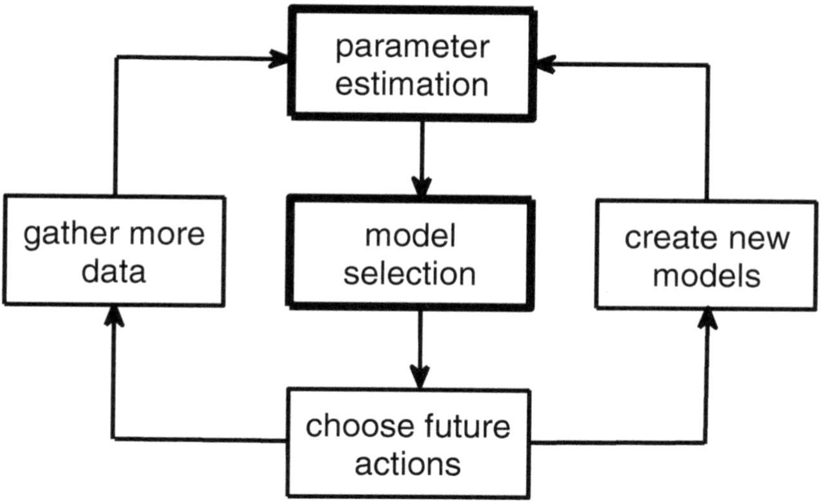

Fig. 9.1 Shows the role played by Bayesian inference in the scientific process.

9.3 Probability Model

The standard procedure for setting up a full probability model, or joint probability distribution, is to write down the likelihood function, i. e. the probability of the observed data given the unknowns and multiply it by the *a priori* distribution of all the unobserved variables and parameters. We assume the observed data, $\mathbf{y}$, comes from a parametric pdf p characterized by the parameters $\boldsymbol{\theta}$. Then the joint pdf can be represented as

$$p(\mathbf{y},\boldsymbol{\theta}|I) = p(\mathbf{y}|\boldsymbol{\theta},I)\pi(\boldsymbol{\theta}|\boldsymbol{\lambda},I) , \tag{9.1}$$

where $\boldsymbol{\theta}$ is assumed to come from some *a priori* distribution, π, with parameters $\boldsymbol{\lambda}$[1] and I denotes all of our relevant background knowledge.

A convenient representation of the model is the graphical model (Fig. 9.2). The input data, $\mathbf{y}$, the missing data, $\mathbf{z}$, and the model parameters, $\boldsymbol{\theta}$, are represented as nodes in a graph. Relationships and influences between the nodes are then represented as straight lines, or edges, which link the nodes.

9.4 A Posteriori Distribution

The Bayesian inference is drawn by examining the probability of all possible values of the parameters, $\boldsymbol{\theta}$, after considering the data, $\mathbf{y}$. If the hyperparameters, $\boldsymbol{\lambda}$, are

[1] Here in after the parameters $\boldsymbol{\lambda}$ are referred to as *hyperparameters* to differentiate them from the parameters $\boldsymbol{\theta}$.

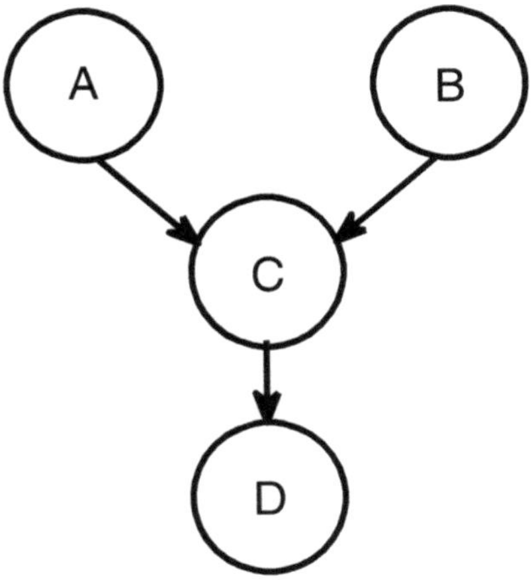

Fig. 9.2 Shows a simple graphical model with nodes $\{A,B,C,D\}$. The set $\{A,B\}$ is the parents of C and C is the child of $\{A,B\}$ and the parent of D. Mathematically we write these relationships as: $\mathrm{pa}(C) = \{A,B\}$, $\mathrm{ch}(A) = C$, $\mathrm{ch}(B) = C$ and $\mathrm{pa}(D) = C$. The parents and child of a node, plus the children's other parents, constitute a *Markov Blanket*. The Markov Blanket of A is $\{B,C\}$. Mathematically we write this as $\mathrm{mk}(A) = \{\mathrm{pa}(A), \mathrm{ch}(A), \mathrm{pa}(\mathrm{ch}(A))\}$ or $\mathrm{mk}(A) = \{B,C\}$.

known (or are estimated), then we obtain the *a posteriori* pdf through the application of Bayes' theorem

$$\begin{aligned} p(\boldsymbol{\theta}|\mathbf{y},I) &= \frac{\overbrace{p(\mathbf{y}|\boldsymbol{\theta},I)}^{\text{likelihood}}\,\overbrace{\pi(\boldsymbol{\theta}|\boldsymbol{\lambda},I)}^{a\ priori\ \text{pdf}}}{\underbrace{p(\mathbf{y}|I)}_{\text{evidence}}}, \\ &= \frac{p(\mathbf{y}|\boldsymbol{\theta},I)\pi(\boldsymbol{\theta}|\boldsymbol{\lambda},I)}{\int p(\mathbf{y}|\boldsymbol{\theta},I)\pi(\boldsymbol{\theta}|\boldsymbol{\lambda},I)d\boldsymbol{\theta}}. \end{aligned} \tag{9.2}$$

The denominator $p(\mathbf{y}|I) = \int p(\mathbf{y}|\boldsymbol{\theta},I)\pi(\boldsymbol{\theta}|\boldsymbol{\lambda},I)d\boldsymbol{\theta}$ is known as the *evidence* and is a normalizing constant which is required so that $p(\boldsymbol{\theta}|\mathbf{y},I)$ will integrate to one. The evidence is obtained by integrating out all the variables, except for the observed data, $\mathbf{y}$, from the joint distribution.

If the hyperparameters, $\boldsymbol{\lambda}$, are not known, then we may also remove them by integration:

$$p(\boldsymbol{\theta}|\mathbf{y},I) = \frac{\int p(\mathbf{y}|\boldsymbol{\theta},I)\pi(\boldsymbol{\theta}|\boldsymbol{\lambda},I)\pi(\boldsymbol{\lambda}|I)d\boldsymbol{\lambda}}{\int\int p(\mathbf{y}|\boldsymbol{\theta},I)\pi(\boldsymbol{\theta}|\boldsymbol{\lambda},I)\pi(\boldsymbol{\lambda}|I)d\boldsymbol{\lambda}d\boldsymbol{\theta}}, \tag{9.3}$$

where $\pi(\boldsymbol{\lambda}|I)$ is the *a priori* pdf for $\boldsymbol{\lambda}$[2].

[2] Here in after $h(\boldsymbol{\lambda})$ is referred to as a *hyperprior* distribution to differentiate it from the *a priori* distribution $\pi(\boldsymbol{\theta}|I)$.

Equation (9.3) tells us how we may systematically update our knowledge of $\boldsymbol{\theta}$ given the data $\mathbf{y}$. For example, if the observations are obtained one at a time, we can update the *a posteriori* distribution as follows:

$$\begin{aligned} p(\boldsymbol{\theta}|\mathbf{y}_1,I) &\sim p(\mathbf{y}_1|\boldsymbol{\theta},I)\pi(\boldsymbol{\theta}|I)\;, \\ p(\boldsymbol{\theta}|\mathbf{y}_1,\mathbf{y}_2,I) &\sim p(\mathbf{y}_2|\boldsymbol{\theta},I)p(\boldsymbol{\theta}|\mathbf{y}_1,I)\;, \\ &\cdots \\ p(\boldsymbol{\theta}|\mathbf{y}_1,\mathbf{y}_2,\ldots,\mathbf{y}_N,I) &\sim p(\mathbf{y}_N|\boldsymbol{\theta},I)p(\boldsymbol{\theta}|\mathbf{y}_1,\mathbf{y}_2,\ldots,\mathbf{y}_{N-1},I)\;. \end{aligned} \tag{9.4}$$

When there is more than one unknown parameter, e. g. $\boldsymbol{\theta} = (\theta_1, \theta_2)$, and we are only interested in one component, say $\boldsymbol{\theta}_1$, those unknown quantities that are not of immediate interest, but are needed by the model, are known as *nuisance parameters*, and are removed by integration:

$$p(\boldsymbol{\theta}_1|\mathbf{y}) = \frac{\int p(\mathbf{y}|\boldsymbol{\theta}_1,\boldsymbol{\theta}_2,I)\pi(\boldsymbol{\theta}_1,\boldsymbol{\theta}_2|I)d\boldsymbol{\theta}_2}{\int p(\mathbf{y}|\boldsymbol{\theta}_1,\boldsymbol{\theta}_2,I)\pi(\boldsymbol{\theta}_1,\boldsymbol{\theta}_2|I)d\boldsymbol{\theta}_1 d\boldsymbol{\theta}_2}\;. \tag{9.5}$$

Despite the deceptively simple-looking form of (9.5), the application of Bayesian statistics can be difficult. The most challenging aspects being: (1) the development of a model, $p(\mathbf{y}|\boldsymbol{\theta},I)\pi(\boldsymbol{\theta},I)$, which must effectively capture the key features of the underlying physical problem; and (2) the necessary computation required for deriving the *a posteriori* distribution.

9.4.1 Standard Probability Distribution Functions

In developing adequate models for the likelihood $p(\boldsymbol{\theta}|\mathbf{y})$ and the *a priori* and *a posteriori* distributions, $p(\mathbf{y}|\boldsymbol{\theta},I)$ and $p(\boldsymbol{\theta}|\mathbf{y},I)$, we often use standard mathematical distributions [35]. Table 9.2 is a list of the most important univariate distributions $f(y)$.

9.4.2 Conjugate Priors

An early effort at making the integration required by (9.2) accessible was the development of the *conjugate priors* [1, 13, 19]. Formally, a conjugate prior is a family of distributions for $\pi(\boldsymbol{\theta}|I)$ that has the same functional form as the likelihood function. As a consequence, when a conjugate prior is used, the functional form of the *a posteriori* distribution is the same as that of the *a priori* distribution. Thus although we can choose any functional form for $\pi(\boldsymbol{\theta}|I)$, the conjugate prior enjoys the greatest mathematical and computational advantage. In Table 9.3 we list the important families of conjugate likelihood and *a priori* distributions [13, 19].

Table 9.2 Important Univariate Distributions $f(y)$

Name	Formula
Beta $\mathscr{B}e(\alpha,\beta)$	$\Gamma(\alpha+\beta)y^{\alpha-1}(1-y)^{\beta-1}/(\Gamma(\alpha)\Gamma(\beta))$; $0 \le y \le 1; \alpha,\beta > 0$.
Binomial $\mathscr{B}in(n,q)$	$C(n,q)q^y(1-q)^{n-y}; n \in N$.
Cauchy $\mathscr{C}a(a,b)$	$1/b\pi \times (1+(y-a)^2/b^2)^{-1}; b > 0$.
Chi-Square $\chi^2(n)$	$\exp-(y/2)y^{(n-2)/2}/(2^{n/2}\Gamma(n/2); y \ge 0, n \in N$.
Exponential $\mathscr{E}(\lambda)$	$\lambda\exp-(\lambda y); y \ge 0; \lambda > 0$.
Gamma $\mathscr{G}a(\alpha,\beta)$	$y^{\alpha-1}\exp-(y/\beta)\beta^\alpha/\Gamma(\alpha); \alpha > 0; \beta > 0$.
Gaussian $\mathscr{N}(\mu,\sigma^2)$	$(2\pi\sigma^2)^{(-1/2)}\exp-\frac{1}{2}((y-\mu)/\sigma)^2; \sigma > 0$.
Inverse Gamma $\mathscr{IG}(\alpha,\beta)$	$y^{-(\alpha+1)}\exp(y/\beta)\beta^\alpha/\Gamma(\alpha);\ \alpha > 0$.
Laplace $\mathscr{L}a(\mu,\sigma)$	$\exp-(\lvert y-\mu\rvert\sigma)/\sigma; \sigma > 0$.
Logistic $\mathscr{L}o(\alpha,\beta)$	$\frac{1}{\beta}(\exp-(y-\alpha)/\beta)/(1+\exp-(y-\alpha/\beta))^2$.
Pareto $\mathscr{P}a(a,\theta)$	$\theta a^\theta/y^{\theta+1}; y \ge a, \theta > 2, a > 0$.
Student-t $\mathscr{S}t(\mu,\sigma,\nu)$	$\Gamma((\nu+1)/2)/(\sigma\sqrt{\nu\pi}\Gamma(\nu/2)) \times [1 + (y-\mu)^2/\nu\sigma^2]^{-(\nu+1)/2}; \nu \in N, \nu > 2, \sigma > 0$.
Uniform $\mathscr{U}(a,b)$	$1/(b-a), 0 \le y \le 1; a,b > 0$.

Table 9.3 Conjugate Distributions

Likelihood $p(y_i\lvert\theta,I)$	*A Priori* $\pi(\theta\lvert I)$	*A Posteriori* $p(\theta\lvert\{y_i\},I)$
$\mathscr{N}(\theta,\sigma^2)$[known σ^2]	$\mathscr{N}(\mu,\tau^2)$	$\mathscr{N}((N\tau^2\bar{y}+\sigma^2\mu)/\Delta,\sigma^2\tau^2/\Delta)$, where $\Delta = \sigma^2 + N\tau^2$.
$\mathscr{N}(\mu,1/\theta)$[known μ]	$\mathscr{G}a(\alpha,\beta)$	$\mathscr{G}a(\alpha+N/2,(2\beta+\sum_{i=1}^N(y_i-\mu)^2)/2)$.
$\mathscr{N}(\theta_1,\theta_2)$		See Ex. 9.3.
$\mathscr{P}ois(\theta)$	$\mathscr{G}a(\alpha,\beta)$	$\mathscr{G}a(\alpha+\sum_{i=1}^N y_i,\beta+N)$.
$\mathscr{G}a(\nu,\theta)$[known ν]	$\mathscr{G}a(\alpha,\beta)$	$\mathscr{G}a(\alpha+N\nu,\beta+\sum_{i=1}^N y_i)$.
$\mathscr{U}(0,\theta)$	$\mathscr{P}a(\theta_0,\alpha)$	$\mathscr{P}a(\max(\theta_0,y),\alpha+1)$.
$\mathscr{G}a(\nu=\frac{1}{2},2\theta)$ [known ν]	$\mathscr{IG}(\alpha,\beta)$	$\mathscr{IG}(\frac{1}{2}+\alpha,1/(\frac{y}{2}+\beta^{-1}))$.
$\mathscr{E}(\theta)$	$\mathscr{G}a(\alpha,\beta)$	$\mathscr{G}a(\alpha+1,\beta+\sum_{i=1}^N y_i)$.
$\mathscr{B}in(n,\theta)$[known n]	$\mathscr{B}e(\alpha,\beta)$	$\mathscr{B}e(\alpha+y,\beta+n-y)$.

The following examples illustrate the use of Table 9.3 in calculating the *a posteriori* pdf for a Gaussian likelihood.

Example 9.2. Conjugate Prior for a Gaussian Distribution with Known Variance [13]. We observe N independent samples $y_i, i \in \{1,2,\ldots,N\}$, which are distributed according to the Gaussian distribution $\mathscr{N}(\mu,\sigma^2)$ with unknown mean μ. The variance σ^2 is assumed known and is set equal to the unbiased maximum likelihood value:

$$\sigma^2_{\mathrm{ML}} = \frac{1}{N-1}\sum_{i=1}^N (y_i-\mu)^2 .$$

According to Table 9.3 the *a posteriori* pdf $p(\mu|\{y_i\},I)$ is

$$p(\mu|\{y_i\},\sigma^2 = \sigma^2_{\mathrm{ML}},I) = \mathcal{N}\left(\frac{\tau^2 \sum_i y_i + \sigma^2_{\mathrm{ML}}\mu}{\sigma^2_{\mathrm{ML}} + N\tau^2}, \frac{\sigma^2_{\mathrm{ML}}\tau^2}{\sigma^2_{\mathrm{ML}} + N\tau^2}\right) .$$

Example 9.3. Conjugate Prior for a Gaussian Distribution with Unknown Mean and Unknown Variance [13]. We reconsider Ex. 9.2 but assume the mean μ and the variance σ^2 are unknown. The *a priori* distribution for μ and σ^2 is

$$\begin{aligned}\pi(\mu,\sigma^2|I) &= \pi(\mu|\sigma^2,I)\pi(\sigma^2|I) ,\\ &= \mathcal{N}(\mu_0,\sigma^2/k_0)\mathcal{IG}(\nu_0,\sigma_0^2) .\end{aligned}$$

The joint *a posteriori* pdf is

$$p(\mu,\sigma^2|\{y_i\},I) = \mathcal{N}\left(\frac{k_0\mu_0 + N\bar{y}}{k_N}, \frac{\sigma^2}{k_N}\right)\mathcal{IG}(\nu_N,\sigma_N^2) ,$$

where $k_N = k_0 + N$, $\nu_N = \nu_0 + N$, $\mu_N = (k_0\mu_0 + N\bar{y})/k_N$ and $\nu_N\sigma_N^2 = \nu_0\sigma_0^2 + \sum_{i=1}^N (y_i - \bar{y})^2 + k_0 N(\bar{y} - \mu_0)^2/k_N$. By integrating out σ^2 and μ we obtain the marginal distributions: $p(\mu|\{y_i\},I)$ and $p(\sigma^2|\{y_i\},I)$:

$$\begin{aligned}p(\mu|\{y_i\},I) &= \int p(\mu,\sigma^2|\{y_i\},I)d\sigma = \mathcal{S}t(\mu_N,\sigma_N^2/k_N,\nu_N) ,\\ p(\sigma^2|\{y_i\},I) &= \int p(\mu,\sigma^2|\{y_i\},I)d\mu = \mathcal{IG}(\nu_N,\sigma_N^2) .\end{aligned}$$

9.4.3 Non-informative Priors

Another class of *a priori* distributions is the *non-informative* or "flat" *a priori* distributions. These distributions are designed for those situations when we essentially we have no knowledge of what the true distributions should be. The simplest non-informative *a priori* distribution for a given variable θ is to assume $\pi(\theta|I)$ is proportional to one: $\pi(\theta|I) \propto 1$. However this choice of *a priori* pdf is not invariant to scale changes.

Considerations of this sort led to the concept of Jeffrey's *a priori* pdfs which are invariant to changes of variables. They are defined as follows:

$$\pi(\theta|I) \propto |I(\theta)|^{1/2}, \tag{9.6}$$

where $I(\theta) = -E(d^2 \log p(\mathbf{y}|\theta)/d\theta^2)$ is the *Fisher information*.

Other non-informative *a priori* distributions are the reference and the maximal data information prior (MDIP) distributions [18, 34]. In Table 9.4 we list the uniform (U), Jeffrey's (J), reference (R) and the MDIP (M) non-informative priors for a Gaussian distribution.

Table 9.4 Non-Informative Priors Distributions

Likelihood $p(y_i\|\theta,I)$	*A Priori* $\pi(\theta\|I)$	*A Posteriori* $p(\theta\|\{y_i\},I)$
$\mathcal{N}(\theta,\sigma^2)$ [σ^2 known]	1 (U,J,R,M)	$\mathcal{N}(\bar{y},\sigma^2/N)$; $\bar{y}=\Sigma_{i=1}^N y_i/N$.
$\mathcal{N}(\mu,\theta)$ [μ known]	1 (U)	$\mathcal{IG}((N-2),2/S^2)$; $S^2=\Sigma_{i=1}^N (y_i-\bar{y})^2$.
	$1/\sigma^2$ (J, R, M)	$\mathcal{IG}(N/2,2/S^2)$.
$\mathcal{N}(\theta_1,\theta_2)$	1 (U)	$\theta_1 \sim \mathcal{S}t(\bar{y},S^2/(N-3),N-3)$.
		$\theta_2 \sim \mathcal{IG}((N-3)/2,2/S^2)$.
	$1/\sigma^4$ (J)	$\theta_1 \sim \mathcal{S}t(\bar{y},S^2/(N(N+1)),N+1)$.
		$\theta_2 \sim \mathcal{IG}((N+1)/2,2/S^2)$.
	$1/\sigma^2$ (R,M)	$\theta_1 \sim \mathcal{S}t(\bar{y},S^2/(N(N-1)),N-1)$.
		$\theta_2 \sim \mathcal{IG}((N-1)/2,2/S^2)$.

Note: The use of non-informative priors is controversal especially their use for multivariate distributions. As we explained not all of the priors given there are location and scale independent [3]. For example, for a Gaussian likelihood in which both μ and σ^2 are unknown, Ref [26] recommends $\pi(\mu,\sigma^2|I) \sim \frac{1}{\sigma^3}$ (this is the only *a priori* pdf which is invariant to changes in location and scale).

Example 9.4. Non-Informative Prior for a Gaussian Distribution with Unknown Mean and Unknown Variance [26]. We reconsider Ex. 9.3 but this time we use non-informative *a priori* distributions. We assume no information regarding μ and σ^2. In this case, a non-informative distribution, which is invariant to translation and scale changes, is

$$\begin{aligned}\pi(\mu,\sigma^2|I) &= \pi(\mu|\sigma^2,I)\pi(\sigma^2|I),\\ &= \frac{1}{\sqrt{2\pi\sigma^2}} \times \frac{1}{\sigma^2}.\end{aligned}$$

By Bayes' rule, the corresponding *a posteriori* pdf is

$$p(\mu,\sigma^2|\{y_i\},I) = \prod_{i=1}^N \frac{\mathcal{N}(y_i|\mu,\sigma^2)\pi(\mu,\sigma^2|I)}{p(y_i|I)} = \frac{1}{\sigma^3\sqrt{2\pi}} \prod_{i=1}^N \frac{\mathcal{N}(y_i|\mu,\sigma^2)}{p(y_i|I)},$$

[3] Nevertheless reasonable results are often obtained with *a priori* distributions which are not location or scale independent.

which we may rewrite [26] as:

$$p(\mu,\sigma^2|\{y_i\},I) = \frac{1}{\Gamma(N/2)\sigma^2}\sqrt{\frac{N}{\pi S}} \times \left(\frac{S}{2\sigma^2}\right)^{(N+1)/2} \exp -\frac{1}{2\sigma^2} \times \left(N(\mu-\bar{y})^2+S\right),$$

where $\bar{y} = \frac{1}{N}\sum_{i=1}^{N} y_i$ and $S = \sum_{i=1}^{N}(y_i-\bar{y})^2$.

9.4.4 Missing Data

In many problems, it is often fruitful to distinguish between two kinds of unknowns: parameters and hyper-parameters on the one hand and *missing* data on the other hand. Although there is no absolute distinction between the two types of unknowns, the missing data is usually directly related to the individual data and its dimensionality tends to increase as more and more data are observed. On the other hand, the parameters and hyperparameters usually characterize the entire population of observations and are fixed in number.

The following example illustrates a case when data is "missing" because it lies below a given detection limit.

Example 9.5. Environmental Data with Below Detection Limit Observations [31]. The processing of environmental samples containing potentially hazardous chemicals is often complicated by the fact that some of these pollutants are present at trace levels, which cannot be measured reliably and therefore, are reported as unknown values which lie below a certain limit of detection t. The corresponding data sets are known as "left-censored", data sets or data sets with "missing" values.

When missing data, $\mathbf{z}$, occurs in a statistical problem, the inference can be achieved by using the "observed-data likelihood", defined as $L = p(\mathbf{y}|\boldsymbol{\theta},I)$, which can be obtained by integrating out the missing data from the "complete-data likelihood", i. e.

$$L = \int p(\mathbf{y},\mathbf{z}|\boldsymbol{\theta},I)d\mathbf{z}\,. \quad (9.7)$$

Bayesian analysis for missing data problems can be achieved coherently through integration:

$$p(\mathbf{y}|I) \sim \int\int p(\mathbf{y},\mathbf{z}|\boldsymbol{\theta},I)p(\boldsymbol{\theta}|I)d\mathbf{z}d\boldsymbol{\theta}\,. \quad (9.8)$$

Examination of (9.8) shows that the **z** are treated as random variables and that the integration required for eliminating missing data is no different from that required for eliminating nuisance parameters and unknown hyperparameters.

Many applications which involve the use of missing data are solved using the *expectation-maximization* (EM) algorithm.

Example 9.6. EM algorithm for left-censored measurements [31]. Let $y_1, y_2, \ldots, y_N$ denote a random sample of N measurements made from an Gaussian distribution, $\mathcal{N}(\mu, \sigma^2)$, where K measurements lie below the threshold limit t. For simplicity we suppose the first K measurements, $y_1, y_2, \ldots, y_K$, are "non-detects", i. e. are below the detection threshold t. Then we may estimate μ and σ^2 using the EM algorithm in which, at each iteration, all the non-detects are replaced by the same conditional expected value. Let ϕ and Φ be the probability density function (pdf) and cumulative density function (cdf) of the Gaussian distribution. Then, the estimates of μ and σ at the $(m+1)$th iteration are:

$$\hat{\mu}^{(m+1)} = \Big[\sum_{i=K+1}^{N} y_i + \sum_{i=1}^{K} E_m(y_i|y_i \leq t)\Big]/N \ ,$$

$$[\hat{\sigma}^{(m+1)}]^2 = \Big[\sum_{i=K+1}^{N} (y_i - \hat{\mu}^{(m)})^2 + \sum_{i=1}^{K} E_m((y_i - \hat{\mu}^{(m)})^2|y_i \leq t)\Big]/(N-1) \ ,$$

where $z = (t - \hat{\mu}^{(m)})/\hat{\sigma}^{(m)}$ and

$$E_m(y_i|y_i \leq t) = \hat{\mu}^{(m)} - \hat{\sigma}^{(m)}\phi(z)/\Phi(z) \ ,$$

$$E_m\big((y_i - \hat{\mu}^{(m)})^2|y_i \leq t\big) = (\hat{\sigma}^{(m)})^2\big(1 - z\phi(z)/\Phi(z)\big) \ .$$

The following example illustrates the use of the EM algorithm to generate a low-dimensional representation.

Example 9.7. PCA Using the EM Algorithm [30]. The aim of principal component analysis (PCA) is to find a L-dimensional 'linear projection that best represents a set of M-dimensional input vectors $\mathbf{y}_i, i \in \{1, 2, \ldots, N\}$ (see Ex. 4.17). The conventional approach to performing PCA is to find the L dominant eigenvectors $\mathbf{u}_l, l \in \{1, 2, \ldots, L\}$, of the sample covariance matrix $\boldsymbol{\Sigma}$, where

$$\boldsymbol{\Sigma}\mathbf{u}_l = \lambda_l \mathbf{u}_l \ ,$$

and $\lambda_1 \geq \lambda_2 \ldots \geq \lambda_M$. This calculation requires $\sim O(NM^2)$ operations whereas the following EM algorithm [30] performs the PCA in $\sim O(LNM)$ operations per iteration.

E-step. Project the $\tilde{\mathbf{y}}_i = \mathbf{y}_i - \boldsymbol{\mu}$ onto an assumed set of principal axes $\mathbf{U}$, where $\mathbf{U} = (\mathbf{u}_1, \mathbf{u}_2, \ldots, \mathbf{u}_L)$. The result is a $L \times N$ projection matrix $\boldsymbol{\theta} = (\boldsymbol{\theta}_1, \boldsymbol{\theta}_2, \ldots, \boldsymbol{\theta}_N)$, where

$$\boldsymbol{\theta} = (\mathbf{U}^T\mathbf{U})^{-1}\mathbf{U}^T\tilde{\mathbf{Y}} ,$$

and $\tilde{\mathbf{Y}} = (\tilde{\mathbf{y}}_1, \tilde{\mathbf{y}}_2, \ldots, \tilde{\mathbf{y}}_N)$ is a $M \times N$ matrix of input measurements with their mean $\boldsymbol{\mu}$ removed.

M-step. Find new principal axes $\mathbf{U}$ which minimize the squared reconstruction error. The new principal axes are

$$\mathbf{U} = \tilde{\mathbf{Y}}\boldsymbol{\theta}^T(\boldsymbol{\theta}\boldsymbol{\theta}^T)^{-1} .$$

We now consider the application of the EM algorithm to create a new representation consisting of K Gaussian densities.

9.5 Gaussian Mixture Model

In the Gaussian mixture model (GMM) we model an arbitrary M-dimensional pdf as a sum of K Gaussian densities, where $K < M$. The following example explains the main steps in forming the GMM.

Example 9.8. Gaussian Mixture Model(GMM) [22]. In the Gaussian Mixture Model (GMM) we model a probability density function, $\boldsymbol{\theta} = f(\mathbf{y})$, as a sum of K Gaussian distributions $\mathcal{N}(\boldsymbol{\mu}_k, \Sigma_k), k \in \{1, 2, \ldots, K\}$:

$$f(\mathbf{y}) \approx \sum_{k=1}^{K} \alpha_k \mathcal{N}(\mathbf{y}|\boldsymbol{\mu}_k, \Sigma_k) ,$$

where $\alpha_k, \boldsymbol{\mu}_k, \Sigma_k, k \in \{1, 2, \ldots, K\}$, are unknown and are to be learnt from a set of training data $(\boldsymbol{\theta}_i, \mathbf{y}_i), i \in \{1, 2, \ldots, N\}$, subject to the constraints

$$\alpha_k > 0 ,$$
$$\sum_{k=1}^{K} \alpha_k = 1 .$$

One way of learning the parameters $\alpha_k, \boldsymbol{\mu}_k, \Sigma_k$, is to reformulate the mixture model in terms of missing data and then use the expectation-maximization (EM) algorithm.

We associate a "hidden" variable z_i with each $\mathbf{y}_i$, where z_i can only take on the values $\{1,2,\ldots,K\}$. Then we suppose that each value $\boldsymbol{\theta}_i = f(\mathbf{y}_i)$ was generated by randomly choosing z_i from $\{1,2,\ldots,K\}$ and drawing $\mathbf{y}_i$ from the Gaussian $\mathscr{N}(\boldsymbol{\mu}_k, \boldsymbol{\Sigma}_k)$, where $k = z_i$. In this case, α_k is given by the proportion of z_i for which $z_i = k$:

$$\alpha_k = \frac{1}{N}\sum_{i=1}^{N} \delta_{ik} ,$$

where δ_{ik} is the indicator function:

$$\delta_{ik} = \begin{cases} 1 \text{ if } z_i = k , \\ 0 \text{ otherwise} . \end{cases}$$

The parameters $(\alpha_k, \boldsymbol{\mu}_k, \boldsymbol{\Sigma}_k), k \in \{1,2,\ldots,K\}$, are calculated using the following EM algorithm by repeating the following steps until convergence.

E-step. Calculate the *a posteriori* probability of the hidden variable, z_i, given the input data $\mathbf{y}_i, i \in \{1,2,\ldots,N\}$, and using the current parameter values $\alpha_k, \boldsymbol{\mu}_k, \boldsymbol{\Sigma}_k$:

$$p(z_i = k|\mathbf{y}_i, \alpha_k, \boldsymbol{\mu}_k, \boldsymbol{\Sigma}_k) \sim p(\mathbf{y}_i|z_i = k, \alpha_k, \boldsymbol{\mu}_k, \boldsymbol{\Sigma}_k) p(\alpha_k, \boldsymbol{\mu}_k, \boldsymbol{\Sigma}_k|z_i = k) .$$

M-step. Update the parameters $\alpha_k, \boldsymbol{\mu}_k, \boldsymbol{\Sigma}_k$:

$$\alpha_k = \frac{1}{N}\sum_{i=1}^{N} w_{ik} ,$$

$$\boldsymbol{\mu}_k = \sum_{i=1}^{N} w_{ik}\mathbf{y}_i \Big/ \sum_{i=1}^{N} w_{ik} ,$$

$$\boldsymbol{\Sigma}_k = \sum_{i=1}^{N} w_{ik}(\mathbf{y}_i - \boldsymbol{\mu}_k)(\mathbf{y}_i - \boldsymbol{\mu}_k)^T \Big/ \sum_{i=1}^{N} w_{ik} ,$$

where

$$w_{ik} = p(z_i = k|\mathbf{y}_i, \alpha_k, \boldsymbol{\mu}_k, \boldsymbol{\Sigma}_k) .$$

Although widely used, the EM/GMM algorithm may converge to a local minimum or may fail as a result of singularities, or degeneracies, in the input data. In addition the GMM suffers from the "curse of dimensionality" (Sect. 13.2): As the dimensionality of the input data increases, the number of input vectors which are required to specify the means $\boldsymbol{\mu}_k, k \in \{1,2,\ldots,K\}$, and the covariance matrices, $\boldsymbol{\Sigma}_k$, increases exponentially [4]. One way to stop this exponential increase in the number of parameters is to constrain the structure of the covariance matrices [5]. A less drastic solution is to use a mixture of probabilistic PCA's instead of a mixture of Gaussians (see Sect. 10.6.3).

The following example shows how spatial constraints may be incorporated into the GMM.

Example 9.9. Neighbourhood EM algorithm (NEM) [3, 17]. We may incorporate spatial constraints in the GMM by adding the following term to the likelihood:

$$G = \frac{1}{2}\sum_{k=1}^{K}\sum_{i=1}^{N}\sum_{j=1}^{N} v_{ik} v_{jk} c_{ij} ,$$

where

$$v_{ij} = \begin{cases} \alpha \text{ if } \mathbf{y}_i \text{ and } \mathbf{y}_j \text{ are neighbours} , \\ 0 \text{ otherwise} , \end{cases}$$

and $\alpha > 0$.

In comparing two GMM's we often use the symmetric Kullback-Leibler (KL) distance d_{symKL}.

Example 9.10. Kullback-Leibler Distance for GMM's. The symmetrical Kullback-Leibler distance is defined as

$$d_{\mathrm{symKL}}(F,G) = \frac{1}{2}\left(d_{\mathrm{KL}}(F|G) + d_{\mathrm{KL}}(G|F)\right) ,$$

where d_{KL} is the (asymmetric) KL distance:

$$d_{\mathrm{KL}}(F|G) = \int F(x)\log\big(F(x)/G(x)\big)dx .$$

For two GMM's: $F = \sum_{k=1}^{K} a_k \mathcal{N}(\mathbf{m}_k, \mathbf{S}_k)$ and $G = \sum_{k=1}^{K} \alpha_k \mathcal{N}(\boldsymbol{\mu}_k, \boldsymbol{\Sigma}_k)$, there is no closed-form expression for $d_{\mathrm{KL}}(F|G)$ except when $K = 1$. In this case, if f and g denote the corresponding M-dimensional Gaussian pdf's, and

$$\begin{aligned} d_{\mathrm{KL}}(f|g) = \frac{1}{2}\Big(&\log\big(\det(\boldsymbol{\Sigma})/\det(\mathbf{S})\big) + \mathrm{trace}(\boldsymbol{\Sigma}^{-1}\mathbf{S}) - M \\ &+ (\mathbf{m}-\boldsymbol{\mu})\boldsymbol{\Sigma}^{-1}(\mathbf{m}-\boldsymbol{\mu})\Big) . \end{aligned}$$

4 Specification of a M-dimensional Gaussian pdf with a full covariance matrix requires $M(M+3)/2$ parameters.

5 Traditionally, the covariance matrix is constrained to be diagonal (see Sect. 13.2). In this case the number of parameters required to specify a M-dimensional Gaussian pdf with a diagonal covariance matrix is $2M$.

If $K > 1$, we may approximate $d_{\mathrm{KL}}(F|G)$ as follows. Let $\hat{f} = \mathcal{N}(\hat{\mathbf{m}}, \hat{\mathbf{S}})$ and $\hat{g} = \mathcal{N}(\hat{\boldsymbol{\mu}}, \hat{\boldsymbol{\Sigma}})$ be two Gaussian pdf's whose means and covariance matrices match, respectively, the means and covariance matrices of F and G:

$$\hat{\mathbf{m}} = \sum_{k=1}^{K} a_k \mathbf{m}_k ,$$

$$\hat{\mathbf{S}} = \sum_{k=1}^{K} a_k \big(\mathbf{S}_k + (\mathbf{m}_k - \hat{\mathbf{m}})(\mathbf{m}_k - \hat{\mathbf{m}})^T\big) ,$$

and

$$\hat{\boldsymbol{\mu}} = \sum_{k=1}^{K} \alpha_k \boldsymbol{\mu}_k ,$$

$$\hat{\boldsymbol{\Sigma}} = \sum_{k=1}^{K} a_k \big(\boldsymbol{\Sigma}_k + (\boldsymbol{\mu}_k - \hat{\boldsymbol{\mu}})(\boldsymbol{\mu}_k - \hat{\boldsymbol{\mu}})^T\big) .$$

Then,

$$\begin{aligned} d_{\mathrm{KL}}(F|G) &\approx d_{\mathrm{KL}}(\hat{f}|\hat{g}) , \\ &\approx \frac{1}{2}\Big(\log\big(\det(\hat{\boldsymbol{\Sigma}})/\det(\hat{\mathbf{S}})\big) + \mathrm{trace}\big(\hat{\boldsymbol{\Sigma}}^{-1}\hat{\mathbf{S}}\big) - M \\ &\qquad + (\hat{\mathbf{m}} - \hat{\boldsymbol{\mu}})\hat{\boldsymbol{\Sigma}}^{-1}(\hat{\mathbf{m}} - \hat{\boldsymbol{\mu}})\Big) . \end{aligned}$$

9.6 Model Selection

At times we may have more than one model, and then our interest may focus on assessing the fitness of each model and conducting model selection. To illustrate the Bayesian model selection procedure, we focus on the comparison between a "null" model, $M = 0$, and an alternative model, $M = 1$. In this case, the corresponding joint pdf is:

$$p(\mathbf{y}, \boldsymbol{\theta}, M = m|I) = p(\mathbf{y}|\boldsymbol{\theta}, M = m, I)\pi(\boldsymbol{\theta}, M = m|I) . \tag{9.9}$$

If we assume that $\mathbf{y}$ depends on the models through their parameters, then (9.9) becomes:

$$p(\mathbf{y}, \boldsymbol{\theta}, M = m|I) = p(\mathbf{y}|\boldsymbol{\theta}_m)\pi(\boldsymbol{\theta}_m|M = m, I)\pi(M = m|I) , \tag{9.10}$$

where $\pi(\boldsymbol{\theta}_m|M = m, I)$ is the *a priori* distribution for the parameters, $\boldsymbol{\theta}_m$, given the model $M = m$ and $\pi(M = m|I)$ is the *a priori* probability of the model $M = m$. The *a posteriori* probability for the model $M = m$ is proportional to the evidence,

$$p(M=m|\mathbf{y},I) \propto \int p(\mathbf{y}|\boldsymbol{\theta}_m,I)\pi(\boldsymbol{\theta}_m|M=m,I)\pi(M=m|I)d\boldsymbol{\theta}_m \; , \tag{9.11}$$

and the optimal model is

$$m_{\mathrm{OPT}} = \arg\max_m \big(p(M=m|\mathbf{y},I)\big) \; . \tag{9.12}$$

The choice of the *a priori* distribution, $\pi(M=m|I)$, is clearly dependent on the application. If both models appear to be equally likely and contain the same number of parameters, then $\pi(M=0|I) = \frac{1}{2} = \pi(M=1|I)$.

In comparing different models we often use a simple approximation to the *a posteriori* pdf, $p(\mathbf{y}|\boldsymbol{\theta},M=m,I)$, which is known as the *Laplace approximation*.

9.6.1 Laplace Approximation

The Laplace approximation assumes that the *a posteriori* distribution, $p(\boldsymbol{\theta}|\mathbf{y}, M=m,I)$, has only one mode, which we approximate with a Gaussian function. We proceed as follows. Let

$$\begin{aligned} t(\boldsymbol{\theta}) &= \ln p(\boldsymbol{\theta}|\mathbf{y},M=m,I) \; , \\ &= \ln p(\mathbf{y}|\boldsymbol{\theta},M=m,I) + \ln \pi(\boldsymbol{\theta}|M=m,I) + \text{const} \, . \end{aligned} \tag{9.13}$$

Then we expand $t(\boldsymbol{\theta})$ about $\hat{\boldsymbol{\theta}}$, its maximum *a posteriori* (MAP) value, i. e. around the value of $\boldsymbol{\theta}$ that maximizes $p(\boldsymbol{\theta}|\mathbf{y},M=m,I)$:

$$\begin{aligned} t(\boldsymbol{\theta}) &= t(\hat{\boldsymbol{\theta}}) + (\boldsymbol{\theta}-\hat{\boldsymbol{\theta}}) \left.\frac{\partial t(\boldsymbol{\theta})}{\partial \boldsymbol{\theta}}\right|_{\boldsymbol{\theta}=\hat{\boldsymbol{\theta}}} + \frac{1}{2}(\boldsymbol{\theta}-\hat{\boldsymbol{\theta}})^T \left.\frac{\partial^2 t(\boldsymbol{\theta})}{\partial \boldsymbol{\theta}\partial \boldsymbol{\theta}^T}\right|_{\boldsymbol{\theta}=\hat{\boldsymbol{\theta}}} (\boldsymbol{\theta}-\hat{\boldsymbol{\theta}}) + \ldots \\ &\approx t(\hat{\boldsymbol{\theta}}) + (\boldsymbol{\theta}-\hat{\boldsymbol{\theta}})H(\boldsymbol{\theta})(\boldsymbol{\theta}-\hat{\boldsymbol{\theta}})^T \; , \end{aligned} \tag{9.14}$$

where $H(\hat{\boldsymbol{\theta}})$ is the Hessian of the natural logarithm of the *a posteriori* distribution and is defined as the matrix of the second derivative of $t(\boldsymbol{\theta})$ evaluated at $\hat{\boldsymbol{\theta}}$,

$$H(\hat{\boldsymbol{\theta}}) = \left.\frac{\partial^2 \ln p(\boldsymbol{\theta}|\mathbf{y},M=m,I)}{\partial \boldsymbol{\theta}\partial \boldsymbol{\theta}^T}\right|_{\boldsymbol{\theta}=\hat{\boldsymbol{\theta}}} \; . \tag{9.15}$$

Substituting (9.14) into the logarithm of (9.9) and integrating over $\boldsymbol{\theta}$, yields

$$\begin{aligned} \ln p(\mathbf{y}|M=m,I) &= \ln \int \exp\big(t(\boldsymbol{\theta})\big) d\boldsymbol{\theta} \; , \\ &\approx t(\hat{\boldsymbol{\theta}}) + \frac{1}{2}\ln|2\pi H^{-1}| \; , \\ &\approx \ln p(\mathbf{y}|\hat{\boldsymbol{\theta}},M=m,I) + \ln \pi(\hat{\boldsymbol{\theta}}|M=m,I) + \\ &\quad \frac{d}{2}\ln 2\pi - \frac{1}{2}\ln|H| \; , \end{aligned} \tag{9.16}$$

where d is the dimensionality of the parameter space $\boldsymbol{\theta}$. The corresponding approximate expression for $p(\mathbf{y}|M=m,I)$ is

$$p(\mathbf{y}|M=m,I) \approx \sqrt{(2\pi)^d|H^{-1}|}\, p(\mathbf{y}|\hat{\boldsymbol{\theta}},M=m,I)\pi(\hat{\boldsymbol{\theta}}|M=m,I)\,. \qquad (9.17)$$

In Table (9.5) we list some approximations to the natural logarithm of the Bayesian evidence which are commonly used in model selection.

Table 9.5 Approximations to the Natural Logarithm of the Bayesian Evidence $\ln p(\mathbf{y}|M=m,I)$

Name	Approximation to $\ln p(\mathbf{y}\vert M=m,I)$
Laplace Approximation	$\ln p(\mathbf{y}\vert\hat{\boldsymbol{\theta}},M=m,I)+\ln\pi(\hat{\boldsymbol{\theta}}\vert M=m,I)+\frac{d}{2}\ln(2\pi)-\frac{1}{2}\ln\vert H\vert$, where $\hat{\boldsymbol{\theta}}=\arg\max(p(\boldsymbol{\theta}\vert\mathbf{y},M=m,I))$, where d is the dimensionality of the parameter space $\boldsymbol{\theta}$.
Bayesian Information Criterion (BIC)	$\ln p(\mathbf{y}\vert\hat{\boldsymbol{\theta}},M=m,I)-\frac{d}{2}\ln N$.
Akaike Information Criterion (AIC)	$\ln p(\mathbf{y}\vert\,\hat{\boldsymbol{\theta}},M=m,I)-d$.
Kullback Information Criterion (KIC)[29]	$\ln p(\mathbf{y}\vert\,\hat{\boldsymbol{\theta}},M=m,I)-3d/2$.
Cheeseman and Stutz Approximation	$\ln p(\mathbf{y},\hat{\mathbf{z}}\vert M=m,I)+\ln p(\mathbf{y}\vert\hat{\boldsymbol{\theta}},M=m,I)-\ln p(\mathbf{y},\hat{\mathbf{z}}\vert\hat{\boldsymbol{\theta}},M=m,I)$, where $\hat{\mathbf{z}}=\arg\max\big(p(\mathbf{z}\vert\mathbf{y},\hat{\boldsymbol{\theta}},M=m,I)\big)$.

In the following example we illustrate the concept of Bayesian model selection by using it to select the optimum number of Gaussians in a Gaussian mixture model (GMM) [22].

Example 9.11. Selecting the Number of Gaussians in a Gaussian Mixture Model [11, 27]. In a Gaussian mixture model (GMM) (see Ex. 9.8) we approximate a function $f(y)$ using a sum of K Gaussian pdf's:

$$f(y) \approx \sum_{k=1}^{K} \alpha_k \mathcal{N}(y|\mu_k,\sigma_k^2)\,,$$

subject to the constraints

$$\alpha_k > 0\,,$$
$$\sum_k \alpha_k = 1\,.$$

Traditionally the user supplies the value of K. However, if the user is unable to supply a value for K, it may be determined by selecting the best GMM from a set of candidate models, each model having a different number of Gaussians. After optimally fitting the parameters of the different models (using the EM algorithm) we find the optimum value for K (i. e. the optimum model) by choosing the model with the largest Bayesian evidence.

The following example illustrates the use of Bayesian model selection to estimate the intrinsic dimensionality of a given data set $\{y_i\}$.

Example 9.12. Intrinsic Data Dimensionality [24]. A central issue in pattern recognition is to estimate the intrinsic dimensionality d of the input data $\{\mathbf{y}_i\}$. Methods for estimating data dimensionality are surveyed in [8]. One approach is to identify d as the optimum number principal components we need to retain in a principal component analysis (PCA) of the $\{\mathbf{y}_i\}$ (see Sect. 4.5). By interpreting PCA within a Bayesian framework, we are able to derive the following BIC approximation for the evidence:

$$p(\{\mathbf{y}_i\}|L) = \Big(\prod_{j=1}^{L} \lambda_j\Big)^{-N/2} \sigma^{-N(D-L)} N^{-(m+L)/2} ,$$

where D is the dimension of the input data and $m = DL - L(L+1)/2$. The intrinsic dimensionality of the input data is then defined as

$$d = L_{\max} = \arg\max\big(p(\{\mathbf{y}_i\}|L)\big) .$$

See [24] for a derivation of this approximation.

Apart from the model selection methods listed in Table 9.5 there are a large number of non-Bayesian model selection methods available. The more popular of these methods are listed in Table 13.8. Recently, Bouchard and Celeux [7] have questioned the use of the Bayesian model selection approximations given in Table 9.5 for applications involving the classification of a given object O. They suggest a new criterion, known as *Bayesian Entropy Criterion*, which is defined as follows. Suppose each measurement $\mathbf{y}_i$ has a label z_i associated with it. Then,

$$\ln p(\mathbf{z}|\mathbf{y}, M = m, I) \approx \ln p(\mathbf{y}, \mathbf{z}|\hat{\boldsymbol{\theta}}, M = m, I) - \ln p(\mathbf{y}|\widetilde{\boldsymbol{\theta}}, M = m, I) , \quad (9.18)$$

where $\widetilde{\boldsymbol{\theta}} = \arg\max_{\boldsymbol{\theta}} \ln p(\mathbf{y}|\boldsymbol{\theta}, M = m, I)$.

9.6.2 Bayesian Model Averaging

An alternative to model selection is Bayesian Model Averaging (BMA) [15]. In BMA we do not select a single "optimal" model. Instead we take the uncertainty involved in selecting an optimal model into consideration by averaging over the *a posteriori* distribution of multiple models. In a BMA framework, the *a posteriori* pdf, $p(\boldsymbol{\theta}|\mathbf{y},I)$, is given by the following formula:

$$p(\boldsymbol{\theta}|\mathbf{y},I) = \sum_m p(\boldsymbol{\theta}|\mathbf{y},M=m,I)p(M=m|\mathbf{y},I) , \\ = \frac{\sum_m p(\boldsymbol{\theta}|\mathbf{y},M=m,I)\pi(M=m|I)p(\mathbf{y}|M=m,I)}{\sum_m \pi(M=m|I)p(\mathbf{y}|M=m,I)} , \quad (9.19)$$

where $\pi(M=m|I)$ is the *a priori* probability of the model $M=m$. In general, BMA has a better performance than any single model that could reasonably have been selected. In practice, we limit the summation in (9.19) to a small number of models[6]. In Sect. 14.2 we illustrate the use of the BMA as a general framework for combining classifiers in a multiple classifier system (MCS).

9.7 Computation *

In many practical problems, the required computation is the main obstacle to applying the Bayesian method. In fact, until recently, this computation was so difficult that practical applications employing Bayesian statistics were very few in number. The introduction of iterative simulation methods, such as data augmentation and the more general Markov chain Monte Carlo (MCMC), which provide Monte Carlo approximations for the required optimizations and integrations, has brought the Bayesian method into mainstream applications.

Table 9.6 lists some of the optimization and integration algorithms which are used in Bayesian inference (see also algorithms given in Table 9.5). *Note*: Apart from the method of conjugate priors all the algorithms in Table 9.6 are approximate.

9.7.1 Markov Chain Monte Carlo

Although the integrals in Bayesian inference are often intractable, sampling methods can be used to numerically evaluate them. There are a wide range of practical sampling techniques which go under the generic name "Monte Carlo". Given an *a posteriori* pdf, $p(\mathbf{y}|I)$, we may approximate the integral $\int f(\mathbf{y})p(\mathbf{y}|I)d\mathbf{y}$ by using the average of N independent samples, $\mathbf{y}_1,\mathbf{y}_2,\ldots,\mathbf{y}_N$, which are drawn from $p(\mathbf{y}|I)$. Mathematically, we have

$$\int f(\mathbf{y})p(\mathbf{y}|I)d\mathbf{y} \approx \frac{1}{N}\sum_{i=1}^{N} f(\mathbf{y}_i) . \quad (9.20)$$

Unfortunately, in many practical applications, $p(\mathbf{y}|I)$ is only non-zero in an extremely small region of the parameter, or $\mathbf{y}$, space. The problem, then, is how to efficiently generate the samples $\mathbf{y}_i$. The Markov chain Monte Carlo method works

[6] In practice we often discard all models whose *a priori* probabilities $\pi(M=m|I)$ are less than 5% of the *a priori* probability of the optimum model.

Table 9.6 Algorithms Used in Bayesian Inference

Name	Description
Conjugate Priors	This is an *exact* method for Bayesian Inference.
Laplace Approximation	This is an approximate method which is often used in point parameter estimation and in model selection.
Variational Algorithms	This uses a set of variational parameters which are iteratively updated so as to minimize the cross-entropy between the approximate and the true probability density.
Belief Propagation (BP)	This involves propagating the probability distributions using a local message passing algorithm.
Expectation-Maximization (EM)	This is an optimization algorithm which is very convenient to use when the problem involves missing data.
Markov Chain Monte Carlo (MCMC)	This is a relatively new method which has quickly established itself as a favourite all-purpose algorithm for Bayesian inference.

by constructing a random walk (Markov chain) in the parameter space such that the probability for being in a region of the space is proportional to the posterior density for that region. We start with some value of $\mathbf{y}$, calculate the *a posteriori* density there, take a small step and then recalculate the density. The step is accepted or rejected according to some rule and the process is repeated. The resulting output is a series of points in the sample space. The various methods differ according to the rules used to make the moves in the parameter space and the rules determining whether or not a step is accepted. It turns out to be relatively simple to invent such rules that are guaranteed to produce a set of steps that correctly sample the *a posteriori* distribution.

9.8 Software

BNT. A matlab toobox for performing Bayesian inference. Author: Kevin Patrick Murphy.

EMGM A matlab routine for fitting a Gaussian mixture model using the em algorithm. Author: Michael Chen

MCMCSTUFF. A matlab toolbox for implementation of the MCMC technique. Author: Aki Vehtari.

NETLAB. A matlab toolbox for performing neural network pattern recognition. Author: Ian Nabney. The toolbox contains m-files on the Gaussian mixture model and the EM algorithm.

STPRTOOL. A matlab toolbox for performing statistical pattern recognition. Authors: Vojtech Franc, Vaclav Hlavac. The toolbox contains m-files on the Gaussian mixture model and the EM algorithm.

9.9 Further Reading

A useful glossary of selected statistical terms is to be found in [28]. Several modern textbooks on Bayesian inference are: [1, 13, 19, 21, 23, 32]. In addition there is a wide choice of specialist review articles on Bayesian inference which includes [2, 9]. Useful books and review articles on approximate methods for Bayesian inference include [5, 10, 25]. Specific references on MCMC are: [12, 14]; on the EM algorithm are: [4, 22, 27] and on the elimination of nuisance parameters is [6]. Full-length discussions on model selection are given in [20, 21]. For a discussion of the philosophy which underlies the Bayesian paradigm we recommend [16].

Problems

9.1. Define the main premise of Bayesian inference. What are its relative advantages and disadvantages?

9.2. Compare and contrast fuzzy logic and Bayesian inference as a framework for multi-sensor data fusion.

9.3. Write down Bayes formula and explain how we may use it to find the a posteriori pdf.

9.4. What are conjugate priors? Give some examples .

9.5. What are non-informative priors? Give some examples.

9.6. Explain the EM algorithm. Write down the main steps in the EM algorithm for modeling a given function as a mixture of Gaussians (Ex. 9.8).

9.7. Describe the use of the EM algorithm for PCA.

9.8. Explain the process of Bayesian model selection.

9.9. Describe Bayesian model averaging? Compare and contrast it with other model selection procedures.

References

1. D'Agostini, G.: Bayesian Reasoning in Data analysis. World Scientific, Singapore (2003)
2. D'Agostini, G.: Bayesian inference in processing experimental data: principles and basic applications. Rep. Prog. Phys. 66, 1383–1419 (2003)
3. Ambroise, C., Dang, M., Govaert, G.: Clustering of spatial data by the em algorithm. In: Geostatistics for Environmental Applications. In: Soares, A., Gomez-Hernandez, J., Froidevaux, R. (eds.), pp. 493–504. Kluwer Academic Press (1997)

4. Archambeau, C.: Probabilistic models in noisy environments. PhD thesis, Universite Catholque de Louvain, Belgium (2005)
5. Beal, M.J.: Variational algorithms for approximate Bayesian inference. PhD thesis, University of London (2003)
6. Berger, J., Liseo, B., Wolpert, R.L.: Integrated likelihood methods for eliminating nuisance parameters. Stat. Sci. 14, 1–28 (1999)
7. Bouchard, G., Celeux, G.: Selection of generative models in classification. IEEE Trans. Patt. Anal. Mach. Intell. 28, 544–554 (2006)
8. Camastra, F.: Kernel methods for unsupervised learning. PhD thesis, University Genova, Italy (2004)
9. Dose, V.: Bayesian inference in physics: case studies. Rep. Prog. Phys. 66, 1421–1461 (2003)
10. Evans, M., Swartz, T.: Methods for approximating integrals in statistics with special emphasis on Bayesian integration problems. Stat. Sci. 10, 254–272 (1995)
11. Figueiredo, M., Jain, A.K.: Unsupervised learning of finite mixture models. IEEE Trans. Patt. Anal. Mach. Intell. 24, 381–396 (2002)
12. Gamerman, D.: Markov chain monte carlo: stochastic simulations for Bayesian inference. Chapman and Hall (1996)
13. Gelman, A., Larlin, J.S., Stern, H.S., Rubin, D.R.: Bayesian Data Analysis. Chapman and Hall (2003)
14. Gilks, W.R., Richardson, S., Spiegelhater, D.J.: Markov Chain Monte Carlo in practice. Chapman and Hall (1996)
15. Hoeting, J.A., Madigan, D., Raftery, A.E., Volinsky, C.T.: Bayesian model averaging: a tutorial. Stat. Sci. 14, 382–417 (1999)
16. Howson, C., Urbach, P.: Scientific Reasoning: the Bayesian approach, 2nd edn. Open Court Publishing Co., Illinois (1996)
17. Hu, T., Sung, S.Y.: A hybrid EM approach to spatial clustering. Comp. Stat. Data. Anal. 50, 1188–1205 (2006)
18. Kass, R.E., Wasserman, L.: Formal rules for selecting prior distributions: A review and annotated bibliography. Technical Report 583, Department of Statistics, Carnegie Mellon University (1994)
19. Lee, P.M.: Bayesian Statistics: An introduction. Oxford University Press (1997)
20. MacKay, D.J.C.: Bayesian methods for adaptive models. PhD thesis, California Institute of Technology, Pasadena, California, USA (1991)
21. MacKay, D.J.C.: Information Theory, Inference and Learning Algorithms. Cambridge University Press (2003)
22. McLachlan, G.J., Krishnan, T.: The EM algorithm and Extensions. John Wiley and Sons (1997)
23. Migon, H.S., Gamerman, D.: Statistical Inference- An Integrated Approach. Arnold Publishers Ltd., London (1999)
24. Minka, T.P.: Automatic choice of dimensionality for PCA. NIPS (2000)
25. Minka, T.P.: A family of algorithms for approximate Bayesian inference. Phd thesis. Masschusetts Institute of Technology, Cambridge, Massachusetts, USA (2001)
26. Minka, T.P.: Inferring a Gaussian distribution. Unpublished article. Available from Minka's Homepage (2005)
27. Paalanen, P., Kristian, J., Kamarainen, J.-K., Ilonen, J., Kalviainen, H.: Feature Representation and Discrimination Based on Gaussian Mixture Model Probability Densities - Practices and Algorithms. Patt. Recogn. 39, 1346–1358
28. Prosper, H.B., Linnemann, J.T., Rolke, W.A.: A glossary of selected statistical terms. In: Proc. Adv. Statist. Tech. Particle Physics, March 18–22, Grey College, Durham (2002)
29. Seghouane, A.-K.: A note on overfitting properties of KIC and KICc. Sig. Process. 86, 3055–3060 (2006)
30. Skocaj, D.: Robust subspaces approaches to learning and recognition. PhD thesis. University Ljubljana (2003)
31. Singh, A., Nocerino, J.: Robust estimation of mean and variance using environmental data sets with below detection limit observations. Chemometrics Intell. Lab. Syst. 60, 69–86 (2002)

32. Sivia, D.S.: Data Analysis: A Bayesian Tutoral. Oxford University Press (1996)
33. Sivia, D.S.: Dealing with duff data. In: Sears, M., Nedeljkovic, N.E., Sibisi, S. (eds.) Proc. Max. Entropy Conf., South Africa, pp. 131–137 (1996)
34. Yang, R., Berger, J.O.: A catalog of noninformative priors. Plann. Infer. 79, 223–235 (1997); Discussion Paper 97-42, ISDS, Duke University, Durham, NC, USA
35. Zwillinger, D., Kokoska, S.: CRC Standard probability and statistics tables and formulae. Chapman and Hall (2000)

Chapter 10
Parameter Estimation

10.1 Introduction

In this chapter we consider *parameter estimation*, which is our first application of Bayesian statistics to the problem of multi-sensor data fusion. Let $\mathbf{y} = (\mathbf{y}_i^T, \mathbf{y}_2^T, \ldots, \mathbf{y}_N^T)^T$ denote a set of N multi-dimensional sensor observations $\mathbf{y}_i, i \in \{1, 2, \ldots, N\}$, where $\mathbf{y}_i = (y_i^{(1)}, y_i^{(2)}, \ldots, y_i^{(M)})^T$. We suppose the $\mathbf{y}_i$ were generated by a parametric distribution $p(\mathbf{y}|\boldsymbol{\theta}, I)$, where $\boldsymbol{\theta} = (\theta_1, \theta_2, \ldots, \theta_K)^T$ are the parameters and I is any background information we may have. Then parameter estimation involves estimating $\boldsymbol{\theta}$ from $\mathbf{y}$.

In the Bayesian framework we assume an *a priori* distribution $\pi(\boldsymbol{\theta}|I)$ for $\boldsymbol{\theta}$, where I denotes any background information we may have regarding $\boldsymbol{\theta}$. The *a posteriori* distribution, $p(\boldsymbol{\theta}|\mathbf{y}, I)$, represents the probability density function of $\boldsymbol{\theta}$ after observing $\mathbf{y}$ and the relation between the *a priori* and the *a posteriori* distributions is given by Bayes' Theorem:

$$p(\boldsymbol{\theta}|\mathbf{y}, I) = \frac{p(\mathbf{y}|\boldsymbol{\theta}, I)\pi(\boldsymbol{\theta}|I)}{p(\mathbf{y}|I)} . \tag{10.1}$$

In many applications, parameter estimation represents the major fusion algorithm in a multi-sensor data fusion system. In Table 10.1

we list some of these applications together with the classification of the type of fusion algorithm involved.

10.2 Parameter Estimation

Once the *a posterior* distribution is available we can use it to estimate the parameters $\boldsymbol{\theta}$. In what follows we shall suppose $\boldsymbol{\theta}$ is a scalar parameter, θ. We define a *loss function*, $L = L(t, \theta)$, which links θ to its estimated value t. By minimizing the expected value of L, we obtain the *optimum* estimated value $\hat{\theta}$, where

Table 10.1 Applications in which Parameter Estimation is the Primary Fusion Algorithm

Class		Application
DaI-DaO	Ex. 2.5	Time-of-Flight Ultrasonic Sensor.
	Ex. 3.2	Image Smoothing.
	Ex. 3.7	Heart Rate Estimation Using a Kalman Filter.
	Ex. 4.15	Debiased Cartesian Coordinates.
	Ex. 8.18	Psychometric Functions.
	Ex. 10.4	Kriging with an External Trend: Fusing Digital Elevation Model and Rainfall Gauge Measurements.
DaI-FeO	Ex. 4.2	Bayesian Triangulation of Direction-Bearing Measurements.
	Ex. 4.17	Principal Component Analysis.
	Ex. 4.19	Recursive LDA algorithm.
	Ex. 4.18	Chemical Sensor Analysis: Principal Discriminant Analysis (PDA) for Small Sample Size Problems.
	Ex. 10.5	Fitting a Finite-Length Straight-Line Segment in Two Dimensions: A Bayesian Solution.
	Ex. 10.6	Non-Intrusive Speech Quality Estimation Using GMM's.
	Ex. 11.15	Target track initialization using the Hough transform
FeI-FeO	Ex. 8.22	Fusing Similarity Coefficients in a Virtual Screening Programme.
	Ex. 8.26	Converting SVM Scores into Probability Estimates.
	Ex. 13.8	Gene selection:An Application of Feature Selection in DNA Microarray Experiments.
	Ex. 13.9	mrMR: Filter Feature Selection.

The designations DaI-DaO, DaI-FeO and FaI-FeO refer, respectively, to the Dasarathy input/output classifications: "Data Input-Data Output", "Data Input-Feature Output" and "Feature Input-Feature Output" (see Sect. 1.3.3).

$$\hat{\theta} = \min_t \int_{-\infty}^{\infty} L(t,\theta)p(\theta|\mathbf{y},I)d\theta \ . \tag{10.2}$$

Clearly the estimates, $\hat{\theta}$, will vary with the choice of loss function. Three commonly used loss functions are:

Quadratic Loss Function. This is defined as $L(t,\theta) = \Delta^2$, where $\Delta = t - \theta$. In this case, the corresponding MMSE (minimum mean square error) estimate, $\hat{\theta}_{\mathrm{MMSE}}$, is equal to the expected value of θ:

$$\hat{\theta}_{\mathrm{MMSE}} = E(\theta) \ . \tag{10.3}$$

Absolute Loss Function. This is defined as $L(t,\theta) = |\Delta|$. In this case, the corresponding MMAE (minimum mean absolute error) estimate, $\hat{\theta}_{\mathrm{MMAE}}$, is equal to the median of the distribution of θ:

$$\hat{\theta}_{\mathrm{MMAE}} = \mathrm{med}\big(p(\theta|\mathbf{y},I)\big) \ . \tag{10.4}$$

Zero-One Loss Function. This is defined as:

$$L(t,\theta) = \begin{cases} 0 & \text{if } |\Delta| \leq b \ , \\ 1 & \text{if } |\Delta| > b \ . \end{cases} \tag{10.5}$$

In this case, the corresponding estimate is the center of the interval of width $2b$ which has the highest probability. If b goes to zero, then the zero-one estimate converges to the maximum *a posteriori* (MAP) estimate:

$$\hat{\theta}_{\mathrm{MAP}} = \arg\max\big(p(\theta|\mathbf{y},I)\big) \ . \qquad (10.6)$$

Assuming $p(\theta|\mathbf{y},I)$ is differentiable, $\hat{\theta}_{\mathrm{MAP}}$ is found from

$$\left.\frac{\partial \ln p(\theta|\mathbf{y},I)}{\partial \theta}\right|_{\theta=\hat{\theta}_{\mathrm{MAP}}} = 0 \ . \qquad (10.7)$$

Although the above loss functions are widely used, in some circumstances it may be better to use a different loss function. The following example illustrates the use of a *robust* version of the quadratic loss function [1].

Example 10.1. Color Constancy and the Minimum Local Mass (MLM) Loss Function [6]. Color constancy is defined as the ability of a visual system to maintain object colour appearance across variations in factors extrinsic to the object. Human vision exhibits approximate color constancy. Let $p(\theta|\mathbf{y},I)$ denote the *a posteriori* distribution for the illuminants and the surface colors in a given scene. By definition, the best estimate of the surface color estimate is

$$\hat{\theta} = \min_t \int_{-\infty}^{\infty} L(t,\theta)p(\theta|\mathbf{y},I)d\theta \ ,$$

where L is an appropriate loss function. For this application, the authors in [6], found that none of the above loss functions were appropriate. Instead they used a "robust" version of the quadratic loss function:

$$L(t,\theta) = \begin{cases} \Delta^2 & \text{if } |\Delta| < \alpha \ , \\ \alpha^2 & \text{otherwise} \ , \end{cases}$$

where $\Delta = t - \theta$ and α is a fixed constant.

In Table 10.2 we list some of the common loss functions along with their Bayesian estimators.

The following example illustrates the calculation of the MAP estimate for the exponential distribution.

Example 10.2. Exponential Distribution. We suppose the likelihood $p(y|\theta,I)$ is exponential:

[1] The concept of a *robust* loss function is explained in more detail in Chapt. 11, where we consider the subject of *robust* statistics.

Table 10.2 Loss Functions $L(t,\theta)$ and their Bayesian Estimators $\hat{\theta}$

Name	$L(t,\theta)$	$\hat{\theta} \equiv \min_t \int_{-\infty}^{\infty} L(t,\theta)p(\theta\|\mathbf{y},I)d\theta$
Quadratic	Δ^2	$\hat{\theta} = E(p(\theta\|\mathbf{y},I))$, i. e. $\hat{\theta}$ is the expectation of $p(\theta\|\mathbf{y})$.
Absolute	$\|\Delta\|$	$\hat{\theta}$ is the median of $p(\theta\|\mathbf{y},I)$, i. e. $\int_{-\infty}^{\hat{\theta}} p(\theta\|\mathbf{y},I)d\theta = \frac{1}{2} = \int_{\hat{\theta}}^{\infty} p(\theta\|\mathbf{y},I)d\theta$.
Zero-One	1 if $(\|\Delta\| - b) \geq 0$, otherwise 0	$\hat{\theta}$ is the center of the interval of width $2b$ which has the highest probability. If $b \to 0$, then $\hat{\theta} = \arg\max\big(p(\theta\|\mathbf{y},I)\big)$.
Asymmetric Quadratic	$a\Delta^2$ if $\theta \geq t$, otherwise $b\Delta^2$	$\hat{\theta} = \min_t \int L(t,\theta)p(\theta\|\mathbf{y},I)d\theta$.
Asymmetric Absolute	$a\|\Delta\|$ if $\theta \geq t$ otherwise $b\|\Delta\|$	$\hat{\theta}$ is the $b/(a+b)$th quantile of $p(\theta\|\mathbf{y},I)$, i. e. $\int_{-\infty}^{\hat{\theta}} p(\theta\|\mathbf{y},I)d\theta = b/(a+b)$ and $\int_{\hat{\theta}}^{\infty} p(\theta\|\mathbf{y},I)d\theta = a/(a+b)$.
Linex	$\beta(\exp(\alpha\Delta) - \alpha\Delta - 1)$	$\hat{\theta} = -\ln\big(E(\exp -\alpha\theta)\big)/\alpha$. If $y \sim \mathcal{N}(\mu,\sigma^2)$, then $\hat{\theta} = \mu - \alpha\sigma^2/2$.
Local Mass	$\min(\Delta^2,\alpha^2)$	$\hat{\theta} = \min_t \int L(t,\theta)p(\theta\|\mathbf{y},I)d\theta$.
Hubert-t	$\min(\|\Delta\|,\alpha)$	$\hat{\theta} = \min_t \int L(t,\theta)p(\theta\|\mathbf{y},I)d\theta$.

$$p(y|\theta,I) = \begin{cases} \theta\exp(-\theta y) & \text{if } y > 0\,, \\ 0 & \text{otherwise}\,. \end{cases}$$

The corresponding prior conjugate distribution is also exponential:

$$\pi(\theta|I) = \begin{cases} \alpha\exp(-\alpha\theta) & \text{if } \theta > 0\,, \\ 0 & \text{otherwise}\,. \end{cases}$$

By definition, the MMSE estimate is given by

$$\hat{\theta}_{\mathrm{MMSE}}(y) = \int_{-\infty}^{\infty} \theta p(\theta|y,I)d\theta\,,$$

where

$$p(\theta|y,I) = \frac{p(y|\theta,I)\pi(\theta|I)}{\int p(y|\theta,I)\pi(\theta|I)d\theta}\,.$$

Substituting the expressions for $p(y|\theta,I)$ and $\pi(\theta|I)$ into $\hat{\theta}_{\mathrm{MMSE}}(y)$ gives

$$\begin{aligned} \hat{\theta}_{\mathrm{MMSE}}(y) &= \int_0^{\infty} (\alpha+y)^2\theta^2\exp(-(\alpha+y)\theta)d\theta\,, \\ &= \frac{2}{\alpha+y}\,. \end{aligned}$$

By using the appropriate loss function we may also find the MMAE and MAP solutions. All three solutions are listed in Table 10.3.

Table 10.3 Bayesian Estimators for the Exponential Distribution

Estimator	Formula
MMSE	$\hat{\theta} = 2/(\alpha + y)$.
MMAE	$\hat{\theta} = T/(\alpha + y)$, where $T \approx 1.68$ is the solution of $(1+T)\exp(-T) = \frac{1}{2}$.
MAP	$\hat{\theta} = 1/(\alpha + y)$.

10.3 Bayesian Curve Fitting

Curve fitting is an important technique in multi-sensor data fusion where it is commonly used to fuse together multiple sensor observations. We may regard curve fitting as a problem in parameter estimation: Given a set of pairs of input measurements $(y_i, u_i), i \in \{1, 2, \ldots, N\}$, we estimate the parameters of an algebraic curve which best fits these data. A classic example of curve fitting is straight-line regression where we fit a straight-line to several sensor measurements $y_i, i \in \{1, 2, \ldots, N\}$.

The following example illustrates the use of multiple straight-line regression algorithms in a multi-target tracking applications.

Example 10.3. Track Initialization Using a Multiple Straight-Line Regression Algorithm [4, 5]. In surveillance applications we often use a Kalman filter (Sect. 12.3) to track all of the objects which move in a given surveillance area. The Kalman filter is a sequential algorithm in which the tracks are updated as new sensor observations are received. To initiate the Kalman filter we must, however, create an initial set of tracks, one track for each object. This is often performed using a Hough transform (see Ex. 11.15).

A curve fitting technique which is widely used in multi-sensor data fusion is Kriging (see Sects. 4.3 and 5.9). The following example illustrates the fusion of a digital elevation model and rain gauge measurements using a Kriging estimator.

Example 10.4. Kriging with an External Trend: Fusing Digital Elevation Model and Rainfall Gauge Measurements [16]. High resolution estimates of the spatial variability of rainfall are important for the identification of locally intense storms which can lead to floods and especially to flash floods. Given a limited number of rainfall gauges we may generate the rainfall at any point $\mathbf{u}_0$ by ordinary Kriging, or curve-fitting the N nearby rain gauge measurements $y(\mathbf{u}_i), i \in \{1,2,\ldots,N\}$ (see Sect. 4.3). We now consider the situation where the rain gauge measurements are supplemented by elevation data which we assume is correlated with the rainfall and which is available for all $\mathbf{u}$.

In Kriging with an external trend (KED) we estimate the rainfall at the point $\mathbf{u}_0$ using a Kriging estimator (4.5):

$$\hat{y}(\mathbf{u}_0) = \mu(\mathbf{u}_0) + \sum_{i=1}^{N} \lambda_i (y_i - \mu(\mathbf{u}_0)) ,$$

in which we model the mean value $\mu(\mathbf{u}_0)$ as a linear function of z. The unknown mean, $\mu_0 = \mu(\mathbf{u}_0)$, and the weights, λ_i, are given by the following matrix equation:

$$\begin{pmatrix} \lambda_1 \\ \vdots \\ \lambda_N \\ \mu_0 \end{pmatrix} = \begin{pmatrix} \Sigma(\mathbf{u}_1,\mathbf{u}_1) & \ldots & \Sigma(\mathbf{u}_1,\mathbf{u}_N) & 1 & z(\mathbf{u}_1) \\ \vdots & \ddots & \vdots & \vdots & \vdots \\ \Sigma(\mathbf{u}_N,\mathbf{u}_1) & \ldots & \Sigma(\mathbf{u}_N,\mathbf{u}_N) & 1 & z(\mathbf{u}_N) \\ 1 & \ldots & 1 & 0 & 0 \\ z(\mathbf{u}_1) & \ldots & z(\mathbf{u}_N) & 0 & 0 \end{pmatrix}^{-1} \begin{pmatrix} \Sigma(\mathbf{u}_0,\mathbf{u}_1) \\ \vdots \\ \Sigma(\mathbf{u}_0,\mathbf{u}_N) \\ 1 \\ z(\mathbf{u}_0) \end{pmatrix} ,$$

where $\Sigma(\mathbf{u}_i,\mathbf{u}_j)$ is the spatial covariance of the rainfall y.

Note: In KED the weights λ_i sum to one. In this case we can write $\hat{y}(\mathbf{u}_0)$ more simply as

$$\hat{y}(\mathbf{u}_0) = \sum_{i=1}^{N} \lambda_i y_i .$$

The following example graphically illustrates the power of *Bayesian* curve fitting. We consider the problem of optimally fitting a straight-line *segment* to a set of two-dimensional sensor measurements $\mathbf{y}_i, i \in \{1,2,\ldots,N\}$. The algorithm calculates both the slope and intercept of the corresponding straight-line and the end points of the straight-line segment.

Example 10.5. Fitting a Straight-Line Segment in Two Dimensions [29]. We consider the problem of fitting a straight-line, $L \equiv L(\boldsymbol{\theta})$, to a set of points,

$\mathbf{z}_i = (x_i, y_i), i \in \{1, 2, \ldots, N\}$, where $\boldsymbol{\theta} = ((x_A, y_A), (x_B, y_B))$ denotes the Cartesian coordinates of the end points, A and B (Fig. 10.1). Let $\mathbf{z} = \{\mathbf{z}_1, \mathbf{z}_2, \ldots, \mathbf{z}_N\}$. Then our problem reduces to calculating the *a posteriori* probability density function $p(\boldsymbol{\theta}|\mathbf{z}, I)$. Using Bayes' theorem and introducing the dummy variable, $\mathbf{P}$, gives

$$\begin{aligned} p(\boldsymbol{\theta}|\mathbf{z}, \mathbf{P}, I) &= \frac{p(\mathbf{z}|\boldsymbol{\theta}, I)\pi(\boldsymbol{\theta}|I)}{p(\mathbf{z}|I)} , \\ &= \frac{p(\mathbf{z}|\mathbf{P}, I)\pi(\mathbf{P}|\boldsymbol{\theta}, I)\pi(\boldsymbol{\theta}|I)}{p(\mathbf{z}|I)} , \end{aligned}$$

where $\mathbf{P} = (x, y)$ is any point on the straight-line $L(\boldsymbol{\theta})$. We treat $\mathbf{P}$ as a nuisance parameter which we eliminate by integration. If the $\mathbf{z}_i$ are independent, and assuming a uniform distribution for $\pi(\boldsymbol{\theta}|I)$, we obtain

$$p(\boldsymbol{\theta}|\mathbf{z}, I) \propto \prod_{i=1}^{N} \int_{L(\boldsymbol{\theta})} p(\mathbf{z}_i|\mathbf{P}, I)\pi(\mathbf{P}|\boldsymbol{\theta}, I) d\mathbf{P} .$$

Assuming a zero-mean, variance σ^2, Gaussian noise model and a uniform distribution for $\pi(\mathbf{P}|\boldsymbol{\theta}, I)$, gives

$$p(\boldsymbol{\theta}|\mathbf{z}, I) \propto \prod_{i=1}^{N} \int_{L(\boldsymbol{\theta})} \frac{1}{\sqrt{2\pi\sigma^2}} \exp\left[-\frac{1}{2}\left(\frac{\mathbf{z}_i - \mathbf{P}}{\sigma}\right)^2\right] d\mathbf{P} .$$

Let $\mathbf{P}_i$ denote the point on $L(\boldsymbol{\theta})$ which is nearest to $\mathbf{P}$, then $|\mathbf{z}_i - \mathbf{P}_i| = \sqrt{s_i^2 + t_i^2}$, where $s_i = |\mathbf{P} - \mathbf{P}_i|$ and $t_i = |\mathbf{z}_i - \mathbf{P}_i|$. We obtain

$$p(\boldsymbol{\theta}|\mathbf{z}, I) \propto \prod_{i=1}^{N} \int_{s_A}^{s_B} \frac{1}{\sqrt{2\pi\sigma^2}} \exp\left[-\left(\frac{s_i^2 + t_i^2}{2\sigma^2}\right)\right] ds_i ,$$

which may be solved numerically.

Special Case. An important special case occurs when we let the line extend to $\pm\infty$. In this case the *a posteriori* probability is

$$p(\boldsymbol{\theta}|\mathbf{z}, I) = \prod_{i=1}^{N} \frac{1}{\sqrt{2\pi\sigma^2}} \exp\left(-\frac{t_i^2}{2\sigma^2}\right) .$$

The corresponding maximum *a posteriori* (MAP) value for θ is

$$\hat{\theta}_{\text{MAP}} = \arg\max \left[\prod_{i=1}^{N} \frac{1}{\sqrt{2\pi\sigma^2}} \exp\left(-\frac{t_i^2}{2\sigma^2}\right)\right] ,$$

or equivalently,

$$\hat{\theta}_{\mathrm{MAP}} = \arg\min\left(\sum_{i=1}^{N} t_i^2\right) .$$

We see that, in this case, the maximum *a posteriori* straight-line, $L(\hat{\theta}_{\mathrm{MAP}})$, is identical to the straight-line found using the standard least square approach (see Sect. 10.5).

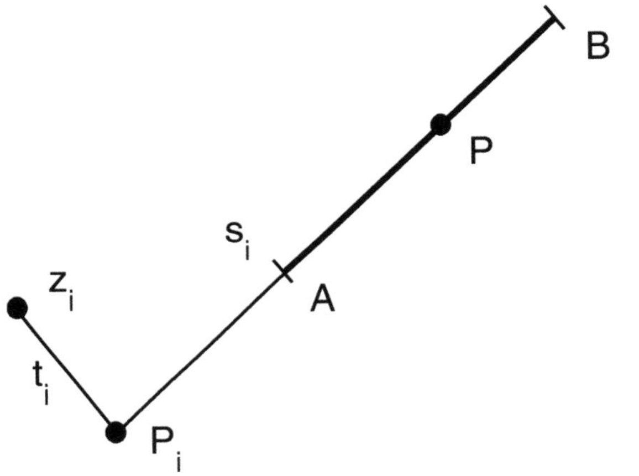

Fig. 10.1 Shows the straight-line segment AB, of finite-length, and a single data point $\mathbf{z}_i$ beside it. $\mathbf{P}_i$ denotes the closest point to $\mathbf{z}_i$, which lies on AB or on its extension. $\mathbf{P}$ denotes any point which lies on the segment AB, s_i is the distance $|\mathbf{P}_i - \mathbf{P}|$ and t_i is the distance $|\mathbf{z}_i - \mathbf{P}_i|$.

10.4 Maximum Likelihood

In some circumstances we do not assume any *a priori* information or, equivalently, we assume a uniform *a priori* distribution $\pi(\boldsymbol{\theta}|I)$. In this case, the *a posteriori* density $p(\boldsymbol{\theta}|\mathbf{y},I)$ is proportional to the likelihood function $p(\mathbf{y}|\boldsymbol{\theta},I)$:

$$p(\boldsymbol{\theta}|\mathbf{y},I) = p(\mathbf{y}|\boldsymbol{\theta},I)\pi(\boldsymbol{\theta}|I)/p(\mathbf{y}|I) , \quad (10.8)$$

$$\propto p(\mathbf{y}|\boldsymbol{\theta},I) , \quad (10.9)$$

and the maximum *a posteriori* solution is identical to the maximum likelihood solution:

$$\hat{\theta}_{\mathrm{MAP}} \equiv \hat{\theta}_{\mathrm{ML}} , \quad (10.10)$$

where

$$\hat{\boldsymbol{\theta}}_{\text{MAP}} = \arg\max_{\boldsymbol{\theta}} \big(p(\boldsymbol{\theta}|\mathbf{y},I)\big) , \quad (10.11)$$

$$\hat{\boldsymbol{\theta}}_{\text{ML}} = \arg\max_{\boldsymbol{\theta}} \big(p(\mathbf{y}|\boldsymbol{\theta},I)\big) . \quad (10.12)$$

Assuming a differentiable function, $\hat{\boldsymbol{\theta}}_{\text{ML}}$ is found from

$$\left.\frac{\partial\big(\ln p(\mathbf{y}|\boldsymbol{\theta},I)\big)}{\partial\boldsymbol{\theta}}\right|_{\boldsymbol{\theta}=\hat{\boldsymbol{\theta}}_{\text{ML}}} = \mathbf{0} . \quad (10.13)$$

If multiple solutions exist, then the one that maximizes the likelihood function is $\hat{\boldsymbol{\theta}}_{\text{ML}}$.

The maximum likelihood estimator is the most widely used estimation method and is often used in combination with other optimization/estimation procedures. For example, the expectation-maximization (EM) algorithm, relies on the maximum-likelihood estimator and for this reason the algorithm is often referred to as the ML-EM algorithm. The following example illustrates the use of the ML-EM algorithm in a speech quality estimation procedure.

Example 10.6. Nonintrusive Speech Quality Estimation Using GMM's [11]. The evaluation and assurance of speech quality is critically important for telephone service providers. A nonintrusive speech quality measurement algorithm is shown in Fig. 10.2. Perceptual features are extracted from test speech. The time segmentation (labeling) module labels the feature vectors of each time frame as belonging to one of three possible classes: voiced (V), unvoiced (U) or inactive (I). Offline, high-quality undistorted speech signals are used to produce a reference model of the behaviour of clean speech features. This is accomplished by modeling the probability distribution of the features for each class $\Lambda \in \{U,V,I\}$ with a mixture model GMM(Λ). Features extracted from the test signal which belong to a class Λ are assessed by calculating a "consistency" measure with respect to GMM(Λ). The consistency measure is defined as

$$c(\Lambda) = \frac{1}{N}\sum_{i=1}^{N} \ln\big(p(y_i|\Lambda)\big) ,$$

where $p(y_i|\Lambda)$ is the likelihood function of the ith test signal, y_i, given the model GMM(Λ). The three consistency values $c(\Lambda), \Lambda \in \{U,V,I\}$ are fused together and then converted to a speech quality Q by passing through an appropriate transfer function (see Ex. 8.18).

Although the ML procedure is widely used in many applications, a certain degree of caution should be exercised when using it. The reason is that the ML has no mechanism for regulating its solution. As a consequence, the solution may prove to be completely wrong, or as in the following example, wildly optimistic.

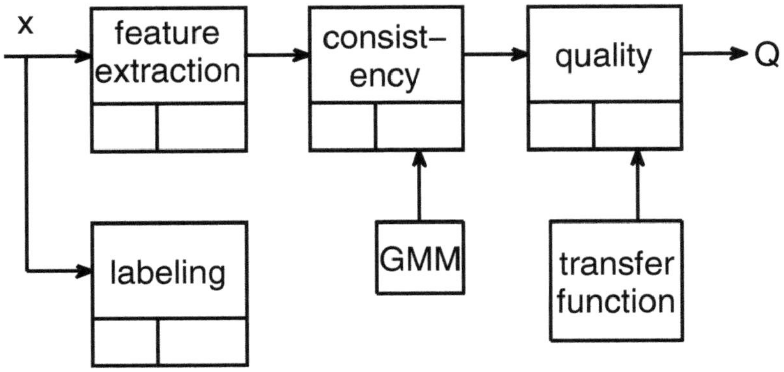

Fig. 10.2 Shows the non-intrusive speech quality estimation algorithm. The main steps in the algorithm are: (1) The segmentation, or labeling, module which labels each time frame of the speech signal as voiced (V), unvoiced (U) or inactive (I). (2) The feature extraction module which extracts perceptual features from each time frame. (3) The consistency module which calculates the consistency measure $c(\Lambda)$ and (4) The quality measure module which calculates the speech quality Q.

Example 10.7. Completion of a Doctoral Thesis: ML vs. MAP [7]. A doctoral student gives his supervisor a *precise* time, T, for the completion of his thesis. If the supervisor belongs to the "maximum likelihood" school then he does not use any *a priori* information and will uncritically accept the students estimate. If, on the other hand, the supervisor belongs to the "Bayesian" school, he will use all the *a priori* information which is available. For example, he may use his knowledge of human psychology and conjecture that a student working under pressure will almost certainly underestimate the time required to complete his PhD thesis. The result is that although the likelihood function is only high for $t \sim T$, the *a priori* probability that T is a reasonable date, given the student is working under pressure, is sufficiently low to make the estimate $t \approx T$ highly unlikely.

10.5 Least Squares

For some applications we cannot define an *a posterior* probability density function $p(\boldsymbol{\theta}|\mathbf{y},I)$, where $\mathbf{y} = (\mathbf{y}_1^T, \mathbf{y}_2^T, \ldots, \mathbf{y}_N^T)^T$. In this case we may optimally determine the parameters $\boldsymbol{\theta}$ by using a deterministic, i. e. *non-probabilistic*, method such as the method of least squares. For simplicity, we shall restrict ourselves to one-dimensional scalar measurements $\mathbf{y} = (y_1, y_2, \ldots, y_N)^T$, which are noisy samples of an "ideal" function $y = f(\boldsymbol{\theta})$. If the y_i are independent, then the least square estimate $\hat{\boldsymbol{\theta}}_{\mathrm{LS}} = (\hat{\theta}_1, \hat{\theta}_2, \ldots, \hat{\theta}_K)^T$, is defined as the set of parameter values which minimize a weighted sum of the square differences $(y_i - f(\boldsymbol{\theta}))^2$. Mathematically,

$$\hat{\boldsymbol{\theta}}_{\rm LS} = \arg\min_{\boldsymbol{\theta}} \left[\sum_{i=1}^{N} \left(\frac{y_i - f(\boldsymbol{\theta})}{\sigma_i} \right)^2 \right] , \tag{10.14}$$

where σ_i^2 is the noise variance in y_i.

If we assume $\big(y_i - f(\boldsymbol{\theta})\big) \sim \mathcal{N}(0, \sigma_i^2)$, then the corresponding likelihood function is

$$p(\mathbf{y}|\boldsymbol{\theta}, I) = \prod_{i=1}^{N} \mathcal{N}(y_i | f(\boldsymbol{\theta}), \sigma_i^2) , \tag{10.15}$$

and the maximum likelihood solution is numerically identical to the least square solution:

$$\hat{\boldsymbol{\theta}}_{\rm ML} = \hat{\boldsymbol{\theta}}_{\rm LS} , \tag{10.16}$$

where

$$\hat{\boldsymbol{\theta}}_{\rm ML} = \arg\max_{\boldsymbol{\theta}} \left(\prod_{i=1}^{N} \mathcal{N}(y_i | f(\boldsymbol{\theta}), \sigma_i^2) \right) , \tag{10.17}$$

$$\hat{\boldsymbol{\theta}}_{\rm LS} = \arg\max_{\boldsymbol{\theta}} \left[\sum_{i=1}^{N} \left(\frac{y_i - f(\boldsymbol{\theta})}{\sigma_i} \right)^2 \right] . \tag{10.18}$$

Additionally, if we assume a uniform *a priori* distribution $\pi(\boldsymbol{\theta}|I)$ for $\boldsymbol{\theta}$, then the maximum *a posteriori* solution becomes equal to the maximum likelihood (and least square) solution.

In general, the MAP, ML and LS solutions are numerically equal, if the *a priori* distribution $\pi(\boldsymbol{\theta}|I)$ is uniform and the input measurement vector $\mathbf{y} = (\mathbf{y}_1^T, \mathbf{y}_2^T, \ldots, \mathbf{y}_N^T)^T$, follows a multivariate Gaussian distribution, i. e. $\mathbf{y} \sim \mathcal{N}(\mathbf{y}|\mathbf{f}(\boldsymbol{\theta}), \boldsymbol{\Sigma})$. Mathematically,

$$\hat{\boldsymbol{\theta}}_{\rm MAP} = \hat{\boldsymbol{\theta}}_{\rm ML} = \hat{\boldsymbol{\theta}}_{\rm LS} \tag{10.19}$$

where

$$\hat{\boldsymbol{\theta}}_{\rm MAP} = \arg\max_{\boldsymbol{\theta}} \Big(\pi(\boldsymbol{\theta}|I) \mathcal{N}\big(\mathbf{y}|\mathbf{f}(\boldsymbol{\theta}), \boldsymbol{\Sigma}\big) \Big) , \tag{10.20}$$

$$\hat{\boldsymbol{\theta}}_{\rm ML} = \arg\max_{\boldsymbol{\theta}} \Big(\mathcal{N}\big(\mathbf{y}|\mathbf{f}(\boldsymbol{\theta}), \boldsymbol{\Sigma}\big) \Big) , \tag{10.21}$$

$$\hat{\boldsymbol{\theta}}_{\rm LS} = \arg\min_{\boldsymbol{\theta}} \Big(\big(\mathbf{y} - \mathbf{f}(\boldsymbol{\theta})\big)^T \boldsymbol{\Sigma}^{-1} \big(\mathbf{y} - \mathbf{f}(\boldsymbol{\theta})\big) \Big) . \tag{10.22}$$

The following example illustrates the least squares procedure by fitting a set of experimental covariance measurements with a parametric covariance model. This procedure is widely used in geostatistical applications including geographical information systems (see Sect. 4.3).

Table 10.4 Representations which use the Linear Gaussian Model

Name	Equivalent Linear Gaussian Model
Sinusoidal	$y_i = \sum_{k=1}^{K}[a_k \sin(\omega_k t_i) + b_k \cos(\omega_k t_i)] + w_i$.
Autoregressive (AR)	$y_i = \sum_{k=1}^{K} a_k y_{i-k} + w_i$.
Autoregressive with External Input (ARX)	$y_i = \sum_{k=1}^{K} a_k y_{i-k} + \sum_{l=0}^{L} b_k u_{i-l} + w_i$.
Nonlinear Autoregressive (NAR)	$y_i = \sum_{k=1}^{K} a_k y_{i-k} + \sum_{l=1}^{L} \sum_{m=1}^{M} b_{lm} y_{i-l} y_{i-m} + \ldots + w_i$.
Volterra	$y_i = \sum_{k=1}^{K} a_k y_{i-k} + \sum_{l=1}^{L} \sum_{m=1}^{M} b_{lm} y_{i-l} y_{i-m} + \ldots + w_i$.
Polynomial	$y_i = \sum_{k=1}^{K} a_k u_i^{k-1}$.

Example 10.8. Fitting Covariance Measurements to a Spatial Covariance Model [8]. In Sect. 4.3.1 we described the construction of a GIS using the Kriging estimators. The calculation of these estimators require a positive-definite spatial covariance matrix $\boldsymbol{\Sigma}(\mathbf{u}_i, \mathbf{u}_j)$. One way of ensuring this is to fit the experimentally measured $\boldsymbol{\Sigma}(\mathbf{u}_i, \mathbf{u}_j)$ values to a parametric model $\boldsymbol{\Sigma}(\mathbf{u}_i, \mathbf{u}_j|\boldsymbol{\theta})$ which has been designed to be positive definite. Traditionally, we determine the parameter vector $\boldsymbol{\theta}$ using the method of least squares in which we suppose all $\mathbf{y}_i$ are equally important. In this case the calculation of $\boldsymbol{\theta}$ reduces to

$$\hat{\boldsymbol{\theta}}_{\mathrm{LS}} = \arg\min_{\boldsymbol{\theta}} \Big((\boldsymbol{\Sigma}(\mathbf{u}_i, \mathbf{u}_j) - \boldsymbol{\Sigma}(\mathbf{u}_i, \mathbf{u}_j|\boldsymbol{\theta}))^T (\boldsymbol{\Sigma}(\mathbf{u}_i, \mathbf{u}_j) - \boldsymbol{\Sigma}(\mathbf{u}_i, \mathbf{u}_j|\boldsymbol{\theta})) \Big) .$$

This is known as the *ordinary* least square procedure.

10.6 Linear Gaussian Model

The *Linear Gaussian Model* describes a linear model between two vectors $\mathbf{y}$ and $\boldsymbol{\theta}$ with Gaussian errors. Mathematically, the model is

$$\mathbf{y} = \mathbf{H}\boldsymbol{\theta} + \mathbf{w} , \tag{10.23}$$

where $\mathbf{y} = (y_1, y_2, \ldots, y_N)^T$ is a $N \times 1$ vector of scalar measurements, $\mathbf{H}$ is a known $N \times K$ matrix, $\boldsymbol{\theta} = (\theta_1, \theta_2, \ldots, \theta_N)^T$ is a $K \times 1$ random vector of linear coefficients with *a priori* pdf, $\mathcal{N}(\boldsymbol{\mu}, \boldsymbol{\Sigma})$, and $\mathbf{w} = (w_1, w_2, \ldots, w_N)^T$ is a $N \times 1$ noise vector with pdf, $\mathcal{N}(\mathbf{0}, \sigma^2\mathbf{I})$, and independent of $\boldsymbol{\theta}$.

The linear Gaussian model forms the basis for much statistical modelling which is used in multi-sensor data fusion. In Table 10.4 we list some common statistical models which may be written in the form of a linear Gaussian model.

The following example shows how we may re-write the polynomial model as a linear Gaussian model.

Example 10.9. Polynomial Model [20]. The polynomial model is defined as

$$y_i = \sum_{k=1}^{K} a_k u_i^{k-1} + w_i ,$$

or, in matrix form, as

$$\begin{pmatrix} y_0 \\ y_1 \\ \vdots \\ y_{N-1} \end{pmatrix} = \begin{pmatrix} 1 & u_0 & u_0^2 & \ldots & u_0^{K-1} \\ 1 & u_1 & u_1^2 & \ldots & u_1^{K-1} \\ \vdots & \vdots & \vdots & \ddots & \vdots \\ 1 & u_{N-1} & u_{N-1}^2 & \ldots & u_{N-1}^{K-1} \end{pmatrix} \begin{pmatrix} a_1 \\ a_2 \\ \vdots \\ a_K \end{pmatrix} + \begin{pmatrix} w_0 \\ w_1 \\ \vdots \\ w_{N-1} \end{pmatrix} .$$

By comparing the matrix equation with (10.23) we make the following identification:

$$\mathbf{H} = \begin{pmatrix} 1 & u_0 & u_0^2 & \ldots & u_0^{K-1} \\ 1 & u_1 & u_1^2 & \ldots & u_1^{K-1} \\ \vdots & \vdots & \vdots & \ddots & \vdots \\ 1 & u_{N-1} & u_{N-1}^2 & \ldots & u_{N-1}^{K-1} \end{pmatrix} ,$$

$$\boldsymbol{\theta} = (a_1, a_2, \ldots, a_N)^T .$$

By Bayes' theorem, the *a posteriori* pdf of the linear Gaussian model parameters is

$$p(\mathbf{H}, \sigma, \boldsymbol{\theta}|\mathbf{y}, I) = \frac{p(\mathbf{y}|\mathbf{H}, \boldsymbol{\theta}, \sigma, I)\pi(\mathbf{H}, \boldsymbol{\theta}, \sigma|I}{p(\mathbf{y}|I)} , \tag{10.24}$$

where

$$p(\mathbf{y}|\mathbf{H}, \sigma, \boldsymbol{\theta}, I) = (2\pi\sigma^2)^{-N/2} \exp\left[-\frac{(\mathbf{y} - \mathbf{H}\boldsymbol{\theta})(\mathbf{y} - \mathbf{H}\boldsymbol{\theta})^T}{\sigma^2}\right] . \tag{10.25}$$

Assuming that $\mathbf{H}$, $\boldsymbol{\theta}$ and σ are statistically independent, we have

$$p(\mathbf{H}, \sigma, \boldsymbol{\theta}|\mathbf{y}, I) = \frac{p(\mathbf{y}|\mathbf{H}, \boldsymbol{\theta}, \sigma, I)\pi(\mathbf{H}|I)\pi(\boldsymbol{\theta}|I)\pi(\sigma|I)}{p(\mathbf{y}|I)} , \tag{10.26}$$

In the case where there is no *a priori* information concerning $\mathbf{H}$, $\boldsymbol{\theta}$ and σ we may use uniform priors for the elements of $\mathbf{H}$ and $\boldsymbol{\theta}$ and Jeffrey's prior for σ. In this case, the joint *a posteriori* pdf for $\mathbf{H}$, $\boldsymbol{\theta}$ and σ becomes

$$p(\mathbf{H},\boldsymbol{\theta},\sigma|\mathbf{y},I) \propto \frac{p(\mathbf{y}|\mathbf{H},\boldsymbol{\theta},\sigma,I)}{p(\mathbf{y}|I)} \times \frac{1}{\sigma}\,,$$
$$\propto \frac{1}{\sigma^{(N+1)}} \exp\left[-\left(\frac{(\mathbf{y}-\mathbf{H}\boldsymbol{\theta})^T(\mathbf{y}-\mathbf{H}\boldsymbol{\theta})}{2\sigma^2}\right)\right]. \qquad (10.27)$$

Until now we assumed the noise components w_i are iid, where $w_i \sim \mathcal{N}(0,\sigma^2)$. However, in some applications the noise components w_i are not independent. In this case $\mathbf{w} \sim \mathcal{N}(\mathbf{0},\boldsymbol{\Sigma})$, where the off-diagonal components of $\boldsymbol{\Sigma}$ are not zero. However, we may always convert the measurements so that $w_i \sim \mathcal{N}(0,\sigma^2)$ by applying a *whitening transformation* $\mathbf{D}$.

Example 10.10. Whitening Transformation [17]. We assume $\boldsymbol{\Sigma}$ is positive definite. In this case it can be factored as $\boldsymbol{\Sigma}^{-1} = \mathbf{D}^T\mathbf{D}$, where $\mathbf{D}$ is an $N \times N$ invertible matrix. The matrix $\mathbf{D}$ acts as a *whitening* transformation when applied to $\mathbf{w}$, since

$$E\left(\mathbf{Dw}(\mathbf{Dw})^T\right) = \mathbf{D}\boldsymbol{\Sigma}\mathbf{D}^T = \mathbf{D}\mathbf{D}^{-1}(\mathbf{D}^T)^{-1}\mathbf{D}^T = \mathbf{I}\,,$$

where $\mathbf{I} = \mathrm{diag}\overbrace{(1,1,\ldots,1)}^{N}$.

As a consequence, if we apply $\mathbf{D}$ to $\mathbf{y} = \mathbf{H}\boldsymbol{\theta} + \mathbf{w}$ we obtain a linear Gaussian model:

$$\mathbf{y}' = \mathbf{Dy} = \mathbf{DH}\boldsymbol{\theta} + \mathbf{Dw} = \mathbf{H}'\boldsymbol{\theta} + \mathbf{w}'\,,$$

where $\mathbf{w}' = \mathbf{Dw} \sim \mathcal{N}(\mathbf{0},\mathbf{I})$.

The linear Gaussian model and the *a posteriori* pdf $p(\mathbf{H},\boldsymbol{\theta},\sigma|\mathbf{y},I)$ are widely used in multi-sensor data fusion. By way of an example, we give three different types of application.

10.6.1 Line Fitting

In curve fitting the matrix $\mathbf{H}$ is completely specified. In this case, our objective is to estimate the vector $\boldsymbol{\theta}$ assuming $\mathbf{H}$ is given but having no *a priori* knowledge regarding $\boldsymbol{\theta}$ or σ. In this case $\mathbf{H}$ does not appear in (10.27). We eliminate σ by marginalization. Then the resulting *a posteriori* pdf is

$$p(\boldsymbol{\theta}|\mathbf{y},I) = \frac{1}{2}\Gamma\left(\frac{N-1}{2}\right)\left(\frac{\mathbf{y}-\mathbf{H}\boldsymbol{\theta})^T(\mathbf{y}-\mathbf{H}\boldsymbol{\theta})}{2}\right)^{-((N-1)/2)}. \qquad (10.28)$$

Maximizing this expression gives us the MAP estimate for the parameter vector $\boldsymbol{\theta}$:

$$\theta_{\mathrm{MAP}} = (\mathbf{H}^T\mathbf{H})^{-1}\mathbf{H}^T\mathbf{y}\,. \qquad (10.29)$$

The following example illustrates the fitting of a straight-line to a set of points $(u_i, y_i), i \in \{1,2,\ldots,N\}$.

Example 10.11. Straight-Line Regression. We consider a set of points $(u_i, y_i), i \in \{1,2,\ldots,N\}$, which we assume to obey the following straight-line equation:

$$y_i = mu_i + c + w_i ,$$

where $w_i \sim \mathcal{N}(0, \sigma^2)$ are independent random noise. The corresponding matrix equation is

$$\begin{pmatrix} y_1 \\ y_2 \\ \vdots \\ y_N \end{pmatrix} = \begin{pmatrix} 1 & u_1 \\ 1 & u_2 \\ \vdots & \vdots \\ 1 & u_N \end{pmatrix} \begin{pmatrix} c \\ m \end{pmatrix} + \begin{pmatrix} w_1 \\ w_2 \\ \vdots \\ w_N \end{pmatrix} .$$

It follows that

$$\mathbf{H}^T\mathbf{H} = \begin{pmatrix} N & \sum_{i=1}^N u_i \\ \sum_{i=1}^N u_i & \sum_{i=1}^N u_i^2 \end{pmatrix} ,$$

$$(\mathbf{H}^T\mathbf{H})^{-1} = \frac{1}{N\sum_{i=1}^N u_i^2 - (\sum_{i=1}^N u_i)^2} \begin{pmatrix} \sum_{i=1}^N u_i^2 & -\sum_{i=1}^N u_i \\ -\sum_{i=1}^N u_i & N \end{pmatrix} ,$$

$$\mathbf{H}^T\mathbf{y} = \begin{pmatrix} \sum_{i=1}^N y_i \\ \sum_{i=1}^N u_i y_i \end{pmatrix} .$$

Substituting all these terms in (10.29) gives the following expression for the map estimate of $(c,m)^T$:

$$\begin{pmatrix} c \\ m \end{pmatrix} = \frac{1}{\sum_{i=1}^N u_i^2 - N\bar{u}^2} \begin{pmatrix} \bar{y}\sum_{i=1}^N u_i^2 - \bar{u}\sum_{i=1}^N u_i y_i \\ \sum_{i=1}^N u_i y_i - \bar{u}\bar{y} \end{pmatrix} ,$$

where $\bar{u} = \sum_{i=1}^N u_i/N$ and $\bar{y} = \sum_{i=1}^N y_i/N$. This solution is identical to the solution obtained using the method of maximum likelihood or least squares. This is not unexpected since we assumed non-informative priors for $\boldsymbol{\theta}$ and σ.

The following example illustrates a novel use of multi-metric fusion using linear regression.

Example 10.12. Multi-Metric Fusion for Visual Quality Assessment [18]. Various image quality metrics have been developed to reflect human visual quality. However, so far, there is no single quality metric that is suitable for all image inputs. In [18] the authors construct a multi-metric fused score (MMF) by non-linearly fusing multiple image quality metrics. Given N training images, let $q_i^{(m)}, m \in \{1,2,\ldots,M\}$, denote the quality score obtained with the mth quality metric on the ith training image. Then we define the MMF score as

$$MMF_i = \mathbf{W}^T \mathbf{Q}_i + b \,,$$

where $\mathbf{Q}_i = (q_i^{(1)}, q_i^{(2)}, \ldots, q_i^{(M)})^T$ and $\mathbf{W} = (w^{(1)}, w^{(2)}, \ldots, w^{(M)})^T$. We determine $\mathbf{W}$ and the bias b by minimizing the difference between MMF_i and the mean opinion score (MOS_i) obtained by human observers on the ith training image.

Experimentally, the mmf is found to outperform all of the individual quality measures $q^{(m)}, m \in \{1,2,\ldots,M\}$.

10.6.2 Change Point Detection

The central concept which underlies a linear change point detector is that an observed time sequence is modeled by different models at different points in time. The models are known but the time instants at which the models change are not known. In this case, our objective is to estimate $\mathbf{H}$. We eliminate θ and σ from (10.27) by marginalization [23]. The resulting *a posteriori* pdf is

$$p(\mathbf{H}|\mathbf{y},I) \propto \frac{\left(\mathbf{y}^T\mathbf{y} - \mathbf{y}\mathbf{H}(\mathbf{H}^T\mathbf{H})^{-1}\mathbf{H}^T\mathbf{y}\right)^{((N-K)/2)}}{\sqrt{|\mathbf{H}^T\mathbf{H}|}} \,. \tag{10.30}$$

The following example illustrates (10.30) given a set of scalar input measurements, $\mathbf{y} = (y_1, y_2, \ldots, y_N)^T$, assuming there is a single (unknown) change point.

Example 10.13. Single Step in Piecewise Constant Time Series [13, 23]. We consider a time series y_i:

$$y_i = \begin{cases} \mu_1 + w_i & \text{if } i \leq m \,, \\ \mu_2 + w_i & \text{if } i > m \,, \end{cases}$$

where m is an unknown integer ($1 \leq m \leq N$) and $w_i \sim \mathcal{N}(0,\sigma^2)$ are independent random noise. The corresponding matrix equation is

$$\begin{pmatrix} y_1 \\ y_2 \\ \vdots \\ y_m \\ y_{m+1} \\ \vdots \\ y_N \end{pmatrix} = \begin{pmatrix} 1 & 0 \\ 1 & 0 \\ \vdots & \vdots \\ 1 & 0 \\ 0 & 1 \\ 0 & 1 \\ \vdots & \vdots \\ 0 & 1 \end{pmatrix} \begin{pmatrix} \mu_1 \\ \mu_2 \end{pmatrix} + \begin{pmatrix} w_1 \\ w_2 \\ \vdots \\ w_m \\ w_{m+1} \\ \vdots \\ w_N \end{pmatrix} .$$

It follows that

$$\mathbf{H}^T\mathbf{H} = \begin{pmatrix} m & 0 \\ 0 & N-m \end{pmatrix} ,$$

$$(\mathbf{H}^T\mathbf{H})^{-1} = \frac{1}{m(N-m)} \begin{pmatrix} N-m & 0 \\ 0 & m \end{pmatrix} ,$$

$$\mathbf{y}^T\mathbf{H} = (\mathbf{H}^T\mathbf{y})^T = \Big(\sum_{i=1}^{m} y_i \ \sum_{i=m+1}^{N} y_i\Big) ,$$

$$\mathbf{y}^T\mathbf{y} = \sum_{i=1}^{N} y_i^2 .$$

Substituting all these terms in (10.30) gives the following expression for the *a posteriori* pdf $p(m|\mathbf{y},I)$:

$$p(m|\mathbf{y},I) \sim \sum_{i=1}^{N} y_i^2 - \Big((N-m)\Big(\sum_{i=1}^{m} y_i\Big)^2\Big) - m\Big(\sum_{i=m+1}^{N} y_i\Big)^2 .$$

The corresponding MAP estimate of m is

$$m_{\mathrm{MAP}} = \arg\max_m \big(p(m|\mathbf{y},I)\big) ,$$

and the associated μ_1 and μ_2 values are:

$$\mu_1 = \frac{1}{m_{\mathrm{MAP}}} \sum_{i=1}^{m_{\mathrm{MAP}}} y_i ,$$

$$\mu_2 = \frac{1}{N - m_{\mathrm{MAP}} + 1} \sum_{i=m_{\mathrm{MAP}}+1}^{N} y_i .$$

Single change point detection with more complicated curves are also easily handled. In the following two examples we derive the matrix $\mathbf{H}$ for a single ramp and a mixture of different order AR processes.

Example 10.14. Single Ramp in a Time Series [13, 23]. We consider a time series y_i:

$$y_i = \begin{cases} \mu + w_i & \text{if } i \le m\,, \\ \mu + r(i-m) + w_i & \text{if } i > m\,, \end{cases}$$

where $w_i \sim \mathcal{N}(0,\sigma^2)$ are independent random noise. The corresponding matrix equation is

$$\begin{pmatrix} y_1 \\ y_2 \\ \vdots \\ y_m \\ y_{m+1} \\ y_{m+2} \\ \vdots \\ y_N \end{pmatrix} = \begin{pmatrix} 1 & 0 \\ 1 & 0 \\ \vdots & \vdots \\ 1 & 0 \\ 1 & 1 \\ 1 & 2 \\ \vdots & \vdots \\ 1 & N-m \end{pmatrix} \begin{pmatrix} \mu \\ r \end{pmatrix} + \begin{pmatrix} w_1 \\ w_2 \\ \vdots \\ w_m \\ w_{m+1} \\ w_{m+2} \\ \vdots \\ w_N \end{pmatrix}.$$

Example 10.15. Mixed Two and Three-Order AR Process [13, 23]. We consider a time series y_i:

$$y_i = \begin{cases} \sum_{j=1}^{2} \alpha_j y_{i-j} + w_i & \text{if } i \le m\,, \\ \sum_{j=1}^{3} \beta_j y_{i-j} + w_i & \text{if } i > m\,, \end{cases}$$

where $w_i \sim \mathcal{N}(0,\sigma^2)$ are independent random noise variables. The corresponding matrix equation is

$$\begin{pmatrix} y_1 \\ y_2 \\ \vdots \\ y_{m-1} \\ y_m \\ y_{m+1} \\ \vdots \\ y_N \end{pmatrix} = \begin{pmatrix} y_0 & y_{-1} & 0 & 0 & 0 \\ y_1 & y_0 & 0 & 0 & 0 \\ y_2 & y_1 & 0 & 0 & 0 \\ \vdots & & \ddots & & \vdots \\ y_{m-2} & y_{m-3} & 0 & 0 & 0 \\ y_{m-1} & y_{m-2} & 0 & 0 & 0 \\ 0 & 0 & y_m & y_{m-1} & y_{m-2} \\ \vdots & & & & \\ 0 & 0 & y_{N-1} & y_{N-2} & y_{N-3} \end{pmatrix} \begin{pmatrix} \alpha_1 \\ \alpha_2 \\ \beta_1 \\ \beta_2 \\ \beta_3 \end{pmatrix} + \begin{pmatrix} w_1 \\ w_2 \\ \vdots \\ w_{m-1} \\ w_m \\ w_{m+1} \\ \vdots \\ w_N \end{pmatrix}.$$

In the most general case the number of change points is unknown. In each segment the y_i are modeled as independent polynomial functions of unknown order. Efficient algorithms for solving the more general case have been developed, including an exact curve fitting algorithm [12].

10.6.3 Probabilistic Subspace

Subspace techniques are a family of techniques which may be used to generate a low dimensional common representational format (see Sect. 4.5). Recently probabilistic versions of several subspace techniques have been described. In this section we concentrate on the *probabilistic principal component analysis* (PPCA) [26, 27] which is a probabilistic version of the principal component analysis (PCA) algorithm (see Ex. 4.17).

In the probabilistic PCA we model the sensor measurements $\mathbf{y}_i$ using a linear Gaussian model of the form

$$\mathbf{y} = \mathbf{H}\boldsymbol{\theta} + \boldsymbol{\mu} + \mathbf{w} , \tag{10.31}$$

where we assume

$$p(\boldsymbol{\theta}|I) \sim \mathcal{N}(\mathbf{0},\mathbf{I}) , \tag{10.32}$$
$$p(\mathbf{w}|I) \sim \mathcal{N}(\mathbf{0},\sigma^2\mathbf{I}) . \tag{10.33}$$

Under these assumptions, we find

$$\begin{aligned} p(\mathbf{y}|\mathbf{H},\boldsymbol{\mu},\sigma^2,I) &= \int p(\mathbf{y}|\mathbf{H},\boldsymbol{\theta},\boldsymbol{\mu},\sigma^2,I)p(\boldsymbol{\theta}|I)d\boldsymbol{\theta} , \\ &\sim \mathcal{N}(\boldsymbol{\mu},\mathbf{H}\mathbf{H}^T + \sigma^2\mathbf{I}) , \end{aligned} \tag{10.34}$$

where

$$p(\mathbf{y}|\mathbf{H},\boldsymbol{\theta},\boldsymbol{\mu},\sigma^2,I) \sim \mathcal{N}(\mathbf{H}\boldsymbol{\theta}+\boldsymbol{\mu},\sigma^2\mathbf{I}) . \tag{10.35}$$

We use (10.32-10.35) in Bayes' rule to give us the *a posteriori* pdf $p(\boldsymbol{\theta}|\mathbf{y},I)$:

$$\begin{aligned} p(\boldsymbol{\theta}|\mathbf{H},\boldsymbol{\mu},\sigma^2,\mathbf{y},I) &= \frac{p(\mathbf{y}|\mathbf{H},\boldsymbol{\theta},\boldsymbol{\mu},\sigma^2,I)p(\boldsymbol{\theta}|I)}{p(\mathbf{y}|\mathbf{H},\boldsymbol{\mu},\sigma^2,I)} , \\ &= \frac{\mathcal{N}(\mathbf{y}|\mathbf{H}\boldsymbol{\theta}+\boldsymbol{\mu},\sigma^2 I)\mathcal{N}(\boldsymbol{\theta}|\mathbf{0},\mathbf{I})}{\mathcal{N}(\mathbf{y}|\boldsymbol{\mu},\mathbf{H}\mathbf{H}^T+\sigma^2\mathbf{I})} , \\ &= \mathcal{N}(\mathbf{H}^T(\mathbf{y}-\boldsymbol{\mu})/\beta,\sigma^2/\beta) , \end{aligned} \tag{10.36}$$

where

$$\beta = \mathbf{H}^T\mathbf{H} + \sigma^2\mathbf{I} . \tag{10.37}$$

Maximum-likelihood estimates of the unknown parameters in (10.36) are given by

$$\boldsymbol{\mu}_{\mathrm{ML}} = \frac{1}{N}\sum_{i=1}^{N}\mathbf{y}_i , \tag{10.38}$$

$$\sigma_{\mathrm{ML}}^2 = \frac{1}{M-L} \sum_{i=L+1}^{M} \lambda_i , \tag{10.39}$$

$$\mathbf{H}_{\mathrm{ML}} = \mathbf{U}(\boldsymbol{\Lambda} - \sigma_{\mathrm{ML}}^2 \mathbf{I})^{1/2} \mathbf{R} , \tag{10.40}$$

where $\lambda_{L+1}, \ldots, \lambda_M$ are the smallest eigenvalues of the sample covariance matrix $\boldsymbol{\Sigma}$, the L columns in the $M \times L$ orthogonal matrix $\mathbf{U}$ are the dominant L eigenvectors of $\boldsymbol{\Sigma}$, $\boldsymbol{\Lambda} = \mathrm{diag}(\lambda_1, \ldots, \lambda_L)$ and $\mathbf{R}$ is an arbitrary $L \times L$ orthogonal matrix [2].

Apart from using the PPCA to create a low-dimensional probabilistic common representational format (Sect. 4.5), we can also model arbitrary pdf's using a mixture of PPCA's instead of a mixture of Gaussians. The main advantage is that by using PPCA's we can help reduce the impact of the "curse of dimensionality" [3] (see Ex. 9.8). The following example compares a mixture of PPCA's with the GMM.

Example 10.16. PPCA vs. GMM [27]. A partially open circle in (x,y)-space is sampled after adding iid (independent and identically distributed) Gaussian noise. In this case $M = 2$. We model the data using three mixtures: (1) a mixture of full covariance Gaussian pdf's, (2) a mixture of diagonal covariance Gaussian pdf's, and (3) a mixture of one component ($L = 1$) PPCA's. Fig. 10.3 shows the results obtained.

The following example illustrates an application which uses a mixture of PPCA's instead of the more usual mixture of Gaussians.

Example 10.17. Voice Conversion Using a Mixture of PPCA's [30]. Voice conversion is the art of transforming one speaker's voice to sound as if it were uttered by another speaker. One application of voice conversion deals with the dubbing of a "original" film sound track. When a popular movie is exported to a foreign country, the sound track is often re-recorded by replacement speakers who speak a translated version of the original performer's part. By using voice conversion we obtain a sound track spoken in a foreign language by a speaker whose voice matches that of the original performer. For this application the authors found it advantageous to use a mixture of PPCA's instead of the more conventional mixture of Gaussians.

---

2 The presence of the matrix $\mathbf{R}$ in (10.40) means that $\mathbf{H}_{\mathrm{ML}}$ is rotationally ambiguous. We may eliminate the ambiguity and obtain the true principal axes by multiplying $\mathbf{H}_{\mathrm{ML}}$ by the rotation matrix $\mathbf{R}_{\mathrm{ML}}$, where the columns of $\mathbf{R}_{\mathrm{ML}}$ are the eigenvectors of $\mathbf{H}_{\mathrm{ML}}^T \mathbf{H}_{\mathrm{ML}}$.

3 The number of free parameters in a mixture of full covariance M-dimensional Gaussian pdf's is $M(M+3)/2$ parameters per pdf. In a mixture of diagonal covariance M-dimensional Gaussian pdf's the number of free parameters is $2M$ per pdf and in a mixture of L component PPCA's the number of free parameters is $M(M+1)+1+L(L-1)/2$ per PPCA. *Note*: The term $L(L-1)/2$ arises from the need to specify the rotation matrix $\mathbf{R}$.

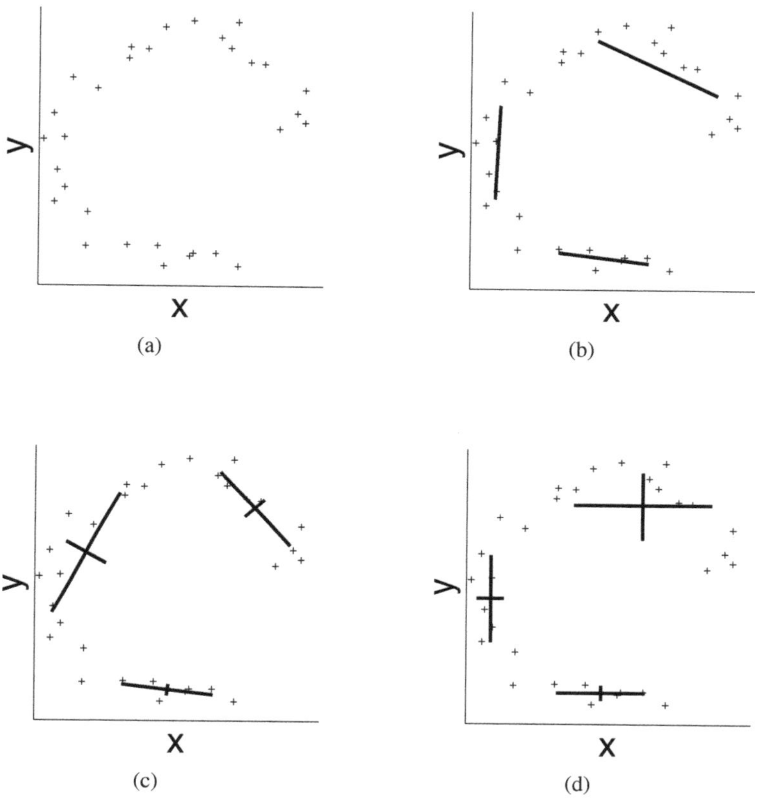

Fig. 10.3 Shows the modeling of a noisy partially open circle in two-dimensions. **(a)** Shows the data points in (x,y) space. **(b)** Shows the three one-component PPCA's we use to model the points. **(c)** Shows the three full-covariance Gaussian mixture we use to model the points. **(d)** Shows the three diagonal covariance Gaussian mixture we use to model the points.

10.7 Generalized Millman Formula *

The *generalized Millman's formula* [24] deals with the optimal linear combination of an arbitrary number of correlated estimates. Suppose we have K local estimates, $\boldsymbol{\mu}_1, \boldsymbol{\mu}_2, \ldots, \boldsymbol{\mu}_K$, of an unknown L-dimensional random vector $\boldsymbol{\theta}$. Then the optimal *linear* estimate of $\boldsymbol{\theta}$ is

$$\widetilde{\boldsymbol{\mu}} = \sum_{k=1}^{K} \mathbf{W}_k \boldsymbol{\mu}_k \,, \tag{10.41}$$

where

$$\sum_{k=1}^{K} \mathbf{W}_k = \mathrm{diag}(\overbrace{1, 1, \ldots, 1}^{L}) \,, \tag{10.42}$$

and the $\mathbf{W}_k$ are $L \times L$ constant weighting matrices determined from the least square error criterion

$$(\mathbf{W}_1, \mathbf{W}_2, \ldots, \mathbf{W}_K) = \min_{\mathbf{W}_k} \Big(|\boldsymbol{\theta} - \sum_{k=1}^{K} \mathbf{W}_k \boldsymbol{\mu}_k|^2 \Big) . \tag{10.43}$$

The solution to this problem is given by the following set of $(K-1)$ linear equations

$$\begin{aligned}
&\sum_{k=1}^{K-1} \mathbf{W}_k(\boldsymbol{\Sigma}_{k,1} - \boldsymbol{\Sigma}_{k,K}) + \mathbf{W}_K(\boldsymbol{\Sigma}_{K,1} - \boldsymbol{\Sigma}_{K,K}) = 0 , \\
&\sum_{k=1}^{K-1} \mathbf{W}_k(\boldsymbol{\Sigma}_{k,2} - \boldsymbol{\Sigma}_{k,K}) + \mathbf{W}_K(\boldsymbol{\Sigma}_{K,2} - \boldsymbol{\Sigma}_{K,K}) = 0 , \\
&\vdots \\
&\sum_{k=1}^{K-1} \mathbf{W}_k(\boldsymbol{\Sigma}_{k,K-1} - \boldsymbol{\Sigma}_{k,K}) + \mathbf{W}_K(\boldsymbol{\Sigma}_{K,K-1} - \boldsymbol{\Sigma}_{K,K}) = 0 ,
\end{aligned} \tag{10.44}$$

subject to (10.41), where $\boldsymbol{\Sigma}_{k,k} \equiv \boldsymbol{\Sigma}_k$ is the $L \times L$ covariance matrix $\boldsymbol{\Sigma}_{k,k} = E\big((\boldsymbol{\theta} - \boldsymbol{\mu}_k)(\boldsymbol{\theta} - \boldsymbol{\mu}_k)^T\big)$ and $\boldsymbol{\Sigma}_{k,l}, k \neq l$ is the $L \times L$ cross-covariance matrix $\boldsymbol{\Sigma}_{k,l} = E\big((\boldsymbol{\theta} - \boldsymbol{\mu}_k)(\boldsymbol{\theta} - \boldsymbol{\mu}_l)^T\big)$. In many applications, only the covariance matrices, $\boldsymbol{\Sigma}_{k,k}, k \in \{1, 2, \ldots, K\}$, are known, while the cross-covariances, $\boldsymbol{\Sigma}_{k,l}, k \neq l$, are unknown. In this case, we often assume $\boldsymbol{\Sigma}_{k,l} = 0, k \neq l$. This may, however, cause problems, especially in decentralized systems (see Sect. 3.4.2).

Example 10.18. Millman's Formula [14]. Given two *uncorrelated* estimates $\boldsymbol{\mu}_1$ and $\boldsymbol{\mu}_2$, the generalized Millman formula reduces to

$$\tilde{\boldsymbol{\mu}} = \boldsymbol{\Sigma}_{22}(\boldsymbol{\Sigma}_{11} + \boldsymbol{\Sigma}_{22})^{-1}\boldsymbol{\mu}_1 + \boldsymbol{\Sigma}_{11}(\boldsymbol{\Sigma}_{11} + \boldsymbol{\Sigma}_{22})^{-1}\boldsymbol{\mu}_2 ,$$

where $\tilde{\boldsymbol{\mu}}$ is the optimum linear estimate.

Example 10.19. Bar-Shalom-Campo Formula [3]. Given two *correlated* estimates $\boldsymbol{\mu}_1$ and $\boldsymbol{\mu}_2$, the generalized Millman formula reduces to the Bar-Shalom-Campo formula:

$$\begin{aligned}
\tilde{\boldsymbol{\mu}} = {} & (\boldsymbol{\Sigma}_{22} - \boldsymbol{\Sigma}_{21})(\boldsymbol{\Sigma}_{11} + \boldsymbol{\Sigma}_{22} - \boldsymbol{\Sigma}_{12} - \boldsymbol{\Sigma}_{21})^{-1}\boldsymbol{\mu}_1 \\
& + (\boldsymbol{\Sigma}_{11} - \boldsymbol{\Sigma}_{12})(\boldsymbol{\Sigma}_{11} + \boldsymbol{\Sigma}_{22} - \boldsymbol{\Sigma}_{12} - \boldsymbol{\Sigma}_{21})^{-1}\boldsymbol{\mu}_2 ,
\end{aligned}$$

where $\tilde{\boldsymbol{\mu}}$ is the optimum linear estimate.

10.8 Software

MATLAB STATISTICS TOOLBOX. The mathworks matlab statistics toolbox. The toolbox contains various m-files for performing parameter eestimation.
MILES. A matlab toolbox for fitting different maximum likelihood models using least squares algorithms.
NETLAB. A matlab toolbox for performing neural network pattern recognition. Author: Ian Nabney.
STPRTOOL. A matlab toolbox for performing statistical pattern recognition. Authors: Vojtech Franc, Vaclav Hlavac.

10.9 Further Reading

Bayesian parameter estimation is discussed in many books and monographs. Good accounts are to be found in [1, 2, 10, 15, 17, 19, 25]. The linear Gaussian models are discussed at length in [21, 22, 31]. A full description of Bayesian curve fitting is given in [9]. Modern references on probabilistic subspace techniques include [27, 28].

Problems

10.1. Describe how we may use Bayesian analysis to find a finite length straight-line segment to a set of noisy points.

10.2. Define the method of maximum likelihood. When should it reduce to MAP (maximum a posteriori)?

10.3. Define the method of least squares. When does it reduce to MAP?

10.4. Define the linear Gaussian model. Describe straight-line regression using the linear Gaussian model.

10.5. Describe the detection of a single step in a piece-wise constant time using the linear Gaussian model.

10.6. Describe the probabilistic PCA.

References

1. D'Agostini, G.: Bayesian Reasoning in Data analysis. World Scientific, Singapore (2003)
2. D'Agostini, G.: Bayesian inference in processing experimental data: principles and basic applications. Rep. Prog. Phys. 66, 1383–1419 (2003)
3. Bar-Shalom, Y., Campo, L.: The effect of the common process noise on the two sensor fused track covariance. IEEE Trans. Aero. Elect. Syst. 22, 803–805 (1986)

4. Bar-Shalom, Y., Li, X.: Multitarget-Multisensor Tracking: Principles and Techniques. YBS Publishing, Storrs (1995)
5. Blackman, S.S., Popoli, R.F.: Design and analysis of modern tracking Systems. Artech House, Norwood (1999)
6. Brainard, D.H., Freeman, W.T.: Bayesian color constancy. J. Opt. Soc. Am. 14A, 1393–1411 (1997)
7. Choudrey, R.A.: Variational methods for Bayesian independent component analysis. PhD thesis, University of Oxford (2002)
8. Cressie, N.A.C.: Statistics for spatial data. John Wiley and Sons (1993)
9. Denison, D.G.T., Holmes, C., Mallick, B.K., Smith, A.F.M.: Bayesian Methods for Nonlinear Classification and Regression. John Wiley and Sons (2002)
10. Dose, V.: Bayesian inference in physics: case studies. Rep. Prog. Phys. 66, 1421–1461 (2003)
11. Falk, T.H., Chan, W.-Y.: Nonintrusive speech quality estimation using Gaussian mixture models. IEEE Sig. Proc. Lett. 13, 108–111 (2006)
12. Fearnhead, P.: Exact Bayesian curve fitting and signal segmentation. IEEE Trans. Sig. Proc. 53, 2160–2166 (2005)
13. Fitzgerald, W.J., Godsill, S.J., Kokaram, A.C., Stark, J.A.: Bayesian methods in signal and image processing. In: Bernardo, J.M., et al. (eds.) Bayesian Statistics, vol. 6, Oxford University Press (1999)
14. Gelb, A.: Applied Optimal estimation. MIT Press (1974)
15. Gelman, A., Larlin, J.S., Stern, H.S., Rubin, D.R.: Bayesian Data Analysis. Chapman and Hall (2003)
16. Goovaerts, P.: Geostatistical approaches for incorporating elevation into the spatial interpolation of rainfall. J. Hydrology 228, 113–129 (2000)
17. Kay, S.M.: Fundamentals of Statistical Signal Processing: Estimation Theory. Prentice-Hall (1993)
18. Liu, T.-J., Lin, W., Jay Kuo, C.-C.: A multi-metric fusion approach to visual quality assessment. (To be published 2011)
19. MacKay, D.J.C.: Information Theory, Inference and Learning Algorithms. Cambridge University Press (2003)
20. Punska, O.: Bayesian approaches to multi-sensor data fusion. MPhil thesis, Signal Processing and Communications Laboratory, Department of Engineering, University of Cambridge (1999)
21. Rosti, A.-V.I., Gales, M.J.F.: Generalized linear Gaussian models. Technical Report CUED/F-INFENG/TR-420, Cambridge University Engineering Dpartment, Cambridge, England (November 23, 2001)
22. Roweis, S., Ghahramani, Z.: A unifying review of linear Gaussian models. Neural Comp. 11, 305–345 (1999)
23. Ruanaidh, J.J.K.O., Fitzgerald, W.J.: Numerical Bayesian methods applied to signal processing. Springer, Heidelberg (1996)
24. Shin, V., Lee, Y., Choi, T.-S.: Generalized Millman's formula and its application for estimation problems. Sig. Proc. 86, 257–266 (2006)
25. Sivia, D.S.: Data Analysis: A Bayesian Tutoral. Oxford University Press (1996)
26. Tipping, M.E., Bishop, C.M.: Probabilistic principal component analysis. J. Roy. Stat. Soc. 61B, 611–622 (1999)
27. Tipping, M.E., Bishop, C.M.: Mixtures of probabilistic principal component analysis. Neural Comp. 11, 443–482 (1999)
28. Wang, C., Wang, W.: Links between ppca and subspace methods for complete gaussian density estimation. IEEE Trans. Neural Networks 17, 789–792 (2006)
29. Werman, M., Keren, D.: A Bayesian method for fitting parametric and nonparametric models to noisy data. IEEE Trans. Patt. Anal. Mach. Intell. 23, 528–534 (2001)
30. Wilde, M.M.: Controlling performance in voice conversion with probabilistic principal component analysis. MSc thesis, Tulane University, USA (2004)
31. Williams, C.K.I., Barber, D.: Bayesian classification with Gaussian processes. IEEE Trans. Patt. Anal. Mach. Intell. 20, 1342–1351 (1998)

Chapter 11
Robust Statistics

11.1 Introduction

In this chapter we shall consider the subject of *robust statistics* and, in particular, *robust parameter estimation*. Robust statistics is defined as the study of statistical methods which are relatively insensitive to the presence of outliers, i. e. input data which is "strange" or "incompatible" with the remaining input data. It might be thought that outliers are a rare event. This is not, however, true. In fact, a common experience in all scientific experiments is that repeated measurements of supposedly one and the same quantity result, occasionally, in values which are in striking disagreement with all the others.

Increasingly, the need to use robust statistics in multi-sensor data fusion is being recognized. The reason for this is the devastating effect outliers may have on standard statistical procedures as illustrated in the following example.

Example 11.1. Fitting a Straight-Line in the Presence of Outliers. We reconsider Ex. 10.5 except this time we suppose that in addition to the N inlier points $(x_i, y_i), i \in \{1, 2, \ldots, N\}$, we have a single outlier (x_{N+1}, y_{N+1}). If the outlier lies sufficiently far from the inliers, then by inspection the optimal straight-line (i. e. the line with the minimum sum of square deviations) will pass through the center of mass $(\bar{x}, \bar{y})$ and through the outlier (x_{N+1}, y_{N+1}) (line BB' in Fig. 11.1). This is in spite of the fact that all the inliers lie on the line AA'.

In many applications, robust parameter estimation represents the major fusion algorithm in a multi-sensor data fusion system. In Table 11.1, we list some of these applications together with the classification of the type of fusion algorithm involved.

Apart from the applications listed in Table 11.1, robust parameter estimation techniques are widely used in forming a common representational format (Chapts. 4-8).

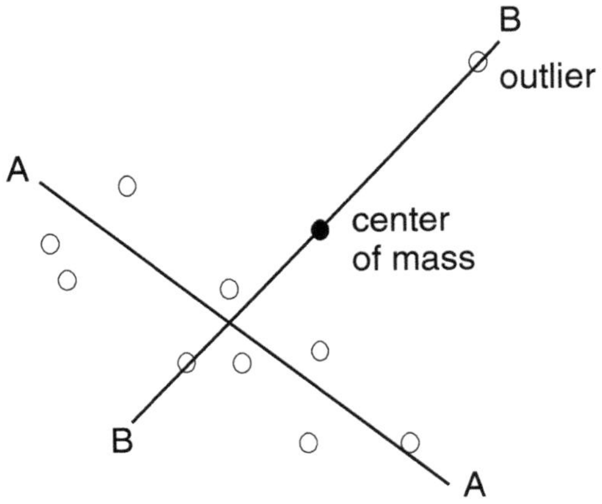

Fig. 11.1 Shows a set of points (including an outlier) in two-dimensions. Without the outlier the optimal straight-line is *AA* while with the outlier the optimal straight-line is *BB*.

Table 11.1 Applications in which Robust Parameter Estimation is the Primary Fusion Algorithm

Class		Application
DaI-DaO	Ex. 2.9	Multi-Sensor Data Fusion with Spurious Data.
	Ex. 10.1	Color Constancy and the Minimum Local Mass (MLM) Loss Function.
	Ex. 11.2	Gating Sensor Observations in Target Tracking.
	Ex. 11.3	Calibration of the SI Unit of the Volt.
	Ex. 11.8	Neutron Fission Cross Sections.
	Ex. 11.9	Estimating Hubble's Constant Using a "Good-and-Bad"Likelihood Function.
DaI-FeO	Ex. 4.18	Chemical Sensor Analysis: Principal Discriminant Analysis (PDA) for Small-Sample Problems.
	Ex. 11.4	Robust Stereo Matching.
	Ex. 11.6	Student Mixture Model.
	Ex. 11.10	Background Estimation in a Powder Diffraction Pattern.
	Ex. 11.7	Robust Probabilistic Principal Component Analysis.
	Ex. 11.13	Robust Discriminant Analysis.
	Ex. 11.15	Target Tracking Initialization Using a Hough Transform.

The designations DaI-DaO and DaI-FeO refer, respectively, to the Dasarathy input/output classifications: "Data Input-Data Output" and "Data Input-Feature Output" (see Sect. 1.3.3).

11.2 Outliers

The simplest approach to dealing with outliers is to eliminate them before parameter estimation takes place. One area in which outliers are routinely eliminated before any parameter estimation takes place is in target tracking when it is known as *gating* and is illustrated in the following example.

*Example 11.2. Gating Sensor Observations in Target Tracking** [5]. Recursive filters (see Chapt. 12) are routinely used to track a moving target. A necessary part of a tracking filter is to eliminate spurious sensor observations which are not associated with the target. In gating this is done by drawing a validation volume, or *gate*, around the predicted sensor observation (Fig. 11.2). The gate is usually formed in such a way that the probability that a true observation (i. e. an observation which is associated with the target) falls within the gate (assuming it is detected) is given by a gating probability P_G [7, 5].

Let $\boldsymbol{\theta}_k$ denote the state of the target at time step k and let $\mathbf{y}_k$ denote the corresponding sensor measurement, where

$$\mathbf{y}_k = \mathbf{H}\boldsymbol{\theta}_k + \mathbf{w}_k ,$$

and $\mathbf{w}_k \sim \mathcal{N}(\mathbf{0}, \mathbf{R})$. If a Kalman filter is used to track the target, then the predicted state of the target at time step k is

$$p(\boldsymbol{\theta}_k|\mathbf{y}_{1:k-1}, I) = \mathcal{N}(\boldsymbol{\mu}_{k|k-1}, \boldsymbol{\Sigma}_{k|k-1}) ,$$

where $\mathbf{y}_{1:k-1} = (\mathbf{y}_1^T, \mathbf{y}_2^T, \ldots, \mathbf{y}_{k-1}^T)^T$ is the sequence of measurements made upto, and including, the time step $k-1$. In this case, the gating region G is defined as the ellipsoid

$$G = \{\mathbf{y}|(\mathbf{y} - \hat{\mathbf{y}}_{k|k-1})^T \hat{\mathbf{S}}_{k|k-1}^{-1}(\mathbf{y} - \hat{\mathbf{y}}_{k|k-1}) \leq \gamma\} ,$$

where $\hat{\mathbf{y}}_{k|k-1} = \mathbf{H}_k\boldsymbol{\mu}_{k|k-1}$ is the predicted sensor measurement and $\hat{\mathbf{S}}_{k|k-1} = \mathbf{H}_k\boldsymbol{\Sigma}_{k|k-1}\mathbf{H}_k^T + \mathbf{R}_k$ is the predicted sensor covariance matrix. The threshold γ maybe chosen to give a specified gating probability P_G. Alternatively, we may choose a threshold such that a true measurement falling within the gate is more likely than a false alarm. In two-dimensions the corresponding threshold is given by

$$\gamma = 2\ln(P_D) - 2\ln(2\pi\lambda(1 - P_D)|\hat{\mathbf{S}}_{k|k-1}|^{1/2}) ,$$

where P_D is the probability of detection of the target and λ is the spatial density of false alarms.

Eliminating spurious observations may, however, cause unforeseen problems as the following example shows.

Example 11.3. Calibration of the SI Unit of the Volt [11]. In 1973 and again in 1986 a committee of the International Council of Scientific Unions performed a least squares adjustment of several independent measurements concerning

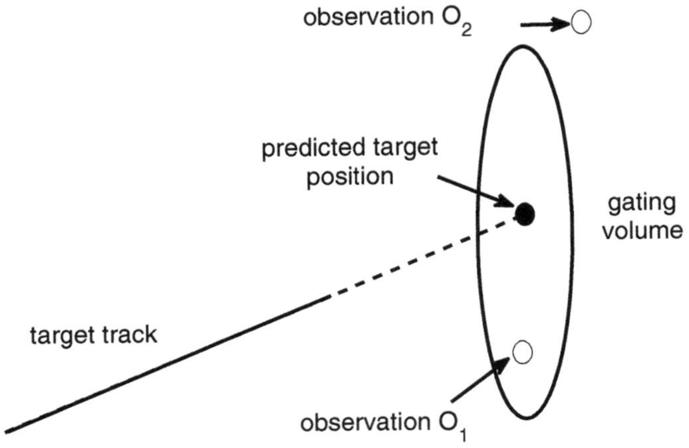

Fig. 11.2 Shows the gating of two sensor observations $O_1 = \langle E_1, *, k, \mathbf{y}_1, * \rangle$ and $O_2 = \langle E_2, *, k, \mathbf{y}_2, * \rangle$ made at time step k. Observation O_1 falls outside the gate while O_2 falls inside the gate. The gate is shown as an ellipse centered on the expected (predicted) sensor observation $\hat{\mathbf{y}}$ for the time step k. *Note*: For an explanation of the notation used in representing O_2 and O_2, see Sect. 2.5.

K_V (a constant involved in the calibration of the SI unit of the Volt [32]). In performing the least square adjustment the committee eliminated measurements which were deemed incompatible with the remaining measurements. A large change was found between K_V(1973) and K_V(1986) although as the committee explain, the large change in K_V could have been avoided if two measurements which seemed to be discrepant with the remaining measurements had *not* been deleted in the 1973 adjustment.

In order to avoid problems of this sort we often prefer to use "robust" techniques which reduce the influence of the outliers without actually identifying the outliers and eliminating them. Given N measurements $x_i, i \in \{1,2,\ldots,N\}$, this often reduces to simply replacing the sample mean $\bar{x}$ by the sample median m and by replacing the sample standard deviation s by the sample mean absolute deviation MAD, where

$$\bar{x} = \frac{1}{N}\sum_{i=1}^{n} x_i , \quad (11.1)$$

$$m = \text{median}(x_1, x_2, \ldots, x_N) , \quad (11.2)$$

$$s^2 = \frac{1}{N-1}\sum_{i=1}^{N}(x_i - \bar{x})^2 , \quad (11.3)$$

$$MAD = \frac{1}{N}\sum_{i=1}^{N}|x_i - m| . \quad (11.4)$$

Example 11.4. Robust Stereo Matching [8]. In stereo matching we match corresponding patches which have been extracted from a pair of stereo images. Intensity distortions, noise, foreshortening, perspective effects and occlusions make this a difficult task in which the choice of similarity metric is critical. Among the measures examined in [8] were the classical zero-mean normalized correlation coefficient (ZNNC) and a robust version of ZNCC ($\mathrm{ZNCC}_{\mathrm{robust}}$).

Let A and B be two stereo images. If a_k and $b_k, k \in \{1,2,\ldots,K\}$, denote K pixel values in two corresponding patches in A and B, then the ZNCC is defined as:

$$\rho = \sum_{k=1}^{K}(a_k-\bar{a})(b_k-\bar{b}) \Bigg/ \sqrt{\sum_{k=1}^{K}(a_k-\bar{a})^2 \sum_{k=1}^{K}(b-\bar{b})^2} \,.$$

The robust version of the ZNCC involves replacing the mean gray-levels $\bar{a} = (\sum_k a_k)/K$ and $\bar{b} = (\sum_k b_k)/K$ with the median gray-levels $m_A = median(a_k)$ and $m_B = median(b_k)$. The robust $\mathrm{ZNCC}_{\mathrm{robust}}$ is:

$$\tilde{\rho} = \sum_{k=1}^{K}(a_k-m_A)(b_k-m_B) \Bigg/ \sqrt{\sum_{k=1}^{K}(a_k-m_A)^2 \sum_{k=1}^{K}(b-m_B)^2} \,.$$

Note: In a detailed evaluation of forty similarity measures, Chambon and Crouzil found the overall performance of ZNCC to be somwhat better than $\mathrm{ZNCC}_{\mathrm{robust}}$, while the best overall performance was obtained with a census transform similarity measure (see Ex. 4.10).

Example 11.5. Median-based Image Thresholding [29]. Given an input image I, two widely used global thresholdings are: Otsu [21] and Kittler-Illingworth [16]:

$$t_{\mathrm{otsu}} = \arg\min_t \left(\omega_{\mathrm{low}}(t)s^2_{\mathrm{low}}(t) + \omega_{\mathrm{high}}(t)s^2_{\mathrm{high}}(t)\right) ,$$

$$t_{\mathrm{KI}} = \arg\min_t \left[\omega_{\mathrm{low}}(t)\log\left(\frac{s_{\mathrm{low}}(t)}{\omega_{\mathrm{low}}(t)}\right) + \omega_{\mathrm{high}}(t)\log\left(\frac{s_{\mathrm{high}}(t)}{\omega_{\mathrm{high}}(t)}\right)\right] ,$$

where the threshold t divides the pixel gray-levels into *low-intensity* and *high-intensity* pixels (see Sect. 8.5). The number of pixels in each group are: $\omega_{\mathrm{low}}(t)$ and $\omega_{\mathrm{high}}(t)$ and the sample standard deviation of each group are: $s_{\mathrm{low}}(t)$ and $s_{\mathrm{high}}(t)$.

Xue and Titterington [29] robustified the Otsu algorithm by replacing the *square* of the sample standard deviations $s^2_{\mathrm{low}}(t)$ and $s^2_{\mathrm{high}}(t)$ with the corresponding MAD values, $MAD_{\text{low}}(t)$ and $MAD_{\text{high}}(t)$. For the KI algorithm they replaced the standard deviations $s_{\text{low}}(t)$ and $s_{\text{high}}(t)$ with $MAD_{\text{low}}(t)$ and $MAD_{\text{high}}(t)$. The corresponding robust thresholds are:

$$t_{\text{otsu-XT}} = \arg\min_t \Big(\omega_{\text{low}}(t) MAD_{\text{low}}(t) + \omega_{\text{high}}(t) MAD_{\text{high}}(t) \Big) ,$$

$$t_{\text{KI-XT}} = \arg\min_t \left[\omega_{\text{low}}(t) \log\left(\frac{MAD_{\text{low}}(t)}{\omega_{\text{low}}(t)}\right) + \omega_{\text{high}}(t) \log\left(\frac{MAD_{\text{high}}(t)}{\omega_{\text{high}}(t)}\right)\right] ,$$

11.3 Robust Parameter Estimation

The aim of robust parameter estimation techniques is to reduce the influence of the outliers without actually identifying and eliminating the outliers. The basic idea is to use a likelihood function, $p(\mathbf{y}|\boldsymbol{\theta},I)$, whose tail size varies in proportion to the number of outliers which are present. Our aim is that as we move away from $\hat{\boldsymbol{\theta}}$, the mean value of $\boldsymbol{\theta}$, the tail will decrease but not too quickly nor too slowly [1].

In Table 11.2 we list some standard functions which may be used to model a robust one-dimensional likelihood function.

Table 11.2 Robust Likelihood Functions

Name	Formula
Student-*t* Function	$\Gamma((\nu+1)/2)/(\sqrt{\nu\pi}\Gamma(\nu/2)\sigma) \times (1+((y-\mu)/\sigma\sqrt{\nu})^2)^{-(\nu+1)/2}$. The parameter ν controls the thickness of the tails, the thickness decreasing with an increase in ν.
"Good-and-Bad" Function	$\alpha\mathcal{N}(y\|\mu,\sigma^2) + (1-\alpha)\mathcal{N}(y\|\mu,\beta^2\sigma^2)$, where $\mathcal{N}(y\|\mu,\sigma^2)$ is the "Good", or "normal", likelihood function which is used to represent the inliers and $\mathcal{N}(y\|\mu,\beta^2\sigma^2)$ is the "Bad", or "aberrant", likelihood function which is used to represent the outliers, where $\beta >> 1$. The parameter α represents the expected proportion of inliers in the input data.
Gaussian + Step Function	$\mathcal{N}(y\|\mu,\sigma^2)+F$, where F is a constant which ensures the probability of an outlier never falls too low.
Uncertain Error Bar	$\text{erf}(\|\mu-y\|/\sqrt{2a^2})/\|\mu-y\|$.

11.3.1 *Student-t Function*

The multivariate Student-*t* likelihood function is

$$p(\boldsymbol{\theta}|\mathbf{y},I) = \mathscr{S}t(\mathbf{y}|\boldsymbol{\mu},\boldsymbol{\Sigma},\nu) , \tag{11.5}$$

[1] *Note*: In this context, the Gaussian distribution, $\mathcal{N}(\boldsymbol{\mu},\boldsymbol{\Sigma})$, is *not* robust since its likelihood drops off very quickly as we move more than a few standard deviations away from $\boldsymbol{\mu}$.

where the Student-t function, $\mathscr{S}t(\boldsymbol{\mu}, \Sigma, \nu)$, is a unimodal function with adjustable tails. In one-dimension it is defined as

$$\mathscr{S}t(y|\mu,\sigma^2,\nu) = \frac{\Gamma((\nu+1)/2)}{\sqrt{\nu\pi}\times\Gamma((\nu/2))\sigma}\left[1+\frac{(y-\mu)^2}{\nu\sigma^2}\right]^{-(\nu+1)/2}, \qquad (11.6)$$

where the parameter, ν, controls the thickness of the tails, the thickness decreasing with an increase in ν. When $\nu > 30$, there is virtually no difference between $\mathscr{S}t(y|\mu,\sigma,\nu)$ and $\mathscr{N}(y|\mu,\sigma^2)$.

The following example illustrates the use of K Student-t likelihood functions in a *robust* finite mixture model.

Example 11.6. Student Mixture Model (SMM) [2]. We reconsider Ex. 9.8. However this time we use a robust mixture of K Student-t functions, instead of a mixture of K one-dimensional Gaussian functions. Our aim is to model a one-dimensional function $\theta = f(y)$:

$$f(y) \simeq \sum_{k=1}^{K} \alpha_k \mathscr{S}t(y|\mu_k,\sigma_k^2,\nu_k) ,$$

where $\alpha_k, \mu_k, \sigma_k, \nu_k, k \in \{1,2,\ldots,K\}$, are unknown and are to be learnt from a set of training data $(y_i, \theta_i), i \in \{1,2,\ldots,N\}$, subject to the constraints

$$\alpha_k > 0 ,$$
$$\sum_{k=1}^{K} \alpha_k = 1 .$$

The maximum likelihood parameter values can be estimated using the expectation-maximization (EM) algorithm (Sect. 9.8). Theoretically the degrees of freedom, ν_k, can also be estimated by the EM algorithm. However, in practice [2] it is better to consider the ν_k as regular hyperparameters which are learnt by an exhaustive search.

We associate a "hidden" variable z_i with each y_i, where z_i can only take on the values $\{1,2,\ldots,K\}$. Then, in one-dimension, the EM algorithm for this problem consists of the following two steps, which are repeated until convergence:

E-Step:

$$p(z_i = k \mid y_i, \phi_k, \nu_k, I) = \frac{p(y_i|z_i = k, \phi_k, \nu_k, I)p(\phi_k|z_i = k, I)}{p(y_i|I)} ,$$
$$Q(z_i = k \mid y_i, \phi_k, \nu_k) = \frac{\nu_k + 1}{\nu_k + ((y_i - \mu_k)^2/(\nu_k\sigma_k^2))} ,$$

where

$$\phi_k = (\alpha_k, \mu_k, \sigma_k) .$$

M-Step:

$$\alpha_k = \frac{1}{N}\sum_{i=1}^{N} w_{ik} ,$$
$$\mu_k = \sum_{i=1}^{N} w_{ik} Q(z_i = k|y_i, \phi_k, \nu_k) y_i / \sum_{i=1}^{N} w_{ik} Q(z_i = k|y_i, \phi_k, \nu_k) ,$$
$$\sigma_k^2 = \sum_{i=1}^{N} p(z_k = k|y_i, \phi_k, \nu_k, I) Q(z_i = k|y_i, \phi_k, \nu_k)(y_i - \mu_j)^2 /$$
$$\sum_{i=1}^{N} p(z_i = k|y_i, \phi_k, \nu_k, I) ,$$

where

$$w_{ik} = p(z_i = k|y_i, \phi_k, \nu_k, I) .$$

Subspace techniques (Sect. 4.5) such as principal component analysis (PCA) and linear discriminant analysis (LDA), are sensitive to outliers and atypical observations. For this reason robust version of the different subspace techniques have been developed. The following example describes a robust probabilistic PCA algorithm.

Example 11.7. Robust Probabilistic Principal Component Analysis [3, 27]. In the robust PPCA we model the sensor measurements y_i using a linear model of the form

$$\mathbf{y} = \mathbf{H}\boldsymbol{\theta} + \boldsymbol{\mu} + \mathbf{w} ,$$

where we use a Student-t distribution for $\mathbf{w}$ and $\boldsymbol{\theta}$:

$$\mathbf{w} \sim \mathscr{S}t(\mathbf{0}, \sigma^2\mathbf{I}, \eta) ,$$
$$\boldsymbol{\theta} \sim \mathscr{S}t(\mathbf{0}, \mathbf{I}, \eta) .$$

The result is a "robust" *a posteriori* pdf:

$$p(\mathbf{y}|\boldsymbol{\theta}, I) \sim \mathscr{S}t(\mathbf{y}|\mathbf{H}\boldsymbol{\theta} + \boldsymbol{\mu}, \tau\mathbf{I}, \eta) ,$$

which may be efficiently solved using an EM algorithm.

11.3.2 "Good-and-Bad" Likelihood Function

The "Good-and-Bad" likelihood function uses a weighted sum of two Gaussian functions:

$$p(\mathbf{y}|\theta,I) = \alpha\mathcal{N}(\mathbf{y}|\mu,\sigma^2) + (1-\alpha)\mathcal{N}(\mathbf{y}|\mu,\beta^2\sigma^2)\ , \tag{11.7}$$

where $\mathcal{N}(\mathbf{y}|\mu,\sigma^2)$ is a "Good" likelihood function and is used to represent the inliers, $\mathcal{N}(\mathbf{y}|\mu,\beta^2\sigma^2)$ is a "Bad" likelihood function and is used to represent the outliers, and $\beta >> 1$. We weight the two Gaussian functions according to the expected proportion of inliers and outliers in the input data.

The "Good-and-Bad" likelihood function is often used to fuse together discordant measurements. The following example illustrates the use of the "Good-and-Bad" likelihood function to fuse five independent neutron fission cross-sections.

Example 11.8. Neutron Fission Cross-Sections [13]. Table 11.4 lists $N = 5$ experimental measurements $\mathbf{y} = (y_1, y_2, \ldots, y_N)^T$, and error bars $\boldsymbol{\sigma} = (\sigma_1, \sigma_2, \ldots, \sigma_N)^T$, obtained in a given neutron fission cross-section experiment. The corresponding (Gaussian) likelihoods, $p(y_i|\theta,I) \sim \mathcal{N}(y_i,\sigma_i^2)$, are shown in Fig. 11.3.

Data point #5 is clearly an outlier being eight standard deviations away from the mean of the other four points. Assuming a flat prior on the cross-section θ, the *a posteriori* probability $p(\theta|\mathbf{y},I)$ is proportional to the product of the individual likelihoods:

$$p(\theta|\mathbf{y},I) \propto \prod_{i=1}^{N} p(y_i|\theta,I)\ .$$

Using "Good" and "Bad" Gaussian likelihood functions, $\mathcal{N}(y_i,\sigma_i^2)$ and $\mathcal{N}(y_i,\beta^2\sigma_i^2)$, the *a posteriori* probability becomes

$$p(\theta|\mathbf{y},I) \propto \int \prod_{i=1}^{5} \Big(\alpha\mathcal{N}(\theta|y_i,\sigma_i^2) + (1-\alpha)\mathcal{N}(\theta|y_i,\beta^2\sigma_i^2)\Big)\pi(\alpha|I)d\alpha\ ,$$

where $\pi(\alpha|I)$ is the *a priori* pdf of α. Assuming a flat distribution for $\pi(\alpha|I)$, we obtain the *a posteriori* probability density function shown in Fig. 11.3(b). This result is close to the product of $\mathcal{N}(\theta|y_i,\sigma_i^2)$ for $i \in \{1,3,4\}$, which is what we might expect if we believe that the points #2 and #5 are outliers that should be discounted.

We may modify the "Good-and-Bad" likelihood function by including the probability that each input value belongs to the inlier distribution. This is illustrated in the following example.

Example 11.9. Estimating Hubble's Constant Using a "Good-and-Bad" Likelihood Function [1, 22, 26]. Table 11.3 lists $N = 13$ experimental measurements $\mathbf{y} = (y_1, y_2, \ldots, y_N)^T$ and error bars $\boldsymbol{\sigma} = (\sigma_1, \sigma_2, \ldots, \sigma_N)^T$ of Hubble's constant θ. The values y_i and the error bars σ_i were read from Fig. 2 in [22]. The values are given in the appropriate units. The measurements include a wide variety of different techniques and are assumed to be independent. We divide the measurements $\{y_i\}$ into "Good" and "Bad" measurements. Altogether there are 2^N different ways, or combinations, of doing this. We use a function $\delta_{ki}, i \in \{1, 2, \ldots, N\}$, to indicate which measurement is "Good" and which measurement is "Bad" in a given combination. Mathematically,

$$\delta_{ki} = \begin{cases} 1 \text{ if } i\text{th measurement in } k\text{th combination is "Good"} \ , \\ 0 \text{ if } i\text{th measurement in } k\text{th combination is "Bad"} \ . \end{cases}$$

Since the measurements are independent, the joint likelihood function for the kth combination is

$$p_k(y|\theta, I) = \prod_{i=1}^{N} (\delta_{ki} \mathcal{N}(y_i|\theta, \sigma_i^2) + (1 - \delta_{ki}) \mathcal{N}(y_i|\theta, \beta^2 \sigma_i^2)) \ ,$$

and the corresponding *a posteriori* probability density function $p(\theta|\mathbf{y}, I)$ is proportional to

$$\sum_{k=1}^{2^N} \pi(k|I) \ p_k(\mathbf{y}|\theta, I) \ ,$$

where $\pi(k|I)$ is the *a priori* probability of the kth combination. By carrying out the summation over all 2^N combinations, we obtain the *a posteriori* distribution shown in Fig. 11.4. The pdf, $p(\theta|\mathbf{y}, I)$, bears a superficial resemblance to a Gaussian function, which is an *outcome* of the calculation and *not* an assumed input.

Table 11.3 Experimental Values y_i and Error bars σ_i for Hubble's Constant

Parameter	Experimental values
y_i	49 51 55 61 89 68 67 80 82 76 81 73 80
σ_i	10 13 25 19 10 22 10 25 18 19 17 10 12

Table 11.4 Experimental Values y_i and Error Bars σ_i for Neutron Fission Cross-Section

Parameter	Experimental values
y_i	2.385 2.521 2.449 2.420 2.670
σ_i	0.027 0.027 0.027 0.027 0.027

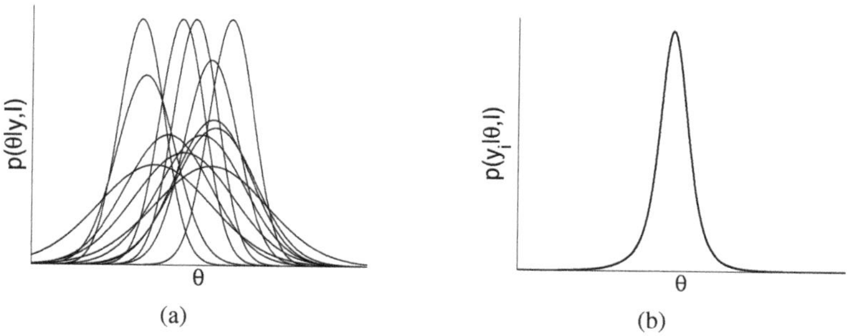

Fig. 11.3 **(a)** Shows the Gaussian likelihoods $p(y_i|\theta,I) = \mathcal{N}(y_i,\sigma_i^2)$ for cross-sections y_i and error bars σ_i. **(b)** Shows the *a posteriori* probability density function $p(\theta|\mathbf{y},I)$ calculated using a "Good-and-Bad" likelihood function with $\beta = 10$. *Note*: The curve was calculated assuming that $\pi(k|I)$ is inversely proportional to the total number of combinations which have the same number of "Good" measurements.

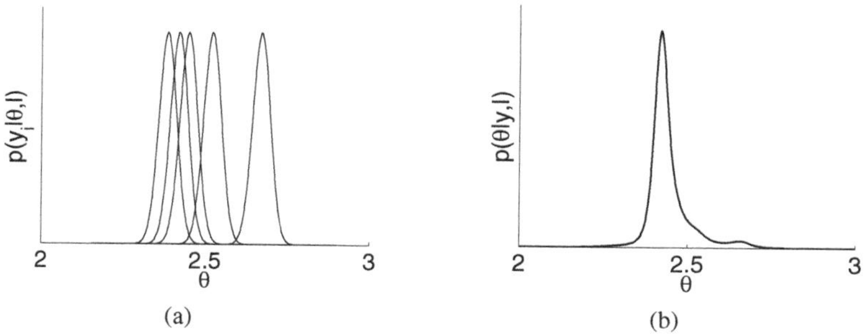

Fig. 11.4 **(a)** Shows the Gaussian likelihood functions $p(y_i|\theta,I) = \mathcal{N}(y_i,\sigma_i^2)$ for the experimental measurements y_i and error bars σ_i. **(b)** Shows the *a posteriori* probability density function $p(\theta|\mathbf{y},I)$ calculated using a "Good-and-Bad" likelihood function with $beta = 10$.

11.3.3 Gaussian Plus Constant

The Gaussian plus constant likelihood function is defined as

$$p(\mathbf{y}|\boldsymbol{\theta},I) = \gamma\mathcal{N}(\mathbf{y}|\boldsymbol{\mu},\boldsymbol{\sigma}^2) + (1-\gamma)\frac{1}{v}, \tag{11.8}$$

where γ is a mixing parameter, $0 \leq \gamma \leq 1$, and v is just a constant over which it is assumed the outlier distribution is uniform. We often write $(1-\gamma)/v$ as a constant F (see Ex. 2.10).

11.3.4 Uncertain Error Bars

In one-dimension, the *uncertain error bar* likelihood function is defined as

$$p(y|\mu,I) = \frac{1}{2|\mu-y|\ln(b/a)}\Big(\mathrm{erf}\big(|\mu-y|/\sqrt{2a^2}\big) - \mathrm{erf}\big(|\mu-y|/\sqrt{2b^2}\big)\Big)\;, \quad (11.9)$$

where the standard deviation σ is assumed to be in the interval $[a,b]$. The uncertain error bar likelihood function is shown in Fig. 11.5. When $a=b$, i. e. the upper and

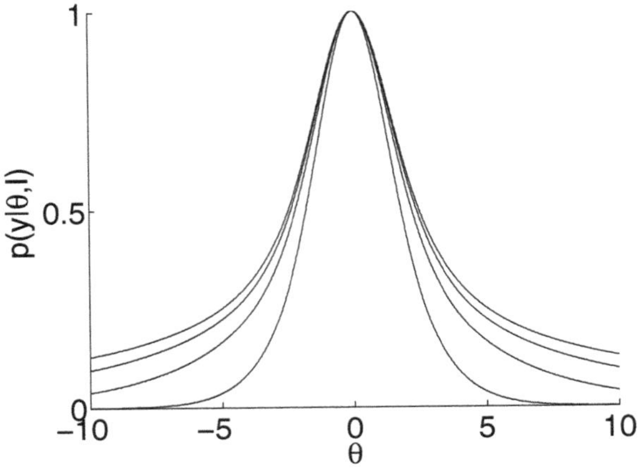

Fig. 11.5 Shows the likelihood $p(y|\mu,a,b,I)$ for $\mu=0$ and $a=1$ vertically scaled to unit amplitude. The curves shown are, in order of increasing width and tail size, for $b=1,3,9,27$ and ∞.

low bounds on σ coincide, (11.9) reduces to $\mathcal{N}(y|\mu,a^2)$, and when $b\rightarrow\infty$, (11.9) reduces to

$$p(y|\mu,I) \propto \frac{1}{|\mu-y|}\mathrm{erf}\Big(|\mu-y|/\sqrt{2a^2}\Big)\;. \quad (11.10)$$

The uncertain error bar likelihood function (11.9) is derived as follows. We assume there are *no* "true" outliers, but rather a y value may give the *appearance* of being an outlier because of a gross error in its uncertainty σ. We may correct for the error in σ by integrating σ out of the conditional pdf $p(y|\mu,\sigma,I)$:

$$p(y|\mu,I) = \int p(y|\mu,\sigma,I)\pi(\sigma|I)d\sigma\;, \quad (11.11)$$

where $\pi(\sigma|I)$ is the *a priori* distribution for σ. A convenient non-informative distribution (see Sect. 9.4.3) for $\pi(\sigma|I)$ is *Jeffrey's* pdf:

$$\pi(\sigma|I) = \begin{cases} \frac{1}{\sigma}\ln(b/a) & \text{if } a\le\sigma\le b\;, \\ 0 & \text{otherwise}\;. \end{cases} \quad (11.12)$$

The following example illustrates the use of the uncertain error bar likelihood function in estimating the background signal B in a powder diffraction pattern experiment.

Example 11.10. Background Estimation in a Powder Diffraction Pattern [9]. We consider a powder diffraction pattern. This consists of an unknown and noisy background signal $B(\lambda)$ which varies with the wavelength λ. Imposed on the background are *positive* high amplitude Bragg diffraction peaks A (Fig. 11.6). We consider a small area of the diffraction pattern near a given wavelength λ. In this case we model the background signal is a constant, but unknown, value B. Let $\mathbf{y} = (y_1, y_2, \ldots, y_N)^T$ denote N diffraction pattern amplitudes taken in the vicinity of λ. Then given a noise variance σ^2 and any background information I (such as the Bragg peak positivity), the probability distribution for the background signal, B, is

$$p(B|\sigma,\mathbf{y},I) = \int p(A,B|\sigma,\mathbf{y},I)dA \,.$$

Invoking Bayes' theorem, assuming the y_i are independent and separating the peak and background distributions gives

$$\begin{aligned} p(B|\sigma,\mathbf{y},I) &\propto \int p(\mathbf{y}|\sigma,A,B,I)\pi(A,B|I)dA \,, \\ &= \pi(B|I) \int p(\mathbf{y}|\sigma,A,B,I)\pi(A|I)dA \,, \\ &= \pi(B|I) \prod_{i=1}^{N} \int p(y_i|\sigma,A,B,I)\pi(A|I)dA \,. \end{aligned}$$

A priori it is difficult to scale the Bragg peak contributions relative to the background. As an alternative we model the probability distribution for A using Jeffrey's pdf (11.12):

$$\pi(A|I) = \frac{1}{A \ln(b/a)} \,.$$

where a and b are, respectively, the lower and upper limits for A. Substituting this *a priori* distribution in $p(B|\sigma,\mathbf{y},I)$ gives:

$$p(B|\sigma,\mathbf{y},I) \propto \pi(B|I) \prod_{i=1}^{N} p(y_i|\sigma,B,I) \,,$$

where

$$p(y|\sigma,B,I) = \int_a^b \frac{1}{A} \exp\left(-\frac{1}{2}\left(-\frac{A+B-y}{\sigma}\right)^2\right) dA \,.$$

denotes the likelihood of y. The natural logarithm of $p(y|\sigma,B,I)$ is shown in Fig. 11.7.

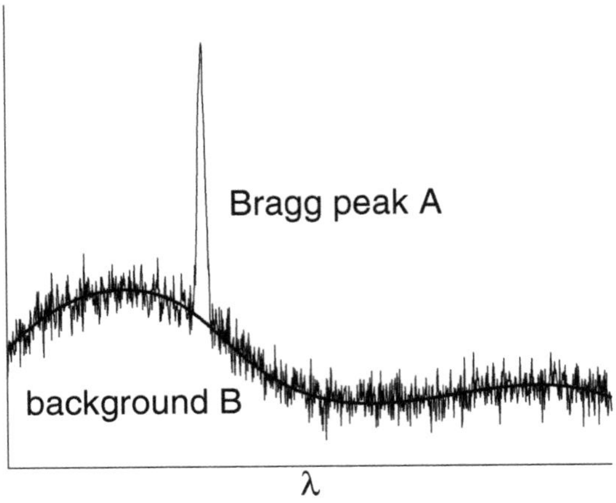

Fig. 11.6 Shows the amplitude of powder diffraction pattern as a function the wavelength λ. Overlaying the noisy diffraction pattern is a thick smooth curve. This is the background signal B. The positive high peak is the Bragg diffraction peak.

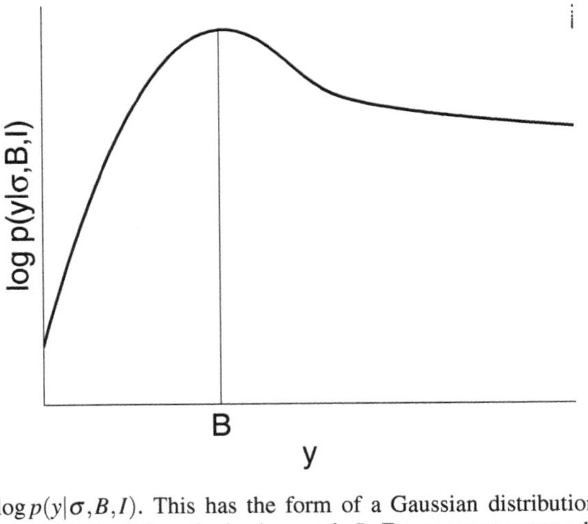

Fig. 11.7 Shows $\log p(y|\sigma,B,I)$. This has the form of a Gaussian distribution for $y < B$, i. e. for measurements y, which lie below the background, B. For measurements which lie above the background, the corresponding robust probability distribution falls off less fast than the Gaussian distribution.

11.4 Classical Robust Estimators

One popular classical robust technique is the M estimators which are a generalization of the maximum likelihood and least squares estimators. The M-estimate of a is defined as

$$\hat{a} = \arg\min_a \left[\sum_i \rho\left(\frac{y_i - a}{\sigma_i}\right) \right] , \tag{11.13}$$

where $\rho(u)$ is a robust loss function that grows subquadratically and is monotonically nondecreasing with increasing $|u|$ and σ_i^2 is the variance, or scale, associated with the residual errors $\varepsilon_i = y_i - a$. The many M-estimators that have been proposed differ in shape of the loss functions $\rho(u)$. Some common loss functions are listed in Table 11.5. Inherent in the different M-estimators is the need to estimate the standard deviation σ of the residual errors ε_i. A robust standard deviation which may be used for this purpose is

$$\sigma = 1.4826\left(1 + \frac{5}{N-1}\right) \mathrm{med}_i |y_i - a| . \tag{11.14}$$

Table 11.5 Robust Loss Functions $L(t,\theta)$

Name	$L(t,\theta)$
L_1	$\|y\|$.
$L_1 - L_2$	$2(\sqrt{1+y^2/2} - 1)$.
L_p	$\|y\|^p/p$.
"Fair"	$c^2((\|y\|/c) - \log(1 + (\|y\|/c)))$.
Geman-McClure	$(y^2/2)/(1+y^2)$.
Beaton and Tukey	$c^2(1-(1-(u/c)^2)^3)/6$ if $\|u\| \leq c$; otherwise $c^2/6$.
Cauchy	$c^2 \log(1+(u/c)^2)/2$.
Huber	$u^2/2$ if $\|u\| \leq k$, otherwise $k(2\|u\| - k/2)$.

11.4.1 Least Median of Squares

A widely used alternative to the M estimators is the *least median of squares* (LMedS) estimator in which we calculate the unknown parameter a by solving the following nonlinear minimization problem:

$$\hat{a} = \arg\left(\min \mathrm{med}_i (y_i - a)^2\right) . \tag{11.15}$$

Increasingly, algorithms based on a least squares (LS) fit are being replaced with the LMedS algorithm.

Example 11.11. Using least median of squares for structural superposition of flexible proteins [17]. Structural comparison of two different conformations of the same protein are often based on a LS fit, which is sensitive to outliers, i. e. to large displacements. In [17] the authors replace the LS algorithm with the LMedS algorithm and obtain substantial improvement in robustness against large displacement and other outliers.

Example 11.12. Robust Image Correlation Measures [4]. Matching two images, or two image patches, A and B is an important task in multi-sensor image fusion. In Table 11.6 (see also Table 8.10) we list some robust correlation measures which may be used for this purpose. In In [4] we perform face recognition by applying N Gabor filters of a different size and orientation to the each image $I_k, k \in \{1,2,\ldots,K\}$, in a given training set D of facial images. The result is an ensemble of N Gabor-filtered images $G_k^{(n)}, n \in \{1,2,\ldots,N\}$. Let $g^{(n)}, n \in \{1,2,\ldots,N\}$, denote the images obtained by applying the Gabor filters to the test image. If $\rho_k^{(n)}$ is the rank correlation between $g^{(n)}$ and $G_k^{(n)}$, then the sum S_k:

$$S_k = \sum_{n=1}^{N} \rho_k^{(n)} ,$$

is a similarity measure which measures the degree to which the test image matches a given training image I_k.

Table 11.6 Robust Image Correlation Measures

Name	Formulae
Median absolute deviation	$\text{med}\lvert(A-B)-\text{med}(A-B)\rvert.$
Least median of power	$\text{med}(\lvert A-B\rvert^p).$
Least trimmed power	$\sum_i \lvert A_i - B_i\rvert^p.$
Smoothed median power deviation	$\sum_i \lvert A_i - B_i - \text{med}(A-B)\rvert^p.$

11.5 Robust Subspace Techniques *

Subspace techniques (see Sect. 4.5), such as principal component analysis (PCA) and linear discriminant analysis (LDA), are sensitive to outliers and to atypical observations. In creating robust subspace techniques, the basic idea is to replace non-robust estimators with robust versions. The following example describes a robust LDA algorithm.

Example 11.13. Robust Discriminant Analysis [15]. One commonly used robust estimator is the minimum covariance determinant (MCD) estimator [15]. In the MCD estimator, we search for a subset containing n input measurements for which the determinant of their covariance matrix is minimal. Then the MCD-estimator of location is given by the mean of these n measurements and the MCD-estimator of covariance is given by their covariance matrix.

The probabilistic subspace techniques use an explicit noise model. For example, the probabilistic PCA and LDA algorithms (see Sect. 10.6.3) use a Gaussian noise model. These algorithms are easily robustified by replacing the Gaussian distribution with a corresponding Student-t distribution. See Ex. 11.7 for an example of a robust probabilistic principal component analysis.

11.6 Robust Statistics in Computer Vision

Special purpose robust statistical techniques are required for computer vision and image analysis applications. The input data in these applications are characterized by multiple structures and models, in which each model accounts for a relatively small percentage of the input data [30]. Under these circumstances, the classical M and LMedS estimators, which require at least 50% inliers, are no longer reliable. In Table 11.7 we list some of the special robust techniques which have been developed for computer vision and image analysis applications [30]. The following example illustrates the RANSAC technique.

Example 11.14. RANSAC in Computer Vision [34]. The RANSAC algorithm works as follows. First we generate a model hypothesis from the input data D by randomly selecting a subset of p points and fitting a given parametric model of dimension p. The quality of the model is evaluated by finding the number of input data which are consistent with the estimated model. Mathematically, the quality is given by

$$Q = \sum_i \rho(e_i^2) ,$$

where $\rho(e_i^2) = 0$ if $e^2 < T^2$, otherwise 1. The random selection of a subset of p points is repeated until a suitable termination criterion is fulfilled. If several solutions are applicable, then the data set with the smallest standard deviation is selected. After the estimation process the outliers are eliminated and the model is recomputed with all the inliers. Alternatively a robust cost function (e. g. a M estimator or a LMedS) can be used to verify the model hypothesis concerning the data D. In contrast to minimization of inconsistent data, the robust error function must be maximized. The main drawback to RANSAC is its high computational load. For data with N points and p unknown parameters, a combination of $N/(p(N-p))$ possibilities must be considered. In an effort to reduce the computational load a genetic algorithm sample consensus (GASAC) may be used [23]. Although very successful one of the problems with the RANSAC is that if the threshold for considering inliers is set too high then the robust estimate can be very poor. This undesirable situation can be remedied by minimizing a new cost function:

$$\widetilde{C} = \sum \widetilde{\rho}(e_i^2) ,$$

where

$$\widetilde{\rho}(e^2) = \begin{cases} e^2 \text{ if } e^2 < T^2 , \\ T^2 \text{ if } e^2 \geq T^2 . \end{cases}$$

We set $T = 1.96\sigma$ so that Gaussian inliers are only incorrect approximately 5 percent of the time.

The following example illustrates the use of the Hough transform [30] to initiate target tracks made by a moving object O.

Example 11.15. Target Tracking Initialization Using the Hough Transform [7]. We consider an object O which starts to move at time step 0 in a two-dimensional, (x,y), space. A sensor, S, detects (with a probability $p_D < 1$) the position of O at time steps $k, k \in \{0,1,2,\ldots\}$. In addition, the sensor may make several false detections at these times. If we assume that initially O moves along a straight-line, we may use the Hough transform to identify the initial track as follows.

Consider a pair of detections $P_i = (x_i, y_i)$ and $P_j = (x_j, y_j)$ made at *different* time steps k_i and k_j. The parameters a_{ij} and b_{ij} of the straight-line $P_i P_j$ is

$$\frac{x}{a_{ij}} + \frac{y}{b_{ij}} = 1 ,$$

where

$$a_{ij} = \frac{x_i y_j - x_j y_i}{y_j - y_i} ,$$

$$b_{ij} = \frac{y_i x_j - y_j x_i}{x_j - x_i} .$$

Table 11.7 Robust Techniques for Computer Vision and Image Analysis

Name	Description
Hough Transform	Hough Transform is a voting technique in which we count the number of data features that are mapped into each cell in quantized parameter space. Its main drawbacks are: excessive storage space and computational complexity and limited parameter precision.
RANSAC	Random Sample Consensus. Subsets of input data are randomly selected and are fitted to a given parametric model. The subsets are known as p-subsets and contain p points where p is the dimension of the parameter space. The quality of model parameters are evaluated on all the input data. Different cost functions may be used: The conventional cost is the number of inliers, i. e. the number of input data points which are consistent with the model.
MUSE	Minimum Unbiased Scale Eestimator. MUSE randomly selects p-subsets and then estimates fits based on these p-subsets. It then calculates the unbiased estimate of the scale for each fit's k smallest smallest residuals where k is set to all possible values and satisfies $1 \leq k \leq N - p$. The smallest scale estimate over all possible k is chosen as th representative value of hypothesized fits. Finally the fit from the p-subset with the smallest scale estimate is chosen as the optimum fit.
ALKS	Adaptive Least kth Order Square. ALKS is based on the least kth order squares (LKS) estimator. The diference between LKS and LMedS is that LKS uses k data points out of N data points $p < k < N$) while LMedS uses $N/2$ data points. ALKS uses a multi-step procedure in which in each step ALKS employs LKS with a different k value. In order to estimate the correct k, a random sampling technique similar to that used in RANSAC is employed. ALKS is robust against multiple structures but it cannot resist the influence of extreme outliers.
RESC	Residual Consensus. The idea of RESC is that if a model is correctly fitted, the residuals of the inliers should be small and at the same time the histogram of the residuals should concentrate within a small range in the lower part of the histogram. The RESC is very robust. However its main disadvantage is that it requires many parameters which must be tuned by the user for optimal performance.

We record each of pair of a_{ij} and b_{ij} values in a Hough space (i. e. the two-dimensional (a,b) space). Then we identify the initial track of the object O by finding the point (a^*,b^*) around which there is a cluster of (a_{ij},b_{ij}) values (see Fig. 11.8). We often only accept (a^*,b^*) if the number of points in the cluster exceeds a given threshold.

The strength of the Hough transform is that we do not specify the number of straight-line tracks. Instead, the Hough transform automatically estimates the number of tracks present.

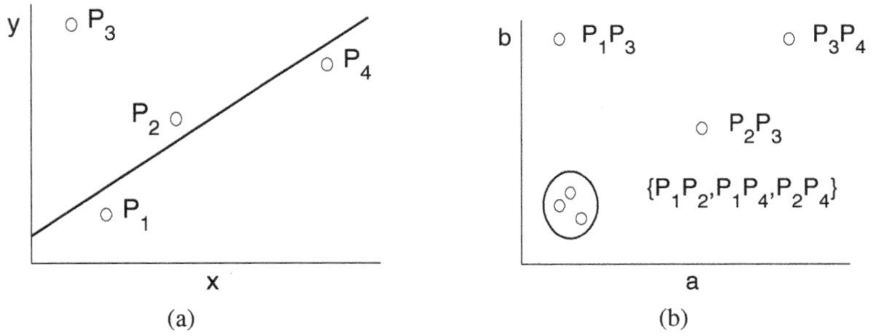

Fig. 11.8 **(a)** Shows four observations $P_i = (x_i, y_i), i \in \{1, 2, \ldots, 4\}$ made at different times t_i by a sensor S. Observations P_1, P_2 and P_4 are associated with the object and observation P_3 is a false alarm. **(b)** Shows the corresponding six points (a_{ij}, b_{ij}) which represent the straight-lines P_iP_j in Hough space. There is a single cluster of three (a_{ij}, b_{ij}) points which represents the straight-lines P_1P_2, P_1P_3 and P_2P_3 and three isolated (a_{ij}, b_{ij}) points which represent the straight-lines P_1P_3, P_2P_3 and P_4P_3.

11.7 Software

LIBRA. A matlab toolbox for performing robust statistical analysis. Authors: Sabine Verboven, Mia Hubert.
MATLAB STATISTICS TOOLBOX. The mathworks matlab statistics toolbox.
RANSAC TOOLBOX. A matlab toolbox for performing RANSAC. Author: Marco Zuliani [33, 34].

11.8 Further Reading

Robustifying the Bayesian framework, i. e. making the Bayesian framework robust against outliers, is a controversial subject which is still being intensively investigated. Apart from the discussion in Sect. 11.3 the reader should also review Sect. 2.7. Some modern references on the subject are: [6, 20, 26]. The development of robust EM algorithms based on Student-t functions are discussed in detail in [2, 18]. A comprehensive review of robust subspace techniques is [27]. Classical references on robust estimation are [12, 14, 24]. Special robust techniques which have been developed for computer vision are described in [19, 28, 30, 31, 33]. A full description of Bayesian curve fitting including robust Bayesian curve fitting is given in [10].

Problems

11.1. Define robust statistics. Give an example showing the importance of using robust statistics.

11.2. Gating is often used in target tracking to eliminate outliers. What is gating? What are the relative advantages and disadvantages of gating?

11.3. Apart from eliminating outliers, robust techniques work by reducing the influence of outliers. Explain how employing student-t function reduces the influence of outliers.

11.4. Write down the "Good-and-Bad" likelihood function. Given N independent measurements $(y_i, \sigma_i), i \in \{1, 2, .., N\}$ of a parameter θ show how we calculate the a posteriori pdf of the parameter θ.

11.5. Derive the uncertain error bar likelihood function. Explain the reasoning behind this likelihood function.

11.6. Describe the Hough transform and its use in tracking initialization.

11.7. Describe the method of RANSAC.

References

1. D'Agostini, G.: Sceptical combinations of experimental results: General considerations and applications to ε'/ε. European Organization for Nuclear Research CERN-EP/99-139 (1999)
2. Archambeau, C.: Probabilistic models in noisy environments. PhD thesis, Departement d'Electricite, Universite Catholique de Louvain, Belgium (2005)
3. Archambeau, C., Delannay, N., Verleysen, M.: Mixture of robust probabilistic principal component analyzers. In: 15th Euro. Symp. Art. Neural Network (ESANN), pp. 229–234 (2007)
4. Ayinde, O., Yang, Y.-H.: Face recognition approach based on rank correlation of Gabor-filtered images. Patt. Recogn. 35, 1275–1289 (2002)
5. Bar-Shalom, Y., Li, X.: Multitarget-Multisensor Tracking: Principles and Techniques. YBS Publishing, Storrs (1995)
6. Berger, J.: An overview of robust Bayesian analysis. Test 3, 5–124 (1994)
7. Blackman, S.S., Popoli, R.F.: Design and analysis of modern tracking Systems. Artech House, MA (1999)
8. Chambon, S., Crouzil, A.: Similarity measures for image matching despite occlusions in stereo vision. Patt. Recogn. 44, 2063–2075 (2011)
9. David, W.I.F., Sivia, D.S.: Background estimation using a robust Bayesian analysis. Appl. Crystallography 34, 318–324 (2001)
10. Denison, D.G.T., Holmes, C., Mallick, B.K., Smith, A.F.M.: Bayesian methods for nonlinear vlassification and regression. John Wiley and Sons (2002)
11. Dose, V., Linden Von der, W.: Outlier tolerant parameter estimation. In: Dose, V., et al. (eds.) Maximum Entropy and Bayesian Methods, Kluwer Academic Publishers, Dordrecht (1999)
12. Hampel, F.: Robust Statistics: The Approach Based on Influence Functions. John Wiley and Sons
13. Hanson, K.M.: Bayesian analysis of inconsistent measurements of neutron cross sections. In: Knuth, K. (ed.) AIP Conf. Proc. of Bayesian Inference and Maximum Entropy Methods in Science and Engineering (2005)
14. Huber, P.J.: Robust Statistics. John Wiley and Sons (1981)
15. Hubert, M., Van Driessen, K.: Fast and robust discriminant analysis. Comp. Stat. Data Anal. 45, 301–320 (2004)
16. Kittler, J., Illingworth, J.: Minimum error thresholding. Patt. Recogn. 19, 41–47 (1986)

17. Liu, Y.-S., Fang, Y., Ramani, K.: Using least median of squares for structural superposition of flexible proteins. BMC Bioinform 10 (2009)
18. McLachlan, G.J., Peel, D.: Robust cluster analysis via mixtures of multivariate t-distributions. In: Amin, A., Pudil, P., Dori, D. (eds.) SPR 1998 and SSPR 1998. LNCS, vol. 1451, pp. 658–666. Springer, Heidelberg (1998)
19. Meer, P.: Robust techniques for computer vision. In: Medioni, G., et al. (eds.) Emerging Topics in Computer Vision. Prentice-Hall (2004)
20. O'Hagen, A., Forster, J.J.: Bayesian Inference - Kendall's Advanced Theory of Statistics, vol. 2B. Arnold Publishers, London (2004)
21. Otsu, N.: A threshold selection method from gray-level histograms. IEEE Trans. Sys. Man Cyber 9, 62–66 (1979)
22. Press, W.H.: Understanding data better with Bayesian and global statistical methods. Unsolved problems in astrophysics. In: Ostriker, J.P. (ed.) Proc. Conf. in Honor of John Bahcall. Princeton University Press (1996)
23. Rodehorst, V., Hellwich, O.: Genetic algorithm sample consensus (GASAC) - a parallel strategy for robust parameter estimation. In: 2006 Conf. Comp. Vis. Patt. Recogn. Workshop (2006)
24. Rousseeuw, P.J., Leroy, A.M.: Robust Regression and Outlier Detection. John Wiley and Sons (1987)
25. Sivia, D.S.: Data Analysis: A Bayesian Tutorial. Oxford University Press (1996a)
26. Sivia, D.S.: Dealing with duff data. In: Sears, M., Nedeljkovic, N.E., Sibisi, S. (eds.) Proc. Maximum Entropy Conf., South Africa, pp. 131–137 (1996b)
27. Skocaj, D.: Robust subspace approaches to visual learning and recognition. PhD thesis. University of Ljubljana (2003)
28. Stewart, C.V.: Robust parameter estimation in computer vision. SIAM Review 41, 513–537 (1999)
29. Xue, J.-H., Titterington, M.: Median-based image thresholding. Preprint (2010)
30. Wang, H.: Robust statistics for computer vision: model fitting, image segmentation and visual motion analysis. PhD thesis. Monash University, Clayton, Australia (2004)
31. Zhang, Z.: Parameter estimation techniques: a tutorial with application to conic fitting. Image Vis. Comp. 15, 59–76 (1997)
32. Zimmerman, N.M.: A primer on electrical units in the Systeme International. Am. Phys. 66, 324–331 (1998)
33. Zuliani, M.: Computational methods for automatic image registration. PhD thesis. University of California (2006)
34. Zuliani, M.: RANSAC for Dummies. Unpublished work available from the authors homepage (2009)

Chapter 12
Sequential Bayesian Inference

12.1 Introduction

The subject of this chapter is *sequential Bayesian inference* in which we consider the Bayesian estimation of a dynamic system which is changing in time. Let $\boldsymbol{\theta}_k$ denote the state of the system, i. e. a vector which contains all relevant information required to describe the system, at some (discrete) time k. Then the goal of sequential Bayesian inference is to estimate the *a posteriori* probability density function (pdf) $p(\boldsymbol{\theta}_k|\mathbf{y}_{1:l},I)$, by fusing together a sequence of sensor measurements $\mathbf{y}_{1:l} = (\mathbf{y}_1,\mathbf{y}_2,\ldots,\mathbf{y}_l)$. In this chapter we shall only consider the calculation of pdf $p(\boldsymbol{\theta}_k|\mathbf{y}_{1:l},I)$ for $k = l$ which is known as (sequential) Bayesian *filtering* or *filtering* for short[1].

We start the chapter by assuming the measurements are all made by the same sensor. In this case we have *single-sensor multi-temporal data fusion*. Then, in Sect. 12.6 we remove this restriction and consider the case when the measurements are made by multiple sensors. This is known as *multi-sensor multi-temporal data fusion*.

For many applications an estimate of $p(\boldsymbol{\theta}_k|\mathbf{y}_{1:k},I)$ is required every time step that a measurement is received. In this case it is common practice to rewrite $p(\boldsymbol{\theta}_k|\mathbf{y}_{1:k},I)$ using a recursive formulation of Bayes' theorem:

$$p(\boldsymbol{\theta}_k|\mathbf{y}_{1:k},I) = \frac{\overbrace{p(\mathbf{y}_k|\boldsymbol{\theta}_k,I)}^{\text{likelihood}}\overbrace{p(\boldsymbol{\theta}_k|\mathbf{y}_{1:k-1},I)}^{\textit{a priori}\text{ pdf}}}{\underbrace{p(\mathbf{y}_k|\mathbf{y}_{1:k-1},I)}_{\text{evidence}}}\ , \tag{12.1}$$

where

$$p(\boldsymbol{\theta}_k|\mathbf{y}_{1:k-1},I) = \int p(\boldsymbol{\theta}_k|\boldsymbol{\theta}_{k-1},I)p(\boldsymbol{\theta}_{k-1}|\mathbf{y}_{1:k-1},I)d\boldsymbol{\theta}_{k-1}\ , \tag{12.2}$$

[1] The calculation of $p(\boldsymbol{\theta}_k|\mathbf{y}_{1:l},I)$ for the cases $k < l$ and $k > l$ are known, respectively, as (sequential) Bayesian *smoothing* and *forecasting*.

$$p(\mathbf{y}_k|\mathbf{y}_{1:k-1},I) = \int p(\mathbf{y}_k|\boldsymbol{\theta}_k,I)p(\boldsymbol{\theta}_k|\mathbf{y}_{1:k-1},I)d\boldsymbol{\theta}_k . \qquad (12.3)$$

Eqs. (12.1)-(12.3) represent a convenient solution to the problem of (sequential) Bayesian filtering: the sensor measurements are processed sequentially and not as a batch. At each time step only the current sensor measurement, $\mathbf{y}_k$, is used and it is not necessary to maintain, or re-process, previous measurements.

The main difficulty in calculating $p(\boldsymbol{\theta}_k|\mathbf{y}_{1:k},I)$ involves solving the integrals defined in (12.2) and (12.3) and this will form the main topic of our discussion in this chapter.

In many applications the recursive filter is the primary fusion algorithm in a single-sensor, or a multi-sensor, multi-temporal data fusion system. In Table 12.1 we list some examples of these applications together with their input/output classification.

Table 12.1 Applications in which the Sequential Filter is the Primary Fusion Algorithm

Class		Application
DaI-DaO	Ex. 3.7	Heart Rate Estimation Using a Kalman Filter.
	Ex. 10.6	Non-Intrusive Speech Quality Estimation Using GMM's.
	Ex. 12.3	Tracking a Satellite's Orbit Around the Earth.
	Ex. 12.4	Tracking a Moving Target.
	Ex. 12.6	Tracking Metrological Features.
	Ex. 12.7	Correcting near surface temperature forecasts using a Kalman filter
	Ex. 12.14	The Lainiotis-Kalman Filter.
	Ex. 12.15	Multi-Sensor Multi-Temporal Data Fusion: Measurement Fusion.
DaI-FeO	Ex. 12.2	Digital Image Stabilizer using Kalman Filter.
	Ex. 12.12	Recursive Parameter Estimation for a First-Order Autoregressive Process.
FeI-FeO	Ex. 12.1	INS/Radar Altimeter: A Hybrid Navigation Algorithm.
	Ex. 12.15	Multi-Sensor Multi-Temporal Data Fusion: Measurement Fusion.
	Ex. 12.16	Multi-Sensor Multi-Temporal Data Fusion: Track-to-Track Fusion.
	Ex. 12.18	Modified Track-to-Track Fusion Algorithm.

The designations DaI-DaO, DaI-FeO and FeI-FeO refer, respectively, to the Dasarathy input/output classifications: "Data Input-Data Output", "Data Input-Feature Output" and "Feature Input-Feature Output" (see Sect. 1.3.3).

12.2 Recursive Filter

The recursive filter is defined in (12.1)-(12.3). These equations may be derived as follows.

Bayes' Theorem. Eq. (12.1) represents a recursive formulation of Bayes' theorem. It is obtained by applying Bayes' theorem to the last measurement in $p(\boldsymbol{\theta}_k|\mathbf{y}_{1:k},I)$:

$$
\begin{aligned}
p(\boldsymbol{\theta}_k|\mathbf{y}_{1:k},I) &= p(\boldsymbol{\theta}_k|\mathbf{y}_k,\mathbf{y}_{1:k-1},I) , \\
&= \frac{p(\boldsymbol{\theta}_k,\mathbf{y}_k,\mathbf{y}_{1:k-1})}{p(\boldsymbol{\theta}_k,\mathbf{y}_{1:k-1},I)} , \\
&= \frac{p(\mathbf{y}_k|\mathbf{y}_{1:k-1},\boldsymbol{\theta}_k,I)p(\mathbf{y}_{1:k-1}|\boldsymbol{\theta}_k,I)p(\boldsymbol{\theta}_k|I)}{p(\mathbf{y}_k|\mathbf{y}_{1:k-1},I)p(\mathbf{y}_{1:k-1}|I)} , \\
&= \frac{p(\mathbf{y}_k|\mathbf{y}_{1:k-1},\boldsymbol{\theta}_k,I)p(\boldsymbol{\theta}_k|\mathbf{y}_{1:k-1},I)p(\mathbf{y}_{1:k-1}|I)}{p(\mathbf{y}_k|\mathbf{y}_{1:k-1},I)p(\mathbf{y}_{1:k-1},I)} , \\
&= \frac{p(\mathbf{y}_k|\boldsymbol{\theta}_k,I)p(\boldsymbol{\theta}_k|\mathbf{y}_{1:k-1},I)}{p(\mathbf{y}_k|\mathbf{y}_{1:k-1},I)} , \qquad (12.4)
\end{aligned}
$$

where the last equation follows from the assumption that the $\mathbf{y}_k$ obey a *Markov measurement model* in which $\mathbf{y}_k$ is independent of the previous states $\boldsymbol{\theta}_{1:k-1}$ and the previous measurements $\mathbf{y}_{1:k-1}$, i. e.

$$
p(\mathbf{y}_k|\boldsymbol{\theta}_k,\mathbf{y}_{1:k-1},I) = p(\mathbf{y}_k|\boldsymbol{\theta}_k,I) . \qquad (12.5)
$$

A Prior Pdf. Eq. (12.2) represents an integral formulation of the *a priori* pdf. It is obtained by marginalizing $\boldsymbol{\theta}_{k-1}$ out of the joint pdf $p(\boldsymbol{\theta}_k,\boldsymbol{\theta}_{k-1}|\mathbf{y}_{1:k-1},I)$:

$$
\begin{aligned}
p(\boldsymbol{\theta}_k|\mathbf{y}_{1:k-1},I) &= \int p(\boldsymbol{\theta}_k,\boldsymbol{\theta}_{k-1}|\mathbf{y}_{1:k-1},I)d\boldsymbol{\theta}_{k-1} , \\
&= \int p(\boldsymbol{\theta}_k|\boldsymbol{\theta}_{k-1},\mathbf{y}_{1:k-1},I)p(\boldsymbol{\theta}_{k-1}|\mathbf{y}_{1:k-1},I)d\boldsymbol{\theta}_{k-1} , \\
&= \int p(\boldsymbol{\theta}_k|\boldsymbol{\theta}_{k-1},I)p(\boldsymbol{\theta}_{k-1}|\mathbf{y}_{1:k-1},I)d\boldsymbol{\theta}_{k-1} , \qquad (12.6)
\end{aligned}
$$

where the last equation follows from the assumption that the $\boldsymbol{\theta}_k$ obey a *Markov process model* in which the $\boldsymbol{\theta}_k$ depends only on $\boldsymbol{\theta}_{k-1}$:

$$
p(\boldsymbol{\theta}_k|\boldsymbol{\theta}_{k-1},\mathbf{y}_{1:k-1},I) = p(\boldsymbol{\theta}_k|\boldsymbol{\theta}_{k-1},I) . \qquad (12.7)
$$

Bayesian Evidence. Eq. (12.3) represents an integral formulation of the Bayesian *evidence*. It follows directly from the definition of evidence which acts as a normalization factor in Bayes' rule.

In order to carry out the integrals in (12.2) and (12.3) we rewrite the Markov process and measurement models as stochastic equations:

$$
\begin{aligned}
\boldsymbol{\theta}_k &= \mathbf{f}_k(\boldsymbol{\theta}_{k-1},\mathbf{v}_{k-1}) , \qquad (12.8) \\
\mathbf{y}_k &= \mathbf{h}_k(\boldsymbol{\theta}_k,\mathbf{w}_k) , \qquad (12.9)
\end{aligned}
$$

where $\mathbf{v}_{k-1}$ is a noise term which represents the stochastic disturbances in the system and $\mathbf{w}_k$ is a noise term which represents the errors in the measurement process. Substituting (12.8) in (12.2) gives

$$p(\boldsymbol{\theta}_k|\boldsymbol{\theta}_{k-1},I) = \int p(\boldsymbol{\theta}_k|\boldsymbol{\theta}_{k-1},\mathbf{v}_{k-1},I)p(\mathbf{v}_{k-1}|\boldsymbol{\theta}_{k-1},I)d\mathbf{v}_{k-1} ,$$
$$= \int \delta(\boldsymbol{\theta}_k - \mathbf{f}_k(\boldsymbol{\theta}_{k-1},\mathbf{v}_{k-1}))p(\mathbf{v}_{k-1}|I)d\mathbf{v}_{k-1} , \qquad (12.10)$$

where the last equation follows from the assumption that the noise is independent of the state. Likewise, substituting (12.9) in (12.3) gives

$$p(\mathbf{y}_{1:k}|\boldsymbol{\theta}_k,I) = \int p(\mathbf{y}_{1:k}|\boldsymbol{\theta}_k,\mathbf{w}_k,I)p(\mathbf{w}_k|\boldsymbol{\theta}_k,I)d\mathbf{w}_k ,$$
$$= \int \delta(\mathbf{y}_k - h_k(\boldsymbol{\theta}_k,\mathbf{w}_k))p(\mathbf{w}_k|I)d\mathbf{w}_k . \qquad (12.11)$$

In Fig. 12.1 we show the main processing steps in the Bayesian recursive filter. The filter is initialized by specifying an *a priori* pdf, $p(\boldsymbol{\theta}_0|\mathbf{y}_0,I)$, where $\mathbf{y}_0$ denotes that no measurements have been made. The filter then operates recursively, performing a single cycle each time a new measurement becomes available. Suppose the *a posteriori* pdf, $p(\boldsymbol{\theta}_{k-1}|\mathbf{y}_{1:k-1},I)$, has been estimated at time step $k-1$ and that a new measurement has become available at time step k. Then we calculate a *predicted* pdf, $p(\boldsymbol{\theta}_k|\mathbf{y}_{1:k-1},I)$, by propagating $p(\boldsymbol{\theta}_{k-1}|\mathbf{y}_{1:k-1},I)$ from the time step $k-1$ to k. The predicted pdf acts as the *a priori* pdf for the time step k and is calculated using a *process model* which describes the evolution of the system. In general the predicted pdf will contain errors due to inaccuracies in the process model and noise in the input measurements. To eliminate these errors we correct $p(\boldsymbol{\theta}_k|\mathbf{y}_{1:k-1},I)$ using

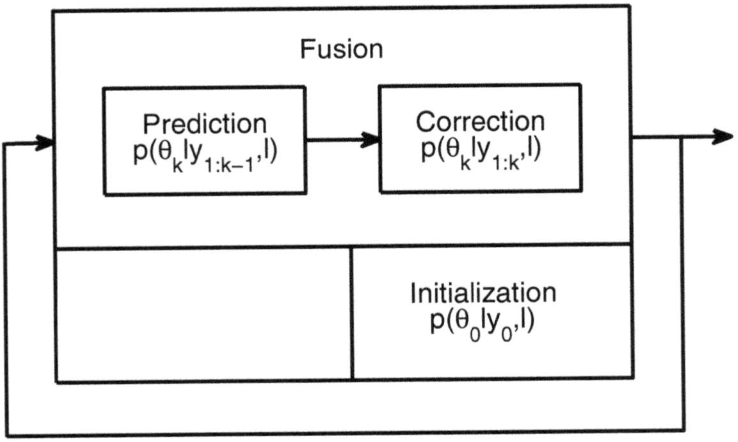

Fig. 12.1 Shows the main processing steps in the recursive filter: (1) Initialization. We give an initial pdf $p(\boldsymbol{\theta}_0|\mathbf{y}_0,I)$. This step is performed only once. (2) Prediction. We calculate a predicted pdf $p(\boldsymbol{\theta}|\mathbf{y}_{1:k-1},I)$ using the process model $\boldsymbol{\theta}_k = \mathbf{f}_{k-1}(\boldsymbol{\theta}_{k-1},\mathbf{v}_{k-1})$ and the *a posteriori* calculated in the previous time step. (2) Correction. We calculate the *a posteriori* pdf $p(\boldsymbol{\theta}_k|\mathbf{y}_{1:k},I)$ by correcting the predicted pdf using the current measurement $\mathbf{y}_k$. The architecture used to represent the filter is the iterative fusion cell (see Sect. 3.3.4).

the latest sensor measurement $\mathbf{y}_k$. This is achieved by using Bayes' theorem which is a mechanism for updating our knowledge in the light of new information (see Chapt. 9).

Fig. 12.2 shows the shapes of the different probability distributions in a generic recursive filter.

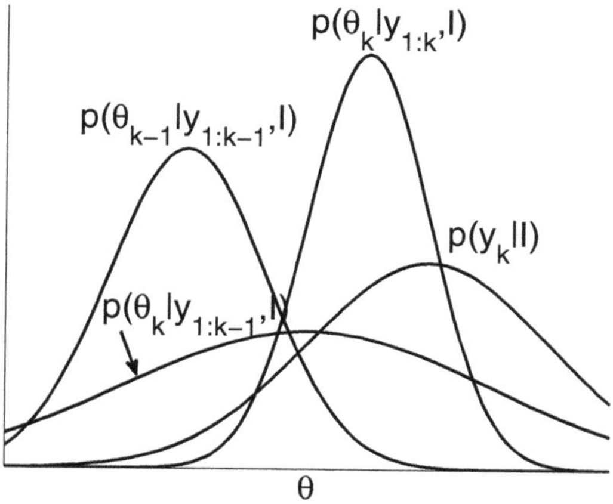

Fig. 12.2 Shows the shapes of the different probability distributions in a generic recursive filter. In the prediction step we use the process model (12.7) to propagate forward the pdf $p(\boldsymbol{\theta}_{k-1}|\mathbf{y}_{1:k-1},I)$ to the next time step k. The result is a predicted pdf which functions as the *a priori* pdf $p(\boldsymbol{\theta}_k|\mathbf{y}_{1:k-1},I)$ for the time step k. Because of the noise $\mathbf{v}_{k-1}$ the *a priori* pdf is a deformed, spread-out version of $p(\boldsymbol{\theta}_{1-k}|\mathbf{y}_{1:k-1},I)$. In the correction step we use the sensor measurement $p(\mathbf{y}_k|I)$ to modify the *a priori* pdf. The result is a sharp *a posteriori* pdf $p(\boldsymbol{\theta}_k|\mathbf{y}_{1:k},I)$ for the time step k.

The following example illustrates the use of (12.1)-(12.3) in terrain-aided navigation.

Example 12.1. INS/Radar Altimeter: A Hybrid Navigation Principle [12]. Terrain-aided navigation is a method for aircraft navigation in which radar altimeter measurements are combined with inertial measurements (INS) to achieve an optimal (minimum mean square error) estimate of the position of the aircraft. Terrain-aided navigation works as follows: At each time step k we use the INS measurements (accelerometer and gyro-meter) to compute $\boldsymbol{\theta}_k$, the INS estimate of the target. We then use sequential Bayesian inference to fuse $\boldsymbol{\theta}_k$ with the radar altimeter measurements y_k. The output is the INS drift $\delta\boldsymbol{\theta}_k$ (i. e. error in $\boldsymbol{\theta}_k$). The INS drift is then used to correct $\boldsymbol{\theta}_k$ as shown in Fig. 12.3. Mathematically, the hybrid navigation algorithm works as follows. We write the INS estimate at time step k as a linear Gaussian equation (see Sect. 10.6)

$$\boldsymbol{\theta}_k = \mathbf{F}_k \boldsymbol{\theta}_{k-1} + \mathbf{v}_{k-1} ,$$

where $\boldsymbol{\theta}_k$ denotes the INS estimates (latitude, longitude, altitude, roll, pitch, yaw and speed) and $\mathbf{v}_{k-1} \sim \mathscr{N}(0, \mathbf{Q}_{k-1})$ is the noise vector in $\boldsymbol{\theta}_{k-1}$.

The scalar radar altimeter measurement equation is given by

$$y_k = z_k - h(l_k, L_k) + w_k ,$$

where l_k, L_k and z_k are, respectively, the longitude, the latitude and the altitude of the aircraft. The function h stands for the terrain profile which is stored on-board the aircraft. The noise $w_k \sim \mathscr{N}(0, R_k)$ represents the measurement error.

Using the above equations for $\boldsymbol{\theta}_k$ and y_k we may compute the *a posteriori* pdf $p(\boldsymbol{\theta}_k | y_{1:k}, I)$ using (12.1) and (12.3). The INS drift is given by

$$\delta\boldsymbol{\theta}_k = \boldsymbol{\mu}_k - \boldsymbol{\theta}_k ,$$

where $\boldsymbol{\mu}_k = \int \boldsymbol{\theta}_k p(\boldsymbol{\theta}_k | y_{1:k}, I) d\boldsymbol{\theta}_k$ is the mean INS value at time step k.

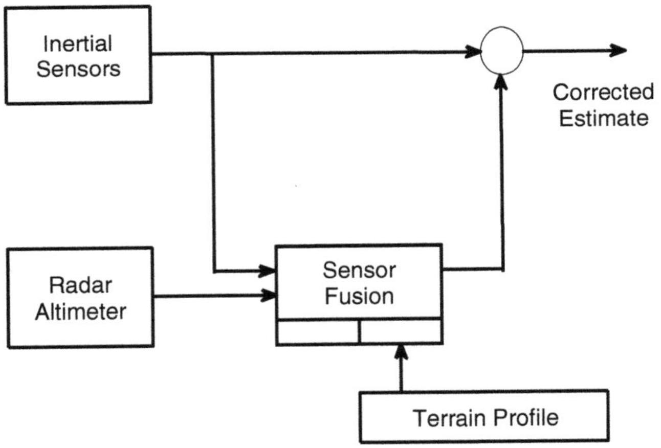

Fig. 12.3 Shows the sensor-fusion filter for terrain-aided navigation. The sensor-fusion filter outputs the INS drift estimate which is then removed from the initial INS estimate to give a corrected INS estimate.

12.3 Kalman Filter

For most applications, (12.2) and (12.3) are analytically intractable and approximate solutions must be used. However, for the special case of linear Gaussian systems (Sect. 10.6), the equations are tractable and a closed-form recursive solution for the sequential Bayesian filter is available. This is the *Kalman* filter (KF) [5] and because of its computational efficiency, it quickly established itself as the favourite algorithm for sequential Bayesian inference.

Mathematically, the KF assumes a linear Gaussian process model:

$$\boldsymbol{\theta}_k = \mathbf{F}_k \boldsymbol{\theta}_{k-1} + \mathbf{v}_{k-1} \ , \tag{12.12}$$

and a linear Gaussian measurement model:

$$\mathbf{y}_k = \mathbf{H}_k \boldsymbol{\theta}_k + \mathbf{w}_k \ , \tag{12.13}$$

where $\mathbf{v}_{k-1} \sim \mathscr{N}(0, \mathbf{Q}_{k-1})$ and $\mathbf{w}_k \sim \mathscr{N}(0, \mathbf{R}_k)$ are independent Gaussian noise sources. The following example illustrates the calculation of $\mathbf{F}_k$ and $\mathbf{H}_k$ for a digital image stabilizer. For more examples, see Sect. 10.6.

Example 12.2. Digital Image Stabilizer Using Kalman Filter [32]. Hand-held cameras usually suffer from image/video instability due to unintentional shaking. Digital image stabilization is a method which aims to remove the camera shakes while keeping the intentional motion of the camera intact. We approximate the camera trajectory as

$$\begin{aligned} \boldsymbol{\theta}_k &= \mathbf{F}\boldsymbol{\theta}_{k-1} + \mathbf{v}_{k-1} \ , \\ \mathbf{y}_k &= \mathbf{H}\boldsymbol{\theta}_k + \mathbf{w}_k \ , \end{aligned}$$

where $\boldsymbol{\theta}_k$ denotes the "state" of the camera at time k which represents the camera position, velocity and acceleration and $\mathbf{y}_k$ denotes the observed camera position at time k. Then, assuming a panning/tilting model for the intentional motion of the camera, we have

$$\begin{pmatrix} \phi_k \\ x_k \\ u_k \\ y_k \\ v_k \end{pmatrix} = \begin{pmatrix} 1&0&0&0&0 \\ 0&1&1&0&0 \\ 0&0&1&0&0 \\ 0&0&0&1&1 \\ 0&0&0&0&1 \end{pmatrix} \begin{pmatrix} \phi_{k-1} \\ x_{k-1} \\ u_{k-1} \\ y_{k-1} \\ v_{k-1} \end{pmatrix} + \begin{pmatrix} m_k \\ 0 \\ n_k \\ 0 \\ n_k \end{pmatrix} ,$$

where ϕ_k, x_k, y_k, u_k and v_k describe the intentional rotation, x and y translation and x and y translational velocities of the camera at time k. The camera rotation is caused by the tilt of the camera and described by

$$\phi_k = \phi_{k-1} + m_{k-1} \ ,$$

where $m_{k-1} \sim N(0, \Sigma_{\mathrm{TILT}})$ is white noise with variance Σ_{TILT}.

The x and y translations are caused by the panning of the camera which is assumed to have a constant velocity (u_k, v_k) with small random permutation over time:

$$
\begin{aligned}
x_k &= x_{k-1} + u_k \,, \\
y_k &= y_{k-1} + v_k \,, \\
u_k &= u_{k-1} + n_k \,, \\
v_k &= v_{k-1} + n_k \,,
\end{aligned}
$$

and $n_k \sim N(0, \Sigma_{\mathrm{PAN}})$ is white noise with variance Σ_{PAN}.

In general, the translational velocities due to the camera tilt, u_k and v_k, are not observed. In this case, the matrix $\mathbf{F}_k$ is a 3×5 matrix:

$$
\begin{pmatrix} \phi_k^0 \\ x_k^0 \\ y_k^0 \end{pmatrix} = \begin{pmatrix} 1\,0\,0\,0\,0 \\ 0\,1\,0\,0\,0 \\ 0\,0\,0\,1\,1 \end{pmatrix} \begin{pmatrix} \phi_k \\ x_k \\ u_k \\ y_k \\ v_k \end{pmatrix} + \begin{pmatrix} m_k^0 \\ n_k^0 \\ n_k^0 \end{pmatrix} ,
$$

where the superscript "0" denotes the variables for observed or global motion; $m_k^0 = N(0, \Sigma_{\mathrm{TILT}}^0)$ and $n_k^0 = N(0, \Sigma_{\mathrm{PAN}}^0)$ are white Gaussian noise with variance Σ_{TILT}^0 and Σ_k^0 standing for measurement noise for rotation and translation respectively.

If we assume the initial pdf is Gaussian, $\pi(\boldsymbol{\theta}_0|\mathbf{y}_0, I) \sim \mathscr{N}(\boldsymbol{\mu}_{0|0}, \boldsymbol{\Sigma}_{0|0})$, then both the predicted pdf (12.2) and the *a posteriori* pdf (12.1) are Gaussian:

$$
\begin{aligned}
p(\boldsymbol{\theta}_k|\mathbf{y}_{1:k-1}, I) &\sim \mathscr{N}(\boldsymbol{\mu}_{k|k-1}, \boldsymbol{\Sigma}_{k|k-1}) \,, && (12.14) \\
p(\boldsymbol{\theta}_k|\mathbf{y}_{1:k}, I) &\sim \mathscr{N}(\boldsymbol{\mu}_{k|k}, \boldsymbol{\Sigma}_{k|k}) \,, && (12.15)
\end{aligned}
$$

where

$$
\begin{aligned}
\boldsymbol{\mu}_{k|k-1} &= \mathbf{F}_k \boldsymbol{\mu}_{k-1|k-1} \,, && (12.16) \\
\boldsymbol{\Sigma}_{k|k-1} &= \mathbf{Q}_{k-1} + \mathbf{F}_k \boldsymbol{\Sigma}_{k-1|k-1} \mathbf{F}_k^T \,, && (12.17) \\
\boldsymbol{\mu}_{k|k} &= \boldsymbol{\mu}_{k|k-1} + \mathbf{K}_k(\mathbf{y}_k - \mathbf{H}_k \boldsymbol{\mu}_{k|k-1}) \,, && (12.18) \\
\boldsymbol{\Sigma}_{k|k} &= \boldsymbol{\Sigma}_{k|k-1} - \mathbf{K}_k \mathbf{H}_k \boldsymbol{\Sigma}_{k|k-1} \,, && (12.19)
\end{aligned}
$$

and $\mathbf{K}_k$ is the *Kalman gain matrix*,

$$
\mathbf{K}_k = \boldsymbol{\Sigma}_{k|k-1} \mathbf{H}_k^T (\mathbf{H}_k \boldsymbol{\Sigma}_{k|k-1} \mathbf{H}_k^T + \mathbf{R}_k)^{-1} \,. \qquad (12.20)
$$

Note. For reasons of numerical stability we often rewrite (12.19) as

$$
\boldsymbol{\Sigma}_{k|k} = (\mathbf{I} - \mathbf{K}_k \mathbf{H}_k) \boldsymbol{\Sigma}_{k|k-1} (\mathbf{I} - \mathbf{K}_k \mathbf{H}_k)^T + \mathbf{K}_k \mathbf{R}_k \mathbf{K}_k^T \,, \qquad (12.21)
$$

where $\mathbf{I}$ is the unit matrix.

Fig. 12.4 is a block diagram which shows the main processing steps in the KF.

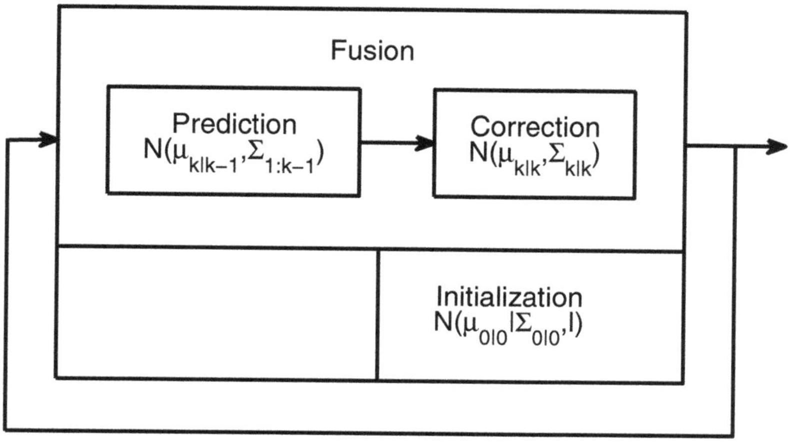

Fig. 12.4 Shows the main processing steps in the Kalman filter: (1) *Initialization*. We give initial parameter values $(\boldsymbol{\mu}_{0|0}|I)$ and $\boldsymbol{\Sigma}_{0|0}|I)$ and the noise covariances $\mathbf{Q}_k$ and $\mathbf{R}_k$. This step is performed only once. (2) *Prediction*. We calculate a predicted pdf $p(\boldsymbol{\theta}|\mathbf{y}_{1:k},I) = \mathcal{N}(\boldsymbol{\mu}_{k|k-1}, \boldsymbol{\Sigma}_{k|k-1})$ using the process model $\boldsymbol{\theta}_k = \mathbf{F}_k(\boldsymbol{\theta}_{k-1}, \mathbf{v}_{k-1})$ and the *a posteriori* calculated in the previous time step. (3) *Correction*. We calculate the *a posteriori* pdf $p(\boldsymbol{\theta}_k|\mathbf{y}_{1:k},I) = \mathcal{N}(\boldsymbol{\mu}_{k|k}, \boldsymbol{\Sigma}_{k|k})$ by correcting the predicted pdf using the current measurement $\mathbf{y}_k$.

The KF is used in a very wide variety of multi-sensor data fusion applications. Ex. 3.7 illustrates the use of a KF to estimate a patients heart rate and the next example illustrates the use of the KF to track a satellite's orbit around the earth.

Example 12.3. Tracking a Satellite's Orbit Around the Earth [24]. The unknown state $\boldsymbol{\theta}_k$ is the position and speed of the satellite at time step k with respect to a spherical coordinate system with origin at the center of the earth. These quantities cannot be measured directly. Instead, from tracking stations around the earth, we obtain measurements of the distance to the satellite and the accompanying angles of the measurement. These form the measurements $\mathbf{y}_k$. The geometrical principles of mapping $\mathbf{y}_k$ into $\boldsymbol{\theta}_k$, are incorporated into $\mathbf{H}_k$, while $\mathbf{w}_k$ reflects the measurement error; $\mathbf{F}_k$ prescribes how the position and speed change in time according to the physical laws governing orbiting bodies, while $\mathbf{v}_k$ allows for deviations from these laws owing to such factors as nonuniformity of the earth's gravitational field etc.

The following example describes the tracking of a moving target which is subject to stochastic accelerations.

Example 12.4. Tracking a Moving Target [3, 4]. We consider a target moving in one-dimension at a nomimal constant velocity while being subjected to *stochastic* accelerations. If θ_k denotes the true position of the target at the kth time step, then the corresponding matrix equation of motion of the target is:

$$\begin{pmatrix} \theta_k \\ \dot{\theta}_k \end{pmatrix} = \begin{pmatrix} 1 & T \\ 0 & 1 \end{pmatrix} \begin{pmatrix} \theta_{k-1} \\ \dot{\theta}_{k-1} \end{pmatrix} + \begin{pmatrix} \frac{T^2}{2} \\ T \end{pmatrix} \ddot{\theta}_{k-1} ,$$

where $\ddot{\theta}_k \sim \mathcal{N}(0, \sigma^2)$ is the (stochastic) acceleration which acts upon the target at time step k and T is the interval between successive time steps. The corresponding process model is

$$\begin{pmatrix} \theta_k \\ \dot{\theta}_k \end{pmatrix} = \begin{pmatrix} 1 & T \\ 0 & 1 \end{pmatrix} \begin{pmatrix} \theta_k \\ \dot{\theta}_k \end{pmatrix} + \mathbf{v}_{k-1} .$$

Equating the matrix equation of motion model with the process model leads to the identification of $\begin{pmatrix} \frac{T^2}{2} \\ T \end{pmatrix} \ddot{\theta}_k$ with the process noise $\mathbf{v}_{k-1} \sim \mathcal{N}(\mathbf{0}, \mathbf{Q})$. In this case,

$$\mathbf{Q} = \sigma^2 \begin{pmatrix} \frac{T^4}{4} & \frac{T^3}{2} \\ \frac{T^3}{2} & T^2 \end{pmatrix} .$$

Ex. 12.3 and 12.4 illustrate a common situation in which we use the KF to track a *single* well-defined object with known process and measurement models. The following examples illustrate the tracking of multiple objects. In the first example, the objects are people in a densely crowded scene. In this case, although the tracked objects are well-defined, their extraction from the input images is very difficult.

Example 12.5. Tracking People in a Crowded Scene [11]. Tracking people in a dense crowd is a challenging problem. One reason for this is the difficulty in consistently and accurately extracting people from a single image. A novel approach to this problem is to use a pair of stereo cameras [11]. In this case it is relatively easy to extract the people since each person is represented by a two-dimensional blob characterized by a given height.

In the second example, the objects are metrological features. In this case, the tracked objects are not well-defined and their extraction from the input images is difficult.

Example 12.6. Tracking Metrological Features [22]. In general metrological features, such as storm centers, are not well-defined: they can change size and intensity; they can split into separate storms; and they can merge with other storms. In addition, individual storms rarely satisfy the linear-Gaussian assumption of the KF. Nevertheless, in practice [22], the KF has been used to track storm centers with reasonable accuracy.

12.3.1 Parameter Estimation

The KF is defined in (12.14)-(12.20). It contains the parameters $\mu_{0|0}$, $\Sigma_{0|0}$, $\mathbf{Q}_k$ and $\mathbf{R}_k$ which must be defined before running the filter. In general the values chosen for $\mu_{0|0}$ and $\Sigma_{0|0}$ do not seriously affect the medium and long-term results obtained with the filter, i. e. the values $\mu_{k|k}$ and $\Sigma_{k|k}$ for $k >> 0$. For this reason we often use simple *a priori* values for $\mu_{0|0}$ and $\Sigma_{0|0}$ or use a track initiation filter (see Ex. 11.15).

Regarding $\mathbf{Q}_k$ and $\mathbf{R}_k$, the way that $\mathbf{Q}_k$ and $\mathbf{R}_k$ are specified crucially affects $\mu_{k|k}$ and $\Sigma_{k|k}$. In mathematical terms, the values of $\mathbf{Q}_k$ and $\mathbf{R}_k$ affect the Kalman gain $\mathbf{K}_k$ which in turn affects the capability of the filter to adjust itself to changing conditions.

The following example illustrates a simple, but very effective, procedure for *adaptively* calculating $\mathbf{Q}_k$ and $\mathbf{R}_k$.

Example 12.7. Correcting near surface temperature forecasts using a Kalman filter [13]. Numerical weather prediction models often exhibit systematic errors in the forecasts of near surface weather parameters. One such parameter is the near surface temperature forecast T which is a commonly biased variable, where the magnitude of the bias depends among other factors, on the geographical location and the season. To correct for the systematic error in T let the measurement y_k be the difference between an observed temperature and the weather prediction model forecast and let the state vector θ_k be the systematic part of this error. We assume the changes in θ_k are random. The corresponding process and measurement models are then

$$\theta_k = \theta_{k-1} + v_{k-1} ,$$
$$y_k = \theta_k + w_k ,$$

where $v_k \sim \mathcal{N}(0, Q_k)$ and $w_k \sim \mathcal{N}(0, R_k)$. In [13] the authors estimate Q_k and R_k using the sample variances of the N latest $v_{k-n} = \theta_{k-n} - \theta_{k-n-1}$ and $w_{k-n} = y_{k-n} - \theta_{k-n}$ values:

$$Q_k = \frac{1}{N-1}\sum_{n=0}^{N-1}((\theta_{k-n}-\theta_{k-n-1})-\bar{Q}_k)^2 ,$$
$$R_k = \frac{1}{N-1}\sum_{n=0}^{N-1}((y_{k-n}-\theta_{k-n})-\bar{R}_k)^2 ,$$

where

$$\bar{Q}_k = \frac{1}{N}\sum_{n=0}^{N-1}(\theta_{k-n}-\theta_{k-n-1}) ,$$
$$\bar{R}_k = \frac{1}{N}\sum_{n=0}^{N-1}(y_{k-n}-\theta_{k-n}) .$$

See [3, 4, 5] for more sophisticated procedures for calculating $\mathbf{Q}_k$ and $\mathbf{R}_k$.

12.3.2 Data Association

In real-world applications spurious measurements are often present (see Sect. 2.5.1). These are called false alarms, or clutter, and their presence creates an ambiguity regarding the origin of the measurements, i. e. it is often not clear which measurement corresponds to the system (a "true"measurement) and which measurement corresponds to clutter (a "false" measurement) [2]. This is known as the *data association* problem [9] and it is clear that it must be solved before we can use the KF. In many multi-sensor data fusion applications the solution of the data association problem is the primary fusion algorithm (see Table 12.2). The following example illustrates the

Table 12.2 Applications in which Solving the Data Association Problem is the Primary Fusion Algorithm

Class		Application
DaI-FeO	Ex. 12.9	The strongest neighbour (SN) filter.
	Ex. 12.10	Covariance Union Filter.
FeI-FeO	Ex. 5.10	Key-point Image Registration.
	Ex. 4.13	Shape Context and Inner Distance.
	Ex. 7.5	Closed contour matching using the shape context.
	Ex. 12.8	Tracking Pedestrians Across Non-Overlapping Cameras.

The designations DaI-FeO and FeI-FeO refer, respectively, to the Dasarathy input/output classifications: "Data Input-Feature Output" and "Feature Input-Feature Output" (see Sect. 1.3.3).

[2] We also allow for the possibility that the sensor will not always return a true measurement.

use of data association in an application involving tracking people in a surveillance area with multiple non-overlapping cameras.

Example 12.8. Tracking Pedestrians Across Non-Overlapping Cameras [20]. In visual surveillance it is not possible to monitor a wide area using a single camera because the field of view (FOV) of one camera is finite and often the structure of the scene limits the visible area in one camera view. For these reasons it is common practice to employ a network of cameras for a surveillance systems. In this case, a major task is to track objects moving across multiple FOV's. If the FOV's do not overlap we use data association to join the tracks from the same objects. The idea is: we estimate the likelihood that ith track T_i observed in the FOV of camera C_i is associated with the jth track T_j observed in the FOV of camera C_j by measuring the similarity $S(T_i, T_j)$ between T_i and T_j.

For a non-tracking applications see Ex. 5.10, Ex. 4.13 and Ex. 7.5.

In general, the solution of the data association problem can be *hard* or *soft*. In a hard solution we assign a given measurement to the system. In this case, the state of the system, $\boldsymbol{\theta}_k$, is updated assuming that the assigned measurement is the true measurement. On the other hand, in a soft solution we assign more than one measurement to the system, each measurement having its own assignment probability. In this case, we use each measurement which has an assignment probability greater than than zero to update $\boldsymbol{\theta}_k$. The collection of updated states is then combined using the assignment probabilities.

In general, different applications will often require different data association algorithms. However, the algorithms given in Table 12.3 have been found useful in a wide-range of applications although they were originally developed for single-target tracking [3, 4].

In general the algorithms listed in Table 12.3 are combined with the KF to create a a new recursive filter. We consider two such filters.

12.3.3 Nearest Neighbour Filter

Let $\tilde{\mathbf{y}}_k = \{\mathbf{y}_k^{(1)}, \mathbf{y}_k^{(2)}, \ldots, \mathbf{y}_k^{(M_k)}\}$, denote the set of M_k measurements received at time step k [3]. In the nearest neighbour (NN) filter we assume the true measurement is

$$\mathbf{y}_k^{(*)} = \arg\min_m (D^{(m)}) \tag{12.22}$$

where

[3] The reader, may find it convenient to regard each measurement $\mathbf{y}_k^{(m)}$ as originating from a separate sensor $S^{(m)}$. In this case, at most, one of the sensors returns a true measurement. Each of the other sensors returning a false measurement.

Table 12.3 Data Association Algorithms

Name	Description
Nearest Neighbour (NN)	Assumes the measurement that is closest to the predicted state of the system is the true measurement. If the true measurement is not detected at time step k, then the nearest neighbour will always be a false measurement which may cause erratic behaviour in the KF. To avoid this we normally restrict ourselves to *validated* measurements. In general, the performance of the NN and the validated NN algorithms is low.
Probabilistic Data Association (PDA)	The state of the system is separately updated using each validated measurement in turn. The collection of updated states are then combined together to form the final state of the system. Each validated measurement $\mathbf{y}_k^{(m)}, m \in \{1,2,\ldots,M_k\}$, is given a weight $\beta^{(m)}$ which is the probability that $\mathbf{y}_k^{(m)}$ is the true measurement. Also computed is a weight $\beta^{(0)}$ which is the probability that none of the validated measurements are the true measurement. These events encompass all possible interpretations of the data, so $\sum_{m=0}^{M_k} \beta^{(m)} = 1$.
Multiple Hypothesis Testing (MHT)	Enumerates all possible hard assignments taken over all time steps. The number of possible assignments increases exponentially with time. As a consequence at each time step the MHT algorithm discards assignments which have a very low probability of being true.

Validated measurements are measurements which fall inside a given region, or validation gate. In general, the validation gate is an area centered on the predicted state of the system. See Ex. 11.2.

$$D^{(m)} = (\mathbf{y}_k^{(m)} - \hat{\mathbf{y}}_{k|k-1})^T \hat{\mathbf{S}}_{k|k-1}^{-1} (\mathbf{y}_k^{(m)} - \hat{\mathbf{y}}_{k|k-1}) , \tag{12.23}$$

is the normalized distance between $\mathbf{y}_k^{(m)}$ and the predicted measurement $\hat{\mathbf{y}}_{k|k-1}$, and $\hat{\mathbf{S}}_{k|k-1}$ is the predicted covariance matrix $\mathbf{H}_k^T \boldsymbol{\Sigma}_{k|k-1} \mathbf{H}_k^T + \mathbf{R}_k$.

In the modified nearest neighbour rule we only use *validated* measurements (i. e. measurements $\mathbf{y}_k^{(*)}$ for which $D^{(*)} \leq \gamma$, where γ is a given threshold [4]). The corresponding *nearest neighbour filter* combines the KF with a validated NN solution of the data association problem:

$$\boldsymbol{\mu}_{k|k-1} = \mathbf{F}_k \boldsymbol{\mu}_{k-1|k-1} , \tag{12.24}$$

$$\boldsymbol{\Sigma}_{k|k-1} = \mathbf{Q}_{k-1} + \mathbf{F}_k \boldsymbol{\Sigma}_{k-1|k-1} \mathbf{F}_k^T , \tag{12.25}$$

$$\boldsymbol{\mu}_{k|k} = \begin{cases} \boldsymbol{\mu}_{k|k-1} + \mathbf{K}_k(\mathbf{y}_k^{(*)} - \mathbf{H}_k \boldsymbol{\mu}_{k|k-1}) & \text{if } m_k \geq 1 , \\ \boldsymbol{\mu}_{k|k-1} & \text{otherwise} , \end{cases} \tag{12.26}$$

$$\boldsymbol{\Sigma}_{k|k} = \begin{cases} \boldsymbol{\Sigma}_{k|k-1} - \mathbf{K}_k \mathbf{H}_k \boldsymbol{\Sigma}_{k|k-1} & \text{if } m_k \geq 1 , \\ \boldsymbol{\Sigma}_{k|k-1} & \text{otherwise} , \end{cases} \tag{12.27}$$

where m_k is the number of validated measurements at time step k, and

$$\mathbf{K}_k = \boldsymbol{\Sigma}_{k|k-1} \mathbf{H}_k^T (\mathbf{H}_k \boldsymbol{\Sigma}_{k|k-1} \mathbf{H}_k^T + \mathbf{R}_k)^{-1} . \tag{12.28}$$

4 See Sect. 11.2.

An important modification of the nearest neighbour (NN) filter is the strongest neighbour (SN) filter.

Example 12.9. The strongest neighbour (SN) filter. The strongest neighbour filter is identical to the NN filter except we identify the true measurement as $\mathbf{y}^{(*)}$, where $\mathbf{y}^{(*)}$ is the measurement with the largest *validated* amplitude. A further refinement of the SN filter is the probabilistic SN filter [19]. In the probabilistic SN filter we weight the strongest neighbour measurement (in the validation gate) with the probability that it is the true measurement. This is in opposition to the SN filter which makes the unrealistic assumption that the SN measurement is always the true measurement. The result is a filter which has substantially better performance than the NN, SN and PDA filters.

In the probabilistic SN filter we take into account the possibility that the measurement with the largest validated amplitude may not be the true measurement. The corresponding *probabilistic strongest neighbour filter* equations are identical to (12.24-12.28) except for the *a posteriori* covariance matrix which is replaced by

$$\boldsymbol{\Sigma}_{k|k} = \begin{cases} \boldsymbol{\Sigma}_{k|k-1} + \mathbf{K}_k(\mathbf{y}_k - \mathbf{H}_k\boldsymbol{\mu}_{k|k-1}) & \text{if } m_k \geq 1\,, \\ \quad \times (\mathbf{y} - \mathbf{H}_k\boldsymbol{\mu}_{k|k-1})^T \mathbf{K}_k^T P_T P_F & \\ \quad + \left(\frac{P_D P_G P_A(1-c)}{1-P_D P_G P_A} P_T - P_F\right) \mathbf{K}\boldsymbol{\Sigma}_{k|k-1}\mathbf{K}_k^T & \\ \boldsymbol{\Sigma}_{k|k-1} + \frac{P_D P_G(1-c)}{1-P_D P_G}\mathbf{K}\boldsymbol{\Sigma}_{k|k-1}\mathbf{K}_k^T & \text{otherwise}\,, \end{cases}$$

where P_D is the probability of the true target being detected and c is a constant determined by the dimensionality of the measurements. For the case of two dimensions, $c = 1 - \frac{1}{2}P_G^{-1}\gamma\exp(-\gamma/2)$.

Another important modification of the NN filter is the covariance union filter which is illustrated next.

Example 12.10. Covariance Union Filter [17]. The *covariance union filter* is similar to the NN and SN filters with one crucial difference: The measurement which is selected by the covariance filter is a "virtual" measurement characterized by a "virtual" value $\mathbf{y}_k^{(*)}$ and a "virtual" covariance matrix $\boldsymbol{\Sigma}_k^{(*)}$, where $\mathbf{y}_k^{(*)}$ and $\boldsymbol{\Sigma}_k^{(*)}$ are chosen to be *consistent* with all validated measurements. For two observations $O^{(1)} = (\mathbf{y}_k^{(1)}, \mathbf{R}_k^{(1)})$ and $O^{(2)} = (\mathbf{y}_k^{(2)}, \mathbf{R}_k^{(2)})$ (Fig. 12.5) we have $\widetilde{O} = (\mathbf{y}_k^{(*)}, \mathbf{R}_k^{(*)})$ as follows [17]:

$$\mathbf{y}_k^{(*)} = \frac{1}{2}(\mathbf{y}_k^{(1)} + \mathbf{y}_k^{(2)})\,,$$
$$\mathbf{R}_k^{(*)} = \mathbf{S}_k^T \mathbf{V}_k \max(\mathbf{D}_k, \mathbf{I})\mathbf{V}_k^T \mathbf{S}_k\,,$$

where $\mathbf{S}_k = |\mathbf{R}_k^{(2)}|^{1/2}$, and $\max(\mathbf{A},\mathbf{B})$ denotes the element-by-element maximum of the matrices $\mathbf{A}$ and $\mathbf{B}$, and $\mathbf{V}_k$ and $\mathbf{D}_k$ are, respectively, the matrices of the eigenvectors and eigenvalues of $(\mathbf{S}_k^{-1})^T \mathbf{R}_k^{(1)} \mathbf{S}_k^{-1}$.

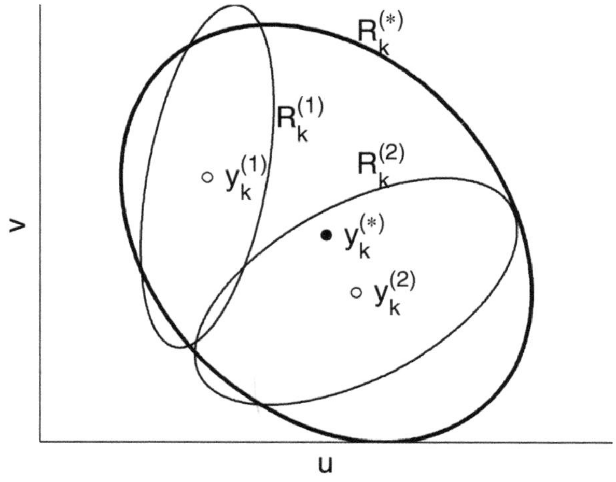

Fig. 12.5 Illustrates the covariance union of two observations $O_1 = (\mathbf{y}_k^{(1)}, \mathbf{R}_k^{(1)})$ and $O_2 = (\mathbf{y}_k^{(2)}, \mathbf{R}_k^{(2)})$ in two dimensions (u,v). The observation means, $\mathbf{y}_k^{(1)}$ and $\mathbf{y}_k^{(2)}$, are shown by small open circles and the observation covariances, $\mathbf{R}_k^{(1)}$ and $\mathbf{R}_k^{(2)}$, are shown by two ellipses drawn with thin solid lines. The mean, $\mathbf{y}_k^{(*)}$, and covariance, $\mathbf{R}_k^{(*)}$, of the covariance union are shown, respectively, by a small filled circle and an ellipse drawn with a thick solid line.

12.3.4 Probabilistic Data Association Filter

The NN, SN and CU filters work by choosing a *single* validated measurement $\mathbf{y}_k^{(*)}$ [5] which is then used to update the state vector $\boldsymbol{\theta}_k$. However, the probabilistic data association (PDA) filter [3, 4] works differently: Let $\beta^{(m)}$ be the probability that $\mathbf{y}_k^{(m)}$ is the true measurement at the kth time step. Then the PDAF uses *all* of the validated measurements, $\mathbf{y}_k^{(m)}, m \in \{1,2,\ldots,M_k\}$, and their probabilities $\beta^{(m)}$ to update $\boldsymbol{\theta}_k$. The corresponding PDAF equations are:

$$\boldsymbol{\mu}_{k|k-1} = \mathbf{F}_k \boldsymbol{\mu}_{k-1|k-1} \ , \tag{12.29}$$

[5] In the NN and SN filters we use a "real" measurement as $\mathbf{y}_k^{(*)}$ while in the CU filter we use a "virtual" measurement for $\mathbf{y}_k^{(*)}$.

$$\boldsymbol{\Sigma}_{k|k-1} = \mathbf{Q}_{k-1} + \mathbf{F}_k \boldsymbol{\Sigma}_{k-1|k-1} \mathbf{F}_k^T , \tag{12.30}$$

$$\boldsymbol{\mu}_{k|k} = \boldsymbol{\mu}_{k|k-1} + \mathbf{K}_k \sum_{m=1}^{m_k} \beta^{(m)} (\mathbf{y}_k^{(m)} - \mathbf{H}_k \boldsymbol{\mu}_{k|k-1}) , \tag{12.31}$$

$$\boldsymbol{\Sigma}_{k|k} = \beta^{(0)} \boldsymbol{\Sigma}_{k|k-1} + (1-\beta^{(0)})(\boldsymbol{\Sigma}_{k|k-1} - \mathbf{K}_k \hat{\mathbf{S}}_{k|k-1} \mathbf{K}_k) + \Delta \boldsymbol{\Sigma}_{k|k-1} , \tag{12.32}$$

where $\beta^{(0)} = 1 - \sum_{m=1}^{m_k} \beta^{(m)}$ is the probability that none of the $\mathbf{y}_k^{(m)}, m \in \{1,2,\ldots,m_k\}$ is the true measurement. The correction term, $\Delta\boldsymbol{\Sigma}_{k|k-1}$, in (12.32) is given by

$$\Delta \boldsymbol{\Sigma}_{k|k-1} = \mathbf{K}_k \Big(\sum_{m=1}^{m_k} \beta^{(m)} \mathbf{v}_k^{(m)} (\mathbf{k}^{(m)})^T - \mathbf{v}_k \mathbf{v}_k^T \Big) \mathbf{K}_k^T - \sum_{m=1}^{m_k} \beta^{(m)} \mathbf{v}_k^{(m)} (\mathbf{v}_k^{(m)})^T - \sum_{m=1}^{m_k} \beta^{(m)} \mathbf{v}_k^{(m)} \Big(\sum_{m=1}^{m_k} \beta^{(m)} (\mathbf{v}_k^{(m)})^T \mathbf{v}_k \mathbf{v}_k^T \Big) \mathbf{K}_k^T , \tag{12.33}$$

where

$$\mathbf{v}_k = \sum_{m=1}^{m_k} \beta^{(m)} \mathbf{v}_k^{(m)} ,$$

and

$$\mathbf{v}_k^{(m)} = \mathbf{y}_k^{(m)} - \mathbf{H}_k \boldsymbol{\mu}_{k|k-1} . \tag{12.34}$$

See [3, 4] for a detailed discussion concerning the PDA filter.

12.3.5 Model Inaccuracies

In the KF we model the system using a linear Gaussian process model (12.12) and a linear Gaussian measurement model (12.13). In practice most systems can never be perfectly modeled and this will set a lower limit on the final performance that can be obtained. However, even when the conditions are far from Gaussian, the KF can often give sensible results by giving more weight to the input measurements, i. e. by increasing the processes noise covariance $\mathbf{Q}$.

The following example illustrates an important case in which there is a clear mismatch between the assumed process model and the true system dynamics.

Example 12.11. Maneuvering Target [3, 4]. We track a target moving in one-dimension with a nominally constant velocity which is subject to stochastic accelerations. The corresponding process model (see Ex. 12.4) is

$$\begin{pmatrix} \theta_k \\ \dot{\theta}_k \end{pmatrix} = \begin{pmatrix} 1 & T \\ 0 & 1 \end{pmatrix} \begin{pmatrix} \theta_{k-1} \\ \dot{\theta}_{k-1} \end{pmatrix} + \mathbf{v}_{k-1} ,$$

where $\mathbf{v}_{k-1} \sim \mathscr{N}(\mathbf{0}, \mathbf{Q})$ and $\mathbf{Q} = \sigma^2 \begin{pmatrix} \frac{T^4}{4} & \frac{T^3}{2} \\ \frac{T^3}{2} & T^2 \end{pmatrix}$. If the process model is correct, the KF will provide optimum estimates of the target's position and velocity. However, if the target initiates and sustains a driver-induced maneuver, then the above model is incorrect. This is because a driver-induced maneuver is not well-modeled as a stochastic random variable.

One solution to this problem is to use a maneuver detector [3]. As long as the maneuver detector has not detected a target maneuver we continue to use the above process model. However, once a maneuver has been detected, we should initialize a *new* recursive filter whose process model describes a target maneuver.

Table 12.4 lists some one-dimensional maneuver models which are used for tracking maneuvering targets [18, 29]. The models given in the table are derived from a general one-dimensional maneuver model:

$$F = \begin{pmatrix} 1 & a & c \\ 0 & b & d \\ 0 & 0 & e \end{pmatrix} , \tag{12.35}$$

Table 12.4 One-Dimensional Maneuver Models

Name	α	β	Maneuver F
Constant Position	∞	∞	$\begin{pmatrix} 1 & 0 & 0 \\ 0 & 0 & 0 \\ 0 & 0 & 0 \end{pmatrix}$
Time Correlated Velocity	>0	∞	$\begin{pmatrix} 1 & a & 0 \\ 0 & b & 0 \\ 0 & 0 & 1 \end{pmatrix}$
Constant Velocity	0	∞	$\begin{pmatrix} 1 & T & 0 \\ 0 & 1 & 0 \\ 0 & 0 & 0 \end{pmatrix}$
Time Correlated Acceleration (Singer Model)	0	>0	$\begin{pmatrix} 1 & T & c \\ 0 & 1 & d \\ 0 & 0 & e \end{pmatrix}$
Constant Acceleration	0	0	$\begin{pmatrix} 1 & T & T^2 \\ 0 & 1 & T \\ 0 & 0 & 1 \end{pmatrix}$
Time Correlated Acceleration and Velocity	>0	>0	$\begin{pmatrix} 1 & a & c \\ 0 & b & d \\ 0 & 0 & e \end{pmatrix}$

in which both the target velocity and acceleration are described by zero-mean first-order Markov processes[6].

12.3.6 Multi-target Tracking

An important application of the recursive filter is the tracking of multiple targets. If the targets are widely separated one from the other then we may track the targets using a separate KF for each target using the NN, SN, CU or PDA filters. In most real-world systems, the targets are not sufficiently separated and modified tracking/data association algorithms must be used. A full discussion of these and other issues is given in [3, 4].

12.4 Extensions of the Kalman Filter *

The KF assumes a single linear Gaussian process model (12.12) and measurement model (12.13). In practice these models may not adequately represent the truth in which case more sophisticated filtering techniques are necessary. Some standard non-linear filter techniques which involve modifications to the KF are listed in Table 12.7.

12.4.1 Robust Kalman Filter

The KF assumes Gaussian process and measurement noise. The filter is therefore sensitive to outliers (see Chapt. 11) which may be present in the measurement $\mathbf{y}_k$. If the $\mathbf{y}_k$ are thought to be contaminated by outliers we may limit their effect by gating (see Sect. 12.3.2) or by using a robust KF which is designed to be insensitive to outliers. Cipra and Romera [8] described a simple robust KF which is identical to the conventional KF except we replace the Kalman gain matrix $\mathbf{K}_k$ with the following expression:

$$\mathbf{K}_k = \Sigma_{k|k-1}\mathbf{H}_k^T(\mathbf{H}_k\Sigma_{k|k-1}\mathbf{H}_k^T + \mathbf{R}_k^{1/2}\mathbf{W}_k\mathbf{R}_k^{1/2})^{-1} . \tag{12.36}$$

In (12.36), $\mathbf{W}_k = \mathrm{diag}(W_k^{(1)}, W_k^{(2)}, \ldots, W_k^{(M)})$ denotes a set of weights given by

$$W_k^{(m)} = \frac{\psi_H\left([\mathbf{R}_k^{-1/2}(\mathbf{y}_k - \mathbf{H}_k\boldsymbol{\mu}_{k|k-1})]^{(m)}\right)}{[\mathbf{R}_k^{-1/2}(\mathbf{y}_k - \mathbf{H}_k\boldsymbol{\mu}_{k|k-1})]^{(m)}} , \tag{12.37}$$

6 The parameters a, b, c, d and e in (12.35) are given by $a = \frac{(1-e^{-\alpha T})}{\alpha}$, $b = e^{-\alpha T}$, $c = \frac{1}{2!}T^2 - \frac{\alpha+\beta}{3!}T^3 + \frac{\alpha^2+\alpha\beta+\beta^2}{4!}T^4 + \ldots$, $d = T - \frac{\alpha+\beta}{2!}T^2 + \frac{\alpha^2+\alpha\beta+\beta^2}{3!}T^3 + \ldots$ and $e = e^{-\beta}T$, where $\alpha = 1/\tau_v$ and $\beta = 1/\tau_a$ are, respectively, the reciprocal of the velocity and acceleration correlation times.

Table 12.5 Kalman Filter and its Modifications

Filter	Description
Kalman Filter (KF)	Assumes linear Gaussian process and measurement models.
Robust Kalman Filter	Assumes linear Gaussian process model and a linear Gaussian measurement model which is contaminated with a small fraction of outliers. Robust estimation techniques (Chapt. 11) are used to reduce influence of outliers [8].
Kriged Kalman Filter	KF is combined with spatial Kriging model [23].
Augmented Kalman Filter	KF in which we use parameterized linear Gaussian process and measurement models. We augment the state vector with the process and measurement model parameters $\alpha,\beta,\ldots,\gamma$. The augmented KF maybe used to include spatial and temporal misalignments between the sensors.
Extended Kalman Filter (EKF)	First-order local linearization of non-linear Gaussian process and measurement models. See Sect. 12.4.2.
Unscented Kalman Filter (UKF)	UKF uses a deterministic set of weighted points $(\omega_k^{(m)},\Omega_k^{(m)},\boldsymbol{\theta}_k^{(m)}), m \in \{1,2,\ldots,M\}$, to represent the *a priori* pdf $p(\boldsymbol{\theta}_{k-1}\|\mathbf{y}_{1:k-1},I)$. See Sect. 12.4.3.
Switching Kalman Filter	Contains multiple KF's but at each time step k we use only one KF. The selected filter is specified by a switching variable Λ_k. See Sect. 12.4.4.
Mixture Kalman Filter	Similar to the switched KF except that at each time step k we use a collection of multiple KF's.
Ensemble Kalman Filter (EnKF)	Used when the dimensions of $\boldsymbol{\theta}_k$ is too large to permit the conventional calculation of the covariance matrix. In the EnKF we estimate the covariance matrices using an ensemble of trajectories.

where $\mathbf{X}^{(m)}$ is the mth component of $\mathbf{X}$ and $\psi_H(z)$ is the Huber influence function (see Ex. 12.12) [15].

Example 12.12. Recursive Parameter Estimation for a First-Order Autoregressive Process [8]. Consider a first-order autoregressive process:

$$y_k = \theta y_{k-1} + w_{k-1} \ .$$

We wish to recursively estimate θ given the sequence of measurements $y_{1:k}$ which may be contaminated by outliers. In this case, the corresponding process and measurement equations are:

$$\theta_k = \theta_{k-1} \ ,$$
$$y_k = y_{k-1}\theta_k + w_k \ ,$$

where $F = 1$, $H = y_{k-1}$, $v_k = 0$ and we use the "Good-and-Bad" model (Sect. 11.3.2) to represent the measurement noise w_k. If α is the expected relative number of outliers, then

$$w_k \sim (1-\alpha)\mathcal{N}(0,\sigma^2)+\alpha\mathcal{N}(0,\beta\sigma^2)\,,$$

where $\beta >> 1$. The *a posteriori* pdf is $p(\theta_k|y_{1:k},I) \sim \mathcal{N}(\mu_{k|k},\Sigma_{k|k})$, where

$$\mu_{k|k} = \mu_{k-1|k-1} + \frac{1}{\sigma}\Sigma_{k|k-1}y_{k-1}\psi_H\left(\frac{\sigma(y_k - y_{k-1}\mu_{k-1|k-1})}{\Sigma_{k|k-1}y_{k-1}^2+\sigma^2}\right)\,,$$

$$\Sigma_{k|k} = \frac{\Sigma_{k|k-1}\sigma^2}{\Sigma_{k|k-1}y_{k-1}^2+\sigma^2}\,,$$

$$\psi_H(z) = \begin{cases} z & \text{if } |z| \leq c\,,\\ cz/|z| & \text{otherwise}\,.\end{cases}$$

The recommended choice of c is the α-quantile of $\mathcal{N}(0,1)$ e. g. $c = 1.645$ for an assumed 5% contamination.

12.4.2 Extended Kalman Filter

In the extended Kalman filter (EKF) we apply the KF to non-linear process and measurement models by linearizing the models. We assume a non-linear Gaussian process model and a non-linear measurement model:

$$\boldsymbol{\theta}_k = \mathbf{f}_k(\boldsymbol{\theta}_{k-1}) + \mathbf{v}_{k-1}\,, \tag{12.38}$$

$$\mathbf{y}_k = \mathbf{h}_k(\boldsymbol{\theta}_k) + \mathbf{w}_k\,, \tag{12.39}$$

where $\mathbf{v}_{k-1} \sim \mathcal{N}(0,\mathbf{Q}_{k-1})$ and $\mathbf{w}_k \sim \mathcal{N}(0,\mathbf{R}_k)$ are independent Gaussian noise sources. We linearize (12.38) and (12.39) by performing the following Taylor series expansions:

$$\boldsymbol{\theta}_k \approx \mathbf{f}_k(\boldsymbol{\mu}_{k-1|k-1}) + \widetilde{\mathbf{F}}_k(\boldsymbol{\theta}_{k-1} - \boldsymbol{\mu}_{k-1|k-1}) + \mathbf{v}_{k-1}\,, \tag{12.40}$$

$$\mathbf{y}_k \approx \mathbf{h}_k(\boldsymbol{\mu}_{k|k-1}) + \widetilde{\mathbf{H}}_k(\boldsymbol{\theta}_k - \boldsymbol{\mu}_{k|k-1}) + \mathbf{w}_k\,, \tag{12.41}$$

where

$$\widetilde{\mathbf{F}}_k = \left.\frac{\partial \mathbf{f}_k(\boldsymbol{\theta})}{\partial \boldsymbol{\theta}}\right|_{\boldsymbol{\theta}=\boldsymbol{\mu}_{k-1|k-1}}\,, \tag{12.42}$$

$$\widetilde{\mathbf{H}}_k = \left.\frac{\partial \mathbf{h}_k(\boldsymbol{\theta})}{\partial \boldsymbol{\theta}}\right|_{\boldsymbol{\theta}=\boldsymbol{\mu}_{k|k-1}}\,. \tag{12.43}$$

By substituting the terms $\boldsymbol{\alpha}_k = \mathbf{f}_k(\boldsymbol{\mu}_{k-1|k-1}) - \widetilde{\mathbf{F}}_k\boldsymbol{\mu}_{k-1|k-1}$ and $\boldsymbol{\beta}_k = \mathbf{h}_k(\boldsymbol{\mu}_{k|k-1}) - \widetilde{\mathbf{H}}_k\boldsymbol{\mu}_{k|k-1}$ into (12.40) and (12.41) we obtain the following linear Gaussian models:

$$\boldsymbol{\theta}_k = \boldsymbol{\alpha}_k + \widetilde{\mathbf{F}}_k\boldsymbol{\theta}_{k-1} + \mathbf{v}_{k-1}\,, \tag{12.44}$$

$$\mathbf{y}_k = \boldsymbol{\beta}_k + \widetilde{\mathbf{H}}_k \boldsymbol{\theta}_k + \mathbf{w}_k \ , \tag{12.45}$$

which may be filtered using the standard KF.

Eqs. (12.46)-(12.52) are the corresponding EKF equations.

$$p(\boldsymbol{\theta}_k|\mathbf{y}_{1:k-1},I) \approx \mathcal{N}(\boldsymbol{\mu}_{k|k-1},\boldsymbol{\Sigma}_{k|k-1}) \ , \tag{12.46}$$
$$p(\boldsymbol{\theta}_k|\mathbf{y}_{1:k},I) \approx \mathcal{N}(\boldsymbol{\mu}_{k|k},\boldsymbol{\Sigma}_{k|k}) \ , \tag{12.47}$$

where

$$\boldsymbol{\mu}_{k|k-1} = \mathbf{f}_k(\boldsymbol{\mu}_{k-1|k-1}) \ , \tag{12.48}$$
$$\boldsymbol{\Sigma}_{k|k-1} = \mathbf{Q}_{k-1} + \widetilde{\mathbf{F}}_k \boldsymbol{\Sigma}_{k-1|k-1} \widetilde{\mathbf{F}}_k^T \ , \tag{12.49}$$
$$\boldsymbol{\mu}_{k|k} = \boldsymbol{\mu}_{k|k-1} + \mathbf{K}_k(\mathbf{y}_k - \mathbf{h}_k(\boldsymbol{\mu}_{k|k-1})) \ , \tag{12.50}$$
$$\boldsymbol{\Sigma}_{k|k} = \boldsymbol{\Sigma}_{k|k-1} - \mathbf{K}_k \widetilde{\mathbf{H}}_k \boldsymbol{\Sigma}_{k|k-1} \ , \tag{12.51}$$

and

$$\mathbf{K}_k = \boldsymbol{\Sigma}_{k|k-1} \widetilde{\mathbf{H}}_k^T (\widetilde{\mathbf{H}}_k \boldsymbol{\Sigma}_{k|k-1} \widetilde{\mathbf{H}}_k^T + \mathbf{R}_k)^{-1} \ . \tag{12.52}$$

The EKF is widely used since it enables us to use the KF framework to perform sequential Bayesian inference on non-linear Gaussian systems. Its main drawbacks are that it requires the *analytical* computation of $\widetilde{\mathbf{F}}_k$ and $\widetilde{\mathbf{H}}_k$ and it is only accurate to first-order.

12.4.3 Unscented Kalman Filter

In the unscented Kalman filter (UKF) we retain the Gaussian approximations in (12.46) and (12.47) but do not perform a Taylor series expansion on the process and measurement models $f_{k-1}(\boldsymbol{\theta}_{k-1})$ and $h_k(\boldsymbol{\theta}_k)$. Instead $\boldsymbol{\theta}_{k-1}$ and $\boldsymbol{\theta}_k$ are specified using a minimal set of carefully chosen sample points. These points (known as sigma-points) completely capture the true mean and covariance of $\boldsymbol{\theta}_{k-1}$ and $\boldsymbol{\theta}_k$ and when propagated, respectively, through the non-linear process and measurement models, they capture the *a posteriori* means and covariances to third order [7]. The unscented

[7] This is true for any non-linear transformation and is the basis of the *unscented transformation* (UT) [16]. The UT is a method for calculating the statistics of a random variable which undergoes a non-linear transformation. Consider propagating a M-dimensional random variable $\mathbf{x} = (x^{(1)}, x^{(2)}, \ldots, x^{(M)})^T$ (mean vector $\bar{\mathbf{x}}$ and covariance matrix $\mathbf{R}$) though a non-linear function $\mathbf{y} = g(\mathbf{x})$. Then the mean and covariance of $\mathbf{y}$ are:

$$\bar{\mathbf{y}} \approx \sum_{m=0}^{2M} \omega^{(m)} y^{(m)} \ ,$$
$$\mathbf{S} \approx \sum_{m=0}^{2M} \Omega^{(m)} (\mathbf{y}^{(m)} - \bar{\mathbf{y}})(\mathbf{y}^{(m)} - \bar{\mathbf{y}})^T \ ,$$

where $\mathbf{x}^{(m)}, m \in \{0, 1, \ldots, 2M\}$ are a set of deterministically selected sigma-points with weights $\omega^{(m)}$ and $\Omega^{(m)}$ and $\mathbf{y}^{(m)} = g(\mathbf{x}^{(m)})$.

Kalman filter is a straightforward extension of the unscented transformation (UT) to the recursive estimation in (12.38) and (12.39), where the state random variable is redefined as the concatenation of the original state $\boldsymbol{\theta}$ and the noise variables $\mathbf{v}$ and $\mathbf{w}$: $\phi_k = (\boldsymbol{\theta}_k^T, \mathbf{v}_k^T, \mathbf{w}_k^T)^T$. The UT sigma point selection is applied to the new augmented state ϕ_{k-1} to calculate the corresponding sigma points. The final UKF equations are:

$$\boldsymbol{\mu}_{k|k-1} = \sum_{m=1}^{2M} \omega_{k-1}^{(m)} \hat{\boldsymbol{\theta}}_k^{(m)} , \tag{12.53}$$

$$\boldsymbol{\Sigma}_{k|k-1} = \sum_{m=0}^{2M} \Omega_{k-1}^{(m)} (\hat{\boldsymbol{\theta}}_k^{(m)} - \boldsymbol{\mu}_{k|k-1})(\hat{\boldsymbol{\theta}}_k^{(m)} - \boldsymbol{\mu}_{k|k-1})^T , \tag{12.54}$$

$$\boldsymbol{\mu}_{k|k} = \boldsymbol{\mu}_{k|k-1} + \mathbf{K}_k(\hat{\mathbf{y}}_k - \bar{\mathbf{y}}_k^{(m)}) , \tag{12.55}$$

$$\boldsymbol{\Sigma}_{k|k} = \boldsymbol{\Sigma}_{k|k-1} + \mathbf{K}_k \mathbf{S}_k \mathbf{K}_k^T , \tag{12.56}$$

where

$$\hat{\boldsymbol{\theta}}_{k-1}^{(m)} = f_{k-1}(\boldsymbol{\theta}_{k-1}^{(m)}) + \mathbf{v}_{k-1} , \tag{12.57}$$

$$\hat{\mathbf{y}}_k^{(m)} = h_k(\boldsymbol{\theta}_{k-1}^{(m)}) + \mathbf{w}_k , \tag{12.58}$$

$$\bar{\mathbf{y}}_k = \sum_{m=0}^{2M} \omega_{k-1}^{(m)} \hat{\mathbf{y}}_k^{(m)} , \tag{12.59}$$

$$\mathbf{S}_k = \sum_{m=0}^{2M} \Omega_{k-1}^{(m)} (\hat{\mathbf{y}}_k^{(m)} - \bar{\mathbf{y}}_k)(\hat{\mathbf{y}}_k^{(m)} - \bar{\mathbf{y}}_k)^T , \tag{12.60}$$

and

$$\mathbf{K}_k = \sum_{m=0}^{2M} \Omega_{k-1}^{(m)} (\hat{\boldsymbol{\theta}}_k^{(m)} - \boldsymbol{\mu}_{k|k-1})(\hat{\mathbf{y}}_k^{(m)} - \bar{\mathbf{y}}_k)^T \mathbf{S}_k^{-1} . \tag{12.61}$$

The *sigma*-points $\boldsymbol{\theta}^{(m)}, m \in \{0,1,\ldots,2M\}$, and their weights, $\omega^{(m)}$ and $\Omega^{(m)}$, are selected in deterministic fashion (see Table 12.6).

Table 12.6 Sigma-Points and their Weights

Variable	Formula
Sigma-Points $\boldsymbol{\theta}^{(m)}$	$\boldsymbol{\theta}^{(0)} = \boldsymbol{\mu}$; $\boldsymbol{\theta}^{(m)} = \boldsymbol{\mu} + [\sqrt{(n+\lambda)\boldsymbol{\Sigma}}]_m$; $\boldsymbol{\theta}^{(m+M)} = \boldsymbol{\mu} - [\sqrt{(n+\lambda)\boldsymbol{\Sigma}}]_m, m \in \{1,2,\ldots,M\}$; where $\lambda = \alpha^2(M+K)$ and $[\sqrt{(n+\lambda)\boldsymbol{\Sigma}}]_m$ is the mth column of the matrix square root of $(n+\lambda)\boldsymbol{\Sigma}$.
Weights $\omega^{(m)}$	$\omega^{(0)} = \lambda/(M+\lambda), \omega^{(m)} = \lambda/(2(M+\lambda)), m \in \{1,2\ldots,2M\}$.
Weights $\Omega^{(m)}$	$\Omega^{(0)} = \lambda/(M+\lambda) + (1-\alpha^2+\beta), \Omega^{(m)} = \lambda/(2(M+\lambda))$, $m \in \{1,2,\ldots,2M\}$.

12.4.4 Switching Kalman Filter

In the switching Kalman filter [22, 25] we generalize (12.12) and (12.13) by assuming that $\mathbf{F}_k$, $\mathbf{H}_k$, $\mathbf{Q}_k$ and $\mathbf{R}_k$ are functions of an indicator, or switching, variable Λ_k: $\mathbf{F}_k = \mathbf{F}(\Lambda_k)$, $\mathbf{H}_k = \mathbf{H}(\Lambda_k)$, $\mathbf{Q}_k = \mathbf{Q}(\Lambda_k)$ and $\mathbf{R}_k = \mathbf{R}(\Lambda_k)$. In this case, (12.12) and (12.13) become

$$\boldsymbol{\theta}_k = \mathbf{F}(\Lambda_k)\boldsymbol{\theta}_{k-1} + \mathbf{v}_{k-1} , \tag{12.62}$$

$$\mathbf{y}_k = \mathbf{H}(\Lambda_k)\boldsymbol{\theta}_k + \mathbf{w}_k , \tag{12.63}$$

where $\mathbf{v}_{k-1} \sim \mathcal{N}(\mathbf{0}, \mathbf{Q}(\Lambda_{k-1}))$ and $\mathbf{w}_k \sim \mathcal{N}(\mathbf{0}, \mathbf{R}(\Lambda_k))$ are independent Gaussian noise sources.

We often suppose that Λ_k obeys a first-order Markov process, where

$$P(\Lambda_k = \lambda_n|I) = \sum_{m=1}^{M} P(\Lambda_k = \lambda_n|\Lambda_{k-1} = \lambda_m, I)P(\Lambda_{k-1} = \lambda_m|I) , \tag{12.64}$$

and Λ_k is limited to M distinct values $\{\lambda_1, \lambda_2, \ldots, \lambda_M\}$. In this case, at a given time step k, the *a posteriori* probability density $p(\boldsymbol{\theta}_k|\mathbf{y}_{1:k}, I)$ is an exponentially growing mixture of M^k Gaussian pdf's, each pdf corresponding to a different possible history of the switching variable [8].

In order to limit the exponential growth in the number of Gaussian pdf's we may proceed as follows. At each time step k we "collapse" the mixture of M^k Gaussians onto a mixture Mr Gaussians where $r < k$. This is called the *Generalized Pseudo Bayesian* (GPB) approximation of order r. The following example illustrates the action of the GPB approximation.

Example 12.13. Generalized Pseudo Bayesian Approximation. We consider the GPB approximation for $r = 1$. In this case, we approximate a mixture of Gaussians with a single Gaussian by matching the mean and covariance of the mixture to that of a single Gaussian, i. e.

$$\begin{aligned} p(\boldsymbol{\theta}) &= \sum_{m=1}^{M} w^{(m)} \mathcal{N}(\boldsymbol{\theta}|\boldsymbol{\mu}^{(m)}, \boldsymbol{\Sigma}^{(m)}) , \\ &\approx \mathcal{N}(\boldsymbol{\theta}|\bar{\mu}, \bar{\Sigma}) , \end{aligned}$$

where

[8] Extensions of the Kalman FilterAt time step $k = 1$ we do not know which KF generated the observation $\mathbf{y}_1$, so a Gaussian pdf for each KF must be maintained. This means we require a mixture of M Gaussian pdf's to describe $\boldsymbol{\theta}_{1|1}$. At time step $k = 2$, which KF generated the observation $\mathbf{y}_2$ is also unknown. Because the KF for time step $k = 1$ is also unknown, we require M^2 Gaussian pdf's to describe $\boldsymbol{\theta}_{2|2}$. We thus conclude that at time step k, we require M^k Gaussian pdf's to describe $\boldsymbol{\theta}_{k|k}$.

$$\bar{\boldsymbol{\mu}} = \sum_{m=1}^{M} w^{(m)} \boldsymbol{\mu}^{(m)} ,$$
$$\bar{\boldsymbol{\Sigma}} = \sum_{m=1}^{M} w^{(m)} \boldsymbol{\Sigma}^{(m)} ,$$

and

$$\boldsymbol{\Sigma}^{(m)} = (\boldsymbol{\mu}^{(m)} - \bar{\boldsymbol{\mu}})(\boldsymbol{\mu}^{(m)} - \bar{\boldsymbol{\mu}})^T .$$

The following example illustrates a switched KF using a GPB approximation of order one.

Example 12.14. The Lainiotis-Kalman Equations [31]. We assume the process and measurement equations:

$$\boldsymbol{\theta}_k = \mathbf{F}_{k-1}\boldsymbol{\theta}_k + \mathbf{v}_{k-1} ,$$
$$\mathbf{y}_k = \mathbf{H}_k(\Lambda_k)\boldsymbol{\theta}_k + \mathbf{w}_k ,$$

where the indicator Λ is limited to a finite set of values $\Lambda \in \{\lambda_1, \lambda_2, \ldots, \lambda_M\}$, and $\mathbf{w}_k \sim \mathcal{N}(0, \mathbf{R}_k(\Lambda))$. Then, we approximate the *a posteriori* pdf, $p(\boldsymbol{\theta}_k|\mathbf{y}_{1:k}, I)$, using a single Gaussian pdf,

$$p(\boldsymbol{\theta}_k|\mathbf{y}_{1:k}, I) \approx \mathcal{N}(\widetilde{\boldsymbol{\mu}}_{k|k}, \widetilde{\boldsymbol{\Sigma}}_{k|k}) ,$$

whose parameters, $\widetilde{\boldsymbol{\mu}}_{k|k}$ and $\widetilde{\boldsymbol{\Sigma}}_{k|k}$, are given by the *Lainiotis-Kalman* equations:

$$\widetilde{\boldsymbol{\mu}}_{k|k} = \sum_{m=1}^{M} p(\Lambda_k = \lambda_m|\mathbf{y}_{1:k}, I)\boldsymbol{\mu}_{k|k}(\Lambda_k = \lambda_m) ,$$
$$\widetilde{\boldsymbol{\Sigma}}_{k|k} = \sum_{m=1}^{M} p(\Lambda_k = \lambda_m|\mathbf{y}_{1:k}, I)[\boldsymbol{\Sigma}_{k|k}(\Lambda_k = \lambda_m)(\boldsymbol{\mu}_{k|k}(\Lambda_k = \lambda_m) - \widetilde{\boldsymbol{\mu}}_{k|k}(\Lambda_k = \lambda_m)) \times (\boldsymbol{\mu}_{k|k}(\Lambda_k = \lambda_m) - \widetilde{\boldsymbol{\mu}}_{k|k}(\Lambda_k = \lambda_m))^T] ,$$

where $\boldsymbol{\mu}_{k|k}^{(m)} \equiv \boldsymbol{\mu}_{k|k}(\Lambda_k = \lambda_m)$ and $\boldsymbol{\Sigma}_{k|k}^{(m)} \equiv \boldsymbol{\Sigma}_{k|k}(\Lambda_k = \lambda_m)$ are the $\boldsymbol{\mu}_{k|k}$ and $\boldsymbol{\Sigma}_{k|k}$ values as computed by the KF assuming $\Lambda_k = \lambda_m$ and

$$P(\Lambda = \lambda_m|\mathbf{y}_{1:k}, I) = \frac{p(\mathbf{y}_{1:k}|\Lambda = \lambda_m, I)P(\Lambda = \lambda_m|I)}{p(\mathbf{y}_{1:k}|I)} ,$$

is the *a posteriori* probability of $\Lambda_k = \lambda_m$ given $\mathbf{y}_{1:k}$.

12.5 Particle Filter *

The techniques discussed in Sect. 12.4 all relax some of the Gaussian linear assumptions which are present in the KF. However, in many practical applications these techniques are still not sufficiently accurate to model all of the aspects of the *a posteriori* pdf $p(\boldsymbol{\theta}_k|\mathbf{y}_{1:k},I)$. If this happens then we may use a different type of filter based on a Monte Carlo approximation. This filter is known as a *particle filter* filter [1].

The key idea underlying the particle filter is to represent the density $p(\boldsymbol{\theta}_k|\mathbf{y}_k,I)$ by a set of N weighted random samples, or particles, which constitutes a discrete approximation of the pdf. In this regard the particle filter is similar to the sigma filter. The key difference being that in the former we may use several thousand particles while in the sigma filter we use a very small number of deterministically selected particles.

12.6 Multi-sensor Multi-temporal Data Fusion

Until now we have considered multi-temporal data fusion in which use a recursive filter to fuse together a sequence of measurements $\mathbf{y}_{1:k}$ made using a single sensor S. We now extend this analysis and consider multi-sensor multi-temporal data fusion in which we fuse together several sequences of measurements $\mathbf{y}_{1:k}^{(m)}, m \in \{1,2,\ldots,M\}$, each sequence $\mathbf{y}_{1:k}^{(m)}$ being made with a different sensor $S^{(m)}$. We start with the simplest method of multi-sensor multi-temporal data fusion which is known as *measurement fusion*.

12.6.1 Measurement Fusion

In measurement fusion we place all the measurements $\mathbf{y}_k^{(m)}, m \in \{1,2,\ldots,M\}$, obtained at time step k into a single augmented measurement vector $\widetilde{\mathbf{y}}_k = ((\mathbf{y}_k^{(1)})^T, (\mathbf{y}_k^{(2)})^T, \ldots, (\mathbf{y}_k^{(M)})^T)^T$. We then fuse the $\widetilde{\mathbf{y}}_{1:k} = (\widetilde{\mathbf{y}}_1, \widetilde{\mathbf{y}}_2, \ldots, \widetilde{\mathbf{y}}_k)^T$ using a single recursive filter. In the following example we illustrate the technique using a Kalman filter.

Example 12.15. Multi-Sensor Multi-Temporal Data Fusion: Measurement Fusion [14]. We consider the tracking of a single target. Let $\boldsymbol{\theta}_k$ denote the filtered state of the target at time step k and $\mathbf{y}_k^{(m)}$ denotes the corresponding measured position of the target as recorded by two sensors $S^{(1)}$ and $S^{(2)}$. The corresponding process and measurement equations are:

$$\begin{aligned}
\boldsymbol{\theta}_k &= \mathbf{F}_k \boldsymbol{\theta}_{k-1} + \mathbf{v}_{k-1} , \\
\mathbf{y}_k^{(1)} &= \mathbf{H}_k^{(1)} \boldsymbol{\theta}_k + \mathbf{w}_k^{(1)} , \\
\mathbf{y}_k^{(2)} &= \mathbf{H}_k^{(2)} \boldsymbol{\theta}_k + \mathbf{w}_k^{(2)} ,
\end{aligned}$$

where $\mathbf{v}_{k-1} \sim \mathcal{N}(0, \mathbf{Q}_k)$, $\mathbf{w}_k^{(1)} \sim \mathcal{N}(0, \mathbf{R}_k^{(1)})$ and $\mathbf{w}_k^{(2)} \sim \mathcal{N}(0, \mathbf{R}_k^{(2)})$.

We estimate $p(\boldsymbol{\theta}_k | \mathbf{y}_{1:k}^{(1:2)}, I) \approx \mathcal{N}(\boldsymbol{\mu}_{k|k}, \boldsymbol{\Sigma}_{k|k})$ using the KF equations:

$$\begin{aligned}
\boldsymbol{\mu}_{k|k-1} &= \mathbf{F}_k \boldsymbol{\mu}_{k-1|k-1} , \\
\boldsymbol{\Sigma}_{k|k-1} &= \mathbf{Q}_{k-1} + \mathbf{F}_k \boldsymbol{\Sigma}_{k-1|k-1} \mathbf{F}_k^T , \\
\boldsymbol{\mu}_{k|k} &= \boldsymbol{\mu}_{k|k-1} + \widetilde{\mathbf{K}}_k (\mathbf{y}_k - \widetilde{\mathbf{H}}_k \boldsymbol{\mu}_{k|k-1}) , \\
\boldsymbol{\Sigma}_{k|k} &= \boldsymbol{\Sigma}_{k|k-1} - \widetilde{\mathbf{K}}_k \widetilde{\mathbf{H}}_k \boldsymbol{\Sigma}_{k|k-1} ,
\end{aligned}$$

where

$$\begin{aligned}
\widetilde{\mathbf{K}}_k &= \boldsymbol{\Sigma}_{k|k-1} \widetilde{\mathbf{H}}_k^T (\widetilde{\mathbf{H}}_k \boldsymbol{\Sigma}_{k|k-1} \widetilde{\mathbf{H}}_k^T + \widetilde{\mathbf{R}}_k)^{-1} , \\
\widetilde{\mathbf{H}}_k &= \begin{pmatrix} \mathbf{H}_k^{(1)} \\ \mathbf{H}_k^{(2)} \end{pmatrix} , \\
\widetilde{\mathbf{w}}_k &= \begin{pmatrix} \mathbf{w}_k^{(1)} \\ \mathbf{w}_k^{(2)} \end{pmatrix} \sim \mathcal{N}(\mathbf{0}, \widetilde{R}_k) ,
\end{aligned}$$

and

$$\widetilde{\mathbf{R}}_k = \begin{pmatrix} \mathbf{R}_k^{(1)} & \mathbf{0} \\ \mathbf{0} & \mathbf{H}_k^{(2)} \end{pmatrix} .$$

Fig. 12.6 is a graphical illustration of the process of measurement fusion for the two sensors $S^{(1)}$ and $S^{(2)}$.

Although theoretically optimum, measurement fusion requires a centralized architecture in which all the measurements $\mathbf{y}_k^{(m)}, m \in \{1, 2, \ldots, M\}$, are transmitted to a single fusion node. The difficulties and drawbacks associated with this architecture were described in Sect. 3.4.1. The most serious drawback associated with the centralized architecture is its sensitivity to errors involved in forming a common representational format, i. e. to the errors in the spatial and temporal alignment of the sensors $S^{(m)}$. An alternative method of multi-sensor multi-temporal data fusion is to separately filter the measurements received from each sensor and then fuse the results together. This is known as track-to-track fusion and is considered in the next section. Its main advantage is that it is relatively insensitive to the errors in the spatial and temporal alignment of the sensors.

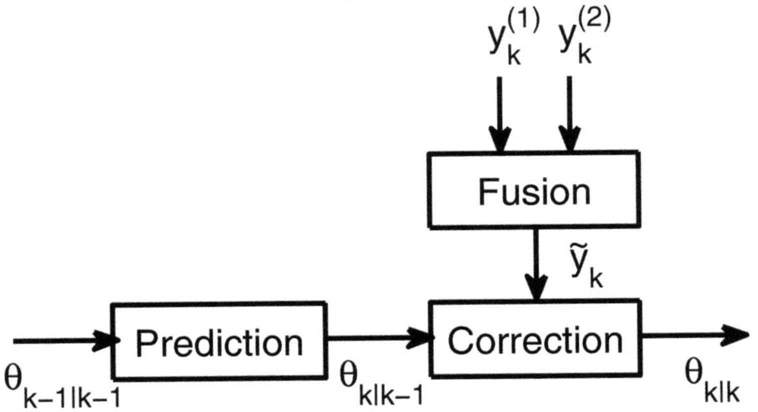

Fig. 12.6 Illustrates the measurement fusion process for two measurement sequencies $\mathbf{y}_{1:k}^{(1)}$ and $\mathbf{y}_{1:k}^{(2)}$. The individual measurements are placed in an augmented measurement sequence $\tilde{\mathbf{y}}_{1:k} = (\tilde{\mathbf{y}}_1, \tilde{\mathbf{y}}_2, \ldots, \tilde{\mathbf{y}}_k)$, where $\tilde{\mathbf{y}}_l = ((\mathbf{y}_l^{(1)})^T, (\mathbf{y}_l^{(2)})^T)^T$. The augmented measurement vector is then fused using a single KF.

12.6.2 Track-to-Track Fusion

In track-to-track fusion we first perform multi-temporal fusion using M recursive filters, in which the mth filter fuses the measurements $\mathbf{y}_{1:k}^{(m)} = \{y_1^{(m)}, y_2^{(m)}, \ldots, y_k^{(m)}\}$ made by the mth sensor $S^{(m)}$. Let $p(\boldsymbol{\theta}_k^{(m)}|\mathbf{y}_{1:k}^{(m)}, I)$ be the corresponding *a posteriori* pdf. Then, at each time step k, we fuse together the M *a posteriori* pdf's $p(\boldsymbol{\theta}_k^{(m)}|\mathbf{y}_{1:k}^{(m)}, I), m \in \{1, 2, \ldots, M\}$, using a linear least squares (minimum mean square error) estimate (see Sect. 10.5). Track-to-track fusion uses a hierarchical architecture (Sect. 3.4.3) containing M local fusion nodes and one central fusion node (Fig. 12.7). In general track-to-track fusion has a lower performance [7, 14] than the corresponding measurement fusion algorthm (Sect. 12.6.1). The reason for the reduction in performance is that the track-to-track fusion is performed using a *linear* estimate.

The following example illustrates track-to-track fusion for two sensors $S^{(1)}$ and $S^{(2)}$ using the KF [2].

Example 12.16. Multi-Sensor Multi-Temporal Data Fusion: Track-to-Track Fusion [2, 31]. We re-analyze Ex. 12.15 using the Bar-Shalom-Campo track-to-track fusion algorithm of [2]. If $\boldsymbol{\theta}_k$ denotes the filtered state of the target at time step k and $\mathbf{y}_k^{(m)}$ denotes the measured position of the target at time step k as recorded by the sensor $S^{(m)}$, then the corresponding process and measurement model equations are:

$$\boldsymbol{\theta}_k = \mathbf{F}_k \boldsymbol{\theta}_{k-1} + \mathbf{v}_{k-1} ,$$

$$\mathbf{y}_k^{(m)} = \mathbf{H}_k^{(m)}\boldsymbol{\theta}_k + \mathbf{w}_k^{(m)} ,$$

where $\mathbf{v}_{k-1} \sim \mathcal{N}(0, \mathbf{Q}_k)$ and $\mathbf{w}_k^{(m)} \sim \mathcal{N}(0, \mathbf{R}_k^{(m)})$.

Let $\boldsymbol{\theta}_k^{(m)} \sim \mathcal{N}(\boldsymbol{\mu}_{k|k}^{(m)}, \boldsymbol{\Sigma}_{k|k}^{(m)})$ denote the KF estimate of $\boldsymbol{\theta}_k$ given the sensor measurements $\mathbf{y}_{1:k}^{(m)}$, where $\boldsymbol{\mu}_{k|k}^{(m)}$ and $\boldsymbol{\Sigma}_{k|k}^{(m)}$ are calculated using the KF equations (12.14-12.20). For $M = 2$ we fuse together the two (correlated) estimates $\boldsymbol{\mu}_{k|k}^{(m)}, m \in \{1,2\}$, using the Bar-Shalom-Campo formula:

$$\begin{aligned}\widetilde{\boldsymbol{\theta}}_{k|k} = {} & \boldsymbol{\mu}_{k|k}^{(1)}(\boldsymbol{\Sigma}_{k|k}^{(22)} - \boldsymbol{\Sigma}_{k|k}^{(21)})(\boldsymbol{\Sigma}_{k|k}^{(11)} + \boldsymbol{\Sigma}_{k|k}^{(22)} - \boldsymbol{\Sigma}_{k|k}^{(12)} - \boldsymbol{\Sigma}_{k|k}^{(21)})^{-1} \\ & + \boldsymbol{\mu}_{k|k}^{(2)}(\boldsymbol{\Sigma}_{k|k}^{(11)} - \boldsymbol{\Sigma}_{k|k}^{(12)})(\boldsymbol{\Sigma}_{k|k}^{(11)} + \boldsymbol{\Sigma}_{k|k}^{(22)} - \boldsymbol{\Sigma}_{k|k}^{(12)} - \boldsymbol{\Sigma}_{k|k}^{(21)})^{-1}\boldsymbol{\mu}_{k|k}^{(2)} ,\end{aligned}$$

where $\widetilde{\boldsymbol{\mu}}_{k|k}$ is the optimum linear estimate, $\boldsymbol{\Sigma}_{k|k}^{(mm)} \equiv \boldsymbol{\Sigma}_{k|k}^{(m)}$ is the covariance matrix of $\mathbf{y}_k^{(m)}$ and $\boldsymbol{\Sigma}_{k|k}^{(mn)}$ is the cross-covariance matrix between $\mathbf{y}_k^{(m)}$ and $\mathbf{y}_k^{(n)}$.

Fig. 12.7 shows the track-to-track fusion of observations received from two sensors $S^{(1)}$ and $S^{(2)}$ (see also Ex. 3.13).

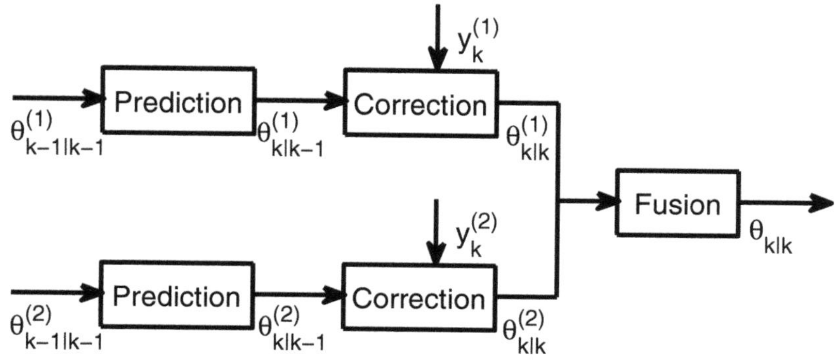

Fig. 12.7 Illustrates the track-to-track fusion process for the measurements $\mathbf{y}_{1:k}^{(1)}$ and $\mathbf{y}_{1:k}^{(2)}$ obtained from two sensors $S^{(1)}$ and $S^{(2)}$. Each sequence of measurements are separately fused together using a KF. At each time step k the outputs $N(\boldsymbol{\mu}_{k|k}^{(1)}, \boldsymbol{\Sigma}_{k|k}^{(1)})$ and $N(\boldsymbol{\mu}_{k|k}^{(2)}, \boldsymbol{\Sigma}_{k|k}^{(2)})$ are then fused together using the Bar-Shalom-Campo formula.

Recently the Bar-Shalom-Campo equations have been extended to the case when $M > 2$ [9]. In the following example we give the equations for track-to-track fusion using $M > 2$ sensors.

[9] See also Sect. 10.7.

Example 12.17. Track-to-Track Fusion using the Generalized Millman Equations [31]. Recently the Bar-Shalom-Campo equations have been extended to more than two sensors. We reconsider Ex. 12.16 except this time we allow $M > 2$. In this case, we contruct the global (subpotimal) estimate $\widetilde{\boldsymbol{\mu}}_{k|k}$ from the local KF estimates $\boldsymbol{\mu}_{k|k}^{(m)}, m \in \{1,2,\ldots,M\}$, using the generalized Millman formula (GMF):

$$\widetilde{\boldsymbol{\mu}}_{k|k} = \sum_{m=1}^{M} \mathbf{C}_k^{(m)} \boldsymbol{\mu}_{k|k}^{(m)} ,$$

where the time-varying matrices $\mathbf{C}_k^{(m)}, m \in \{1,2,\ldots,M\}$, are given by

$$\sum_{m=1}^{M-1} \mathbf{C}_k^{(m)} (\mathbf{S}_{k|k}^{(m,1)} - \mathbf{S}_{k|k}^{(m,M)}) + \mathbf{C}_k^{(M)} (\mathbf{S}_{k|k}^{(M,1)} - \boldsymbol{\Sigma}_{k|k}^{(M,M)}) = 0 ,$$

$$\sum_{m=1}^{M-1} \mathbf{C}_k^{(m)} (\mathbf{S}_{k|k}^{(m,2)} - \mathbf{S}_{k|k}^{(m,M)}) + \mathbf{C}_k^{(M)} (\mathbf{S}_{k|k}^{(M,2)} - \boldsymbol{\Sigma}_{k|k}^{(M,M)}) = 0 ,$$

$$\vdots$$

$$\sum_{m=1}^{M-1} \mathbf{C}_k^{(m)} (\mathbf{S}_{k|k}^{(m,M-1)} - \mathbf{S}_{k|k}^{(m,M)}) + \mathbf{C}_k^{(M)} (\mathbf{S}_{k|k}^{(M,M-1)} - \mathbf{S}_{k|k}^{(M,M)}) = 0 ,$$

where

$$\sum_{m=1}^{M} \mathbf{C}_k^{(m)} = \mathbf{I} ,$$

and

$$\mathbf{S}_{k|k}^{(m,n)} = \begin{cases} (\mathbf{I} - \mathbf{K}_k^{(m)} \mathbf{H}_k^{(m)}) \boldsymbol{\Sigma}_{k|k-1}^{(m)} & \text{if } n = m , \\ (\mathbf{I} - \mathbf{K}_k^{(m)} \mathbf{H}_k^{(m)}) (\mathbf{F}_k \mathbf{S}_{k-1|k-1}^{(m,n)} \mathbf{F}_k^T + \mathbf{Q}_k) & \\ \quad \times (\mathbf{I} - \mathbf{K}_k^{(n)} \mathbf{H}_k^{(n)})^T & \text{otherwise} . \end{cases}$$

Track-to-track fusion is widely used and many different variants of the basic scheme have been described. The following example illustrates one such variant.

Example 12.18. Modified track-to-track fusion algorithm [14]. In the modified track-to-track fusion algorithm of Gao and Harris [14] we calculate a single predicted pdf $p(\boldsymbol{\theta}_k|\mathbf{y}_{1:k-1}^{(1:M)},I)$ using the *fused* pdf $p(\boldsymbol{\theta}_k|\mathbf{y}_{1:k}^{(1:M)},I)$. Fig. 12.8 shows a schematic implementation of the Gao-Harris track-to-track fusion algorithm.

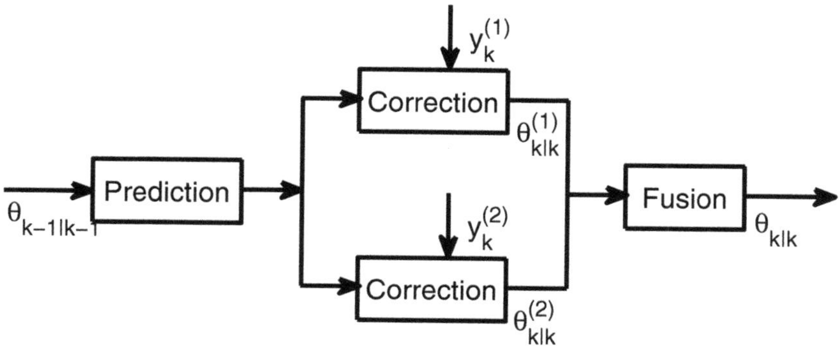

Fig. 12.8 Illustrates the modified track-to-track fusion process for the measurements $\mathbf{y}_{1:k}^{(1)}$ and $\mathbf{y}_{1:k}^{(2)}$ obtained from two sensors $S^{(1)}$ and $S^{(2)}$. As in track-to-track fusion each sequence of measurements are separately fused together using a KF. However, the KF's use a *common* predictor which is based on the fused output $N(\boldsymbol{\mu}_{k-1|k-1},\boldsymbol{\Sigma}_{k-1|k-1})$ obtained at the previous time step.

12.7 Software

COPAP. This is a matlab toolbox for cyclic order-preserving assignment problem. Authors: C. Scott and R. Nowak.

KFT. This is a matlab toolbox for simulating the Kalman filter. Author: Kevin Patrick Murphy.

UPF. This is a matlab toolbox for the unscented particle filter. Author: Nando de Freitas.

12.8 Further Reading

Chen [6] has provided a very readable, comprehensive survey of Bayesian filtering. Additional modern references on Bayesian filters include [1, 21]. References on specific topics in sequential Bayesian inference include: [5] (Kalman filters), [27] (unscented Kalman and particle filters), [1] (particle filters), [3, 4, 28] (multi-target tracking). Ref. [10] is an up-to-date survey of multisensor fusion in target tracking. A useful taxonomy on multi-target tracking is [30].

Table 12.7 Kalman Filter and its Variants

Filter	Description
Kalman Filter (KF)	Assumes linear Gaussian process and measurement models.
Robust Kalman Filter	Assumes linear Gaussian process model and a linear Gaussian measurement model which is contaminated with a small fraction of outliers. Robust estimation techniques (Chapt. 10) are used to reduce influence of outliers [8].
Kriged Kalman Filter	KF is combined with spatial Kriging model [23].
Augmented Kalman Filter	KF in which we use parameterized linear Gaussian process and measurement models. We augment the state vector with the process and measurement model parameters $\alpha,\beta,\ldots,\gamma$. The AKF maybe used to include spatial and temporal misalignments between the sensors.
Extended Kalman Filter (EKF)	First-order local linearization of non-linear Gaussian process and measurement models. See Sect. 12.4.2.
Unscented Kalman Filter (UKF)	UKF uses a deterministic set of weighted points $(\omega_k^{(m)},\Omega_k^{(m)},\boldsymbol{\theta}_k^{(m)}), m \in \{1,2,\ldots,M\}$, to represent the *a priori* pdf $p(\boldsymbol{\theta}_{k-1}\|\mathbf{y}_{1:k-1},I)$. See Sect. 12.4.3.
Switching Kalman Filter	Contains multiple KF's but at each time step k we use only one KF. The selected filter is specified by a switching variable Λ_k. See Sect. 12.4.4.
Mixture Kalman Filter	Similar to the switched KF except that at each time step k we use a collection of multiple KF's.
Ensemble Kalman Filter (EnKF)	Used when the dimensions of $\boldsymbol{\theta}_k$ is too large to permit the conventional calculation of the covariance matrix. In the EnKF we estimate the covariance matrices using an ensemble of trajectories.

Problems

12.1. Define sequential Bayesian inference. Differentiate between smoothing, filtering and forecasting.

12.2. The Kalman filter assumes a linear Gaussian process and measurement model. Explain what these models are.

12.3. Describe the Kalman filter equations.

12.4. Describe some maneuver models.

12.5. Compare and contrast the Robust, Kriged, Augmented, Extended, Unscented and Switching Kalman filters.

12.6. Compare and contrast measurement and track-to-track fusion.

12.7. Define the data association problem. Compare and contrast the following data association algorithms: nearest neighbour, strongest neighbour, covariance union and probabilistic data association.

References

1. Arulampalam, M.S., Maskell, S., Gordon, N., Clapp, T.: A tutorial on particle filters for online nonlinear/non-Gaussian Bayesian tracking. IEEE Trans. Sig. Process. 50, 174–188 (2002)
2. Bar-Shalom, Y., Campo, L.: The effect of the common process noise on the two sensor fused track covariance. IEEE Trans. Aero. Elect. Sys. 22, 803–805 (1986)
3. Bar-Shalom, Y., Li, X.: Multitarget-Multisensor Tracking: Principles and Techniques. YBS Publishing, Storrs (1995)
4. Blackman, S.S., Popoli, R.F.: Design and analysis of modern tracking Systems. Artech House (1999)
5. Brown, R.G., Hwang, P.Y.C.: Introduction to random signal analysis and Kalman filtering. John Wiley and Sons (1997)
6. Chen, Z.: Bayesian filtering: from Kalman filters to particle filters, and beyond. Adaptive Systems Laboratory, Communications Research Laboratory. McMaster University,Canada (2003)
7. Chang, K., Saha, R., Bar-Shalom, Y.: On optimal track-to-track fusion. IEEE Trans. Aero. Elect. Sys. 33, 1271–1276 (1997)
8. Cipra, T., Romera, R.: Kalman filter with outliers and missing observations. Test 6, 379–395 (1997)
9. Cox, I.J.: A review of statistical data association techniques for motion correspondence. Int. J. Comp. Vis. 10, 53–66 (1993)
10. Duncan, S., Singh, S.: Approaches to multisensor data fusion in target tracking: a survey. IEEE Trans. Know Data Engng. 18, 1696–1710 (2006)
11. Eshel, R., Moses, Y.: Tracking people in a crowded scene. Int. J. Comp. Vis. 88, 129–143 (2011)
12. Flament, M., Fleury, G., Davoust, M.-E.: Particle filter and Gaussian-mixture filter efficiency evaluation for terrain-aided navigation. In: 12th Euro. Sig. Proc. Conf. EUSIPCO 2004, Vienna, Austria (2004)
13. Galanis, G., Anadranistakis, M.: A one-dimensional Kalman filter vfor the correction of near surface temperature forecasts. Meteorological Appl. 9, 437–441 (2002)
14. Gao, J.B., Harris, C.J.: Some remarks on Kalman filters for the multisensor fusion. Inf. Fusion 3, 191–201 (2002)
15. Huber, P.J.: Robust Statistics. John Wiley and Sons (1981)
16. Julier, S., Uhlmann, J.: General decentralized data fusion with covariance intersection (CI). In: Hall, D., Llians, J. (eds.) Handbook of Multidimensional Data Fusion, ch. 12. CRC Press (2001)
17. Julier, S.J., Uhlmann, J.K., Nicholson, D.N.: A method for dealing with assignment ambiguity. In: Proc. 2004 Am. Control Conf., Boston, Mass (2004)
18. Li, X.R., Jilkov, V.P.: A survey of maneuvering target tracking. Part I: Dynamic models. IIEEE Trans. Aero. Elec. Syst. 39, 1333–1363 (2003)
19. Li, X., Zhi, X.: PSNR: a refined strongest neighbor filter for tracking in clutter. In: Proc. 35th IEEE Conf. Dec. Control. (1996)
20. Lian, G., Lai, J., Zheng, W.-S.: Spatial-temporal consistent labeling of tracked pedestrians across non-overlapping camera views. Patt. Recogn. 44, 1121–1136 (2011)
21. Maskell, S.: Sequentially structured Bayesian solutions. PhD thesis. Cambridge University (2004)
22. Manfredi, V., Mahadevan, S., Kurose, J.: Switching Kalman filters for prediction and tracking in an adaptive meteorological sensing network. In: 2nd Annual IEEE Commun. Soc. Conf. Sensor and Ad Hoc Commun. and Networks, SECON 2005, Santa Clara, California, USA (2005)
23. Mardia, K.V., Goodall, C., Redfern, E.J., Alonso, F.J.: The Kriged Kalman filter. Test 7, 217–284 (1998)
24. Meinhold, R.J., Singpurwalla, N.D.: Understanding the Kalman filter. Am. Stat. 37, 123–127 (1983)

25. Murphy, K.P.: Switching Kalman Filters. Unpublished Rept. Available from homepage of K. P. Murphy (1998)
26. Murphy, K.P.: Dynamic Bayesian networks: representation, inference and learning, PhD thesis. University of California, Berkeley, USA (2002)
27. van der Merwe, R.: Sigma-point Kalman filters for probabilistic inference in dynamic state-space models. PhD thesis. Oregon Health and Science University (2004)
28. Moore, J.R., Blair, W.D.: Practical aspects of multisensor tracking. In: Bar-Shalom, Y., Blair, W.D. (eds.) Multitarget-Multisensor Tracking: Applications and Advances, vol. III, pp. 1–78. Artech House, USA (2000)
29. Moore, M., Wang, J.: An extended dynamic model for kinematic positioning. J. Navig. 56, 79–88 (2003)
30. Pulford, G.W.: Taxonomy of multiple target tracking methods. IEEE Proc-Radar, Sonar, Navig. 152, 291–304 (2005)
31. Shin, V., Lee, Y., Choi, T.-S.: Generalized Millman's formula and its application for estimation problems. Sig. Proc. 86, 257–266 (2006)
32. Wang, C., Kim, J.-H., Byun, K.-Y., Ni, J., Ko, S.-J.: Robust digital image stabilization using the Kalman filter. IEEE Trans. Consum. Elec. 55, 6–14 (2009)

Chapter 13
Bayesian Decision Theory

13.1 Introduction

The subject of this chapter, and the one that follows it, is *Bayesian decision theory* and its use in multi-sensor data fusion. To make our discussion concrete we shall concentrate on the pattern recognition problem [19] in which an unknown pattern, or object, O is to be assigned to one of K possible classes $\{c_1, c_2, \ldots, c_K\}$. In this chapter we shall limit ourselves to using a single logical sensor, or classifier, S. We shall then remove this restriction in Chapt. 14 where we consider multiple classifier systems.

In many applications Bayesian decision theory represents the primary fusion algorithm in a multi-sensor data fusion system. In Table 13.1

we list some of these applications together with their Dasararthy classification.

13.2 Pattern Recognition

Formally, the pattern recognition problem deals with the optimal assignment of an object O to one of K possible classes, $\{c_1, c_1, \ldots, c_K\}$. We suppose O is characterized by L measurements, or features, $y^{(l)}, l \in \{1, 2, \ldots, L\}$. Then, mathematically, the classifier defines a mapping between a class variable C and the feature vector $\mathbf{y} = (y^{(1)}, y^{(2)}, \ldots, y^{(L)})^T$:

$$\mathbf{y} \rightarrow C \in \{c_1, c_2, \ldots, c_K\} . \tag{13.1}$$

If we wish to minimize the number of classification errors, we use a zero-one loss function (see Sect. 10.2) and assign O to the class with the largest *a posteriori* probability:

$$C = c_{MAP} \qquad \text{if } p(C = c_{MAP}|\mathbf{y}, I) = \max_k(p(C = c_k|\mathbf{y}, I)) , \tag{13.2}$$

Table 13.1 Applications in which Bayesian Decision Theory is the Primary Fusion Algorithm

Class		Description
FeI-DeO	Ex. 1.13	Automatic Speech Recognition: Preventing Catastrophic Fusion.
	Ex. 3.5	Tire Pressure Monitoring Device.
	Ex. 3.6	Multi-Modal Biometric Identification Scheme.
	Ex. 6.4	Speech-Recognition in an Embedded-Speech Recognition System.
	Ex. 6.5	Word-Spotting in Historical Manuscripts.
	Ex. 13.3	Plug-In MAP Bayes' Classifier vs. Naive Bayes' Classifier.
	Ex. 13.4	Naive Bayes' Classifier: Correlated Features.
	Ex. 13.8	Gene Selection: An application of feature selection in DNA microarray experiments.
	Ex. 13.9	mRMR: A Filter Feature Selection Algorithm.
	Ex. 13.10	Wrapper Feature Selection.
	Ex. 13.12	Adjusted Probability Model.
	Ex. 13.13	Homologous Naive Bayes' Classifier.
	Ex. 13.7	Selective Neighbourhood Naive Bayes' Classifier.
DeI-DeO	Ex. 13.14	Naive Bayes' Classifier Committee (NBCC).
	Ex. 13.15	Multi-Class Naive Bayes' Classifier.
	Ex. 14.1	Bayesian Model Averaging of the Naive Bayes' Classifier
	Ex. 14.17	Boosting a Classifier.

The designations FeI-DeO and DeI-DeO refer, respectively, to the Dasarathy input/output classifications: "Feature Input-Decision Output" and "Decision Input-Decision Output" (Sect. 1.3.3).

where I denotes all relevant background information we have regarding O and its measurement vector $\mathbf{y}$.

Example 13.1. Confusion Matrix. A useful tool for recording the performance of a given classifier is the *confusion matrix* Ω. This is a $K \times K$ matrix in which the (l,k)th element $\Omega^{(l,k)}$ records the *a posteriori* probability $P(C_{\text{TRUE}} = c_l | C = c_k)$ where C_{TRUE} and C are, respectively, the true and predicted class variables.

In practice the elements $\Omega^{(l,k)}$ are approximated as follows. Let D denote a training set of N objects $O_i, i \in \{1,2,\ldots,N\}$, then

$$\Omega^{(l,k)} = \frac{n(l,k)}{\sum_{l=1}^{K} n(l,k)} ,$$

where $n(l,k)$ is the number of objects O_i which belong to the class $C_{\text{TRUE}} = c_l$ but which are assigned by the classifier, S, to the class $C = c_k$.

Equation (13.2) is known as the maximum *a posteriori* (MAP) rule. The computation of the *a posteriori* pdf, $p(C = c_{\text{MAP}}|\mathbf{y},I)$, is performed by applying Bayes' Theorem:

$$c_{\text{MAP}} = \arg\max_k \big(p(C = c_k|\mathbf{y},I)\big) ,$$
$$= \arg\max_k \left(\frac{\pi(C = c_k|I)p(\mathbf{y}|C = c_k,I)}{p(\mathbf{y}|I)}\right) , \tag{13.3}$$

where $p(\mathbf{y}|I)$ serves as a normalization constant. This constant can be ignored for classification purposes, giving,

$$c_{\text{MAP}} = \arg\max_k \big(\pi(C = c_k|I)p(\mathbf{y}|C = c_k,I)\big) . \tag{13.4}$$

For *known* probability densities the MAP rule is the best any classifier can do with a zero-one loss function. However, in practice the pdf's in (13.4) are unknown. In this case, we use the following empirical, or *plug-in*, MAP rule:

$$c_{\text{PLUG-IN}} = \arg\max_k \big(\hat{\pi}(C = c_k|I)\hat{p}(\mathbf{y}|C = c_k,I)\big) , \tag{13.5}$$

where $\hat{\pi}(C = c_k|I)$ and $\hat{p}(\mathbf{y}|C = c_k,I)$ denote *estimated* pdf's [45, 46].

Direct calculation of the *a priori* probability $\hat{\pi}(C = c_k|I)$ is straightforward: Let D denote a training set containing N objects $O_i, i \in \{1,2,\ldots,N\}$, of which N_k belong to the class $C = c_k$. If D is a representative sample of the underlying true population, then an unbiased maximum likelihood estimate of the true *a priori* probability is

$$\hat{\pi}(C = c_k|I) = N_k/N . \tag{13.6}$$

However, in many applications, D is not a random sample of the true population. In this case, and in the absence of any specific information, we often simply let $\hat{\pi}(C = c_k|I) = 1/K$.

Direct calculation of the likelihood function, $\hat{p}(\mathbf{y}|C = c_k,I)$, is more difficult, especially for high-dimensional data [41]. The main difficulty is that the required size of D increases *exponentially* with L [38]. This is known as the *curse of dimensionality* [7] and it affects all aspects of classification theory. For example, the curse of dimensionality implies that for a given number of objects O_i in D, there is a maximum number of features above which the performance of Bayes' classifier will degrade rather than improve. This is illustrated in the following example.

Example 13.2. The Small Sample Problem [47]. We consider a two-class classification problem with equal *a priori* probabilities. Each class is characterized by an L-dimensional Gaussian distribution: $\mathcal{N}(\mathbf{y}|\boldsymbol{\mu},\boldsymbol{\Sigma})$ and $\mathcal{N}(\mathbf{y}|-\boldsymbol{\mu},\boldsymbol{\Sigma})$, where

$$\boldsymbol{\mu} = \left(1, 1/\sqrt{2}, 1/\sqrt{3}, \ldots, 1/\sqrt{L}\right)^T .$$

and

$$\Sigma = \text{diag}(\overbrace{1, 1, \ldots, 1}^{L}) .$$

The features are statistically independent and the discriminating power of the successive features decrease monotonically with the first feature providing the maximum discrimination between the two classes. If μ is *known* then the probability of a misclassification error $P_\varepsilon(L)$ decreases with L as shown in Fig. 13.1. However, if μ is *unknown*, then we estimate μ and Σ from a training set D containing N labelled objects $O_i, i \in \{1,2,\ldots,N\}$, with instantiated measurements $\mathbf{y}_i$. In this case, the probability of misclassification, $P_\varepsilon(N,L)$, is a function of both L and N: For a given value of N, $P_\varepsilon(N,L)$ initially decreases with L but eventually it increases and takes on the value of $\frac{1}{2}$.

An explanation of this phenomenom is as follows. For a fixed value of N, the reliability of the likelihood functions decrease with an increase in L, and as a consequence, the performance of the classifiers also fall with an increase in L.

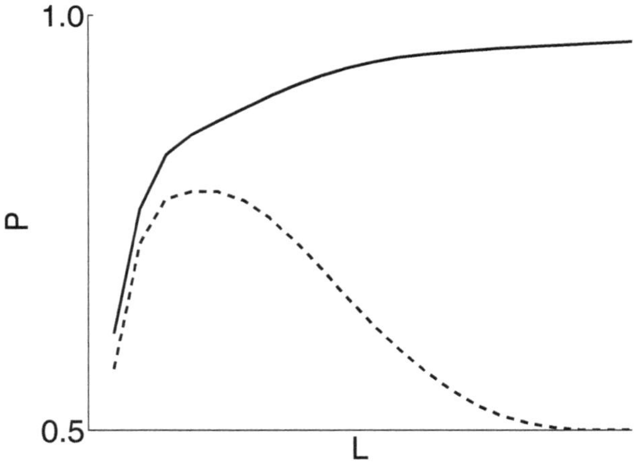

Fig. 13.1 Shows the probability of correct classification P (where P is defined as $1 - P_\varepsilon$) as a function of the number of features L. The solid curve shows P calculated assuming μ is known. The curve increases montonically to one as the number of features increases. The dashed curve shows P calculated assuming μ is estimated from a training set of N labelled objects. This curve initially increases with L but eventually it decreases and takes on the value of $\frac{1}{2}$.

From this perspective,the main effort in Bayesian classification theory may be understood as an attempt to make accurate estimates of the likelihoods, $\hat{p}(\mathbf{y}|C = c_k, I)$, while using a small number of labeled objects O_i. One way of doing this is to impose a given structure on $\hat{p}(\mathbf{y}|C=c_k,I)$. This idea forms the basis of the naive Bayes' classifier which is probably the most widely used classifier used today.

Note. To keep notation simple we shall not, in general, differentiate between the estimated pdf $\hat{p}(\mathbf{y}|C=c_k,I)$ and the true pdf $p(\mathbf{y}|C=c_k,I)$. The reader should observe, however, that in most practical cases, the pdf's are estimated using a maximum likelihood or maximum *a posteriori* estimator.

13.3 Naive Bayes' Classifier

In the naive Bayes' classifier (NBC) we reduce the impact of the curse of dimensionality, by supposing we may write $p(\mathbf{y}|C=c_k,I)$ as a product of L one-dimensional likelihood functions $p(y^{(l)}|C=c_k,I)$:

$$p(\mathbf{y}|C=c_k,I)=\prod_{l=1}^{L}p(y^{(l)}|C=c_k,I)\ . \tag{13.7}$$

In this case, the required number of objects O_i in D increases only *linearly* with L.

Example 13.3. Plug-in MAP Classifier vs. Naive Bayes' Classifier. Consider a classifier with L features $y^{(1)},y^{(2)},\ldots,y^{(L)}$, whose instantiated values are limited to a discrete set of M_l values $\alpha_m^{(l)}, m\in\{1,2,\ldots,M_l\}$. If we assume a minimum of one object O_i for each combination of feature values, then the plug-in MAP classifier requires $\prod_{l=1}^{L}M_l$ objects, while the NBC only requires $\sum_{l=1}^{L}M_l$ objects.

Equation (13.7) is clearly not the only approximation which can be used and in some applications the following approximations [23] are used instead:

$$p(\mathbf{y}|C=c_k,I)=\begin{cases}\min_l\big(p(y^{(l)}|C=c_k,I)\big)\ ,\\ \frac{1}{L}\sum p(y^{(l)}|C=c_k,I)\ ,\\ \mathrm{median}_l\big(p(y^{(l)}|C=c_k,I)\big)\ ,\\ \max_l\big(p(y^{(l)}|C=c_k,I)\big)\ .\end{cases} \tag{13.8}$$

However, in general, (13.7) is the preferred approximation [1].

[1] Equation 13.7 may be "justified" as follows: Let $\{\pi(C=c_k|I),p(y^{(1)}|C=c_k,I),\ldots,p(y^{(L)}|C=c_k,I)\}$ denote the set of "low-order" estimates used by the NBC. Then the product approximation contains the smallest amount of information (i. e. the maximum entropy) of all possible approximations to $p(\mathbf{y}|C=c_k,I)$ that use the same set of low-order estimates [49].

13.3.1 Representation

The NBC is a highly restricted Bayesian network in which all of the features $y^{(l)}, l \in \{1,2,\ldots,L\}$, have one and the same independent class variable C. When depicted graphically, a NBC has the form shown in Fig. (13.2(a)), in which all arcs are directed from the class variable C to the features $y^{(l)}, l \in \{1,2,\ldots,L\}$. If $pa(y^{(l)})$ represents the *parents* of a given feature $y^{(l)}$, then by definition the NBC has $pa(C) = 0$ and $pa(y^{(l)}) = C$.

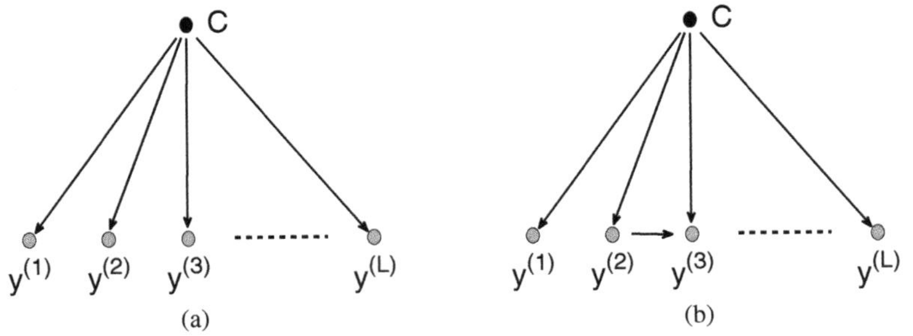

Fig. 13.2 (**a**)Shows a NBC. It contains one class variable C and L features $y^{(l)}, l \in \{1,2,\ldots,L\}$. The NBC has $pa(C) = 0$ and $pa(y^{(l)}) = C$. (**b**) Shows a Tree Augmented Naive (TAN) Bayes' classifier containing one class variable C and four features $y^{(l)}, l \in \{1,2,3,4\}$. The parents of $y^{(1)}, y^{(2)}$ and $y^{(4)}$ are C. The parents of $y^{(3)}$ are C and $y^{(2)}$.

13.3.2 Performance

The NBC rule corresponding to (13.7) is

$$c_{NBC} = \arg\max_k \left(\pi(C = c_k|I) \prod_{l=1}^{L} p_l(y^{(l)}|C = c_k, I)\right) , \tag{13.9}$$

This equation assumes the features $y^{(l)}$ are independent. At first sight, we may believe that the NBC is optimal, i. e. has the smallest number of classification errors, when the $y^{(l)}$ are independent, and perhaps close to optimal when the $y^{(l)}$ are weakly dependent. However, in practice the classifier has a good performance in a wide variety of domains, including many where there are clear dependencies between the $y^{(l)}$ [40]. The following example provides a plausible explanation of this phenomenom.

Example 13.4. Naive Bayes' Classifier: Correlated Features [5]. We consider the classification of an object O into two classes c_1 and c_2. The object O is described by the feature vector $\mathbf{y} = (y^{(1)}, y^{(2)}, y^{(3)})^T$, where we assume $y^{(1)}$

and $y^{(3)}$ are independent, and $y^{(1)}$ and $y^{(2)}$ are completely dependent (i. e. $y^{(1)} = y^{(2)}$). For simplicity we shall further assume that $\pi(C = c_1|I) = \frac{1}{2} = \pi(C = c_2|I)$. In this case, the plug-in MAP and the NBC, classification rules are, respectively,

$$\begin{aligned} c_{\text{PLUG-IN}} &= \arg\max_k \big(p(y^{(1)}|C = c_k, I)p(y^{(3)}|C = c_k, I)\big) , \\ &= \begin{cases} c_1 \text{ if } B > 1 - A , \\ c_2 \text{ otherwise} , \end{cases} \end{aligned}$$

and

$$\begin{aligned} c_{NBC} &= \arg\max_k \big(p(y^{(1)}|C = c_k, I)^2 p(y^{(3)}|C = c_k, I)\big) . \\ &= \begin{cases} c_1 \text{ if } B > (1-A)^2/(A^2 + (1-A)^2) , \\ c_2 \text{ otherwise} , \end{cases} \end{aligned}$$

where $A = p(y^{(l)}|C = c_1, I)$ and $B = p(y^{(l)}|C = c_2, I)$. The two curves, $B = 1 - A$ and $B = (1-A)^2/(A^2 + (1-A)^2)$, are shown in Fig. 13.3. We observe that, although the independence assumption is decisively violated ($y^{(2)} = y^{(1)}$), the NBC disagrees with the plug-in MAP classifier only in the two narrow regions that are above one of the curves and below the other; everywhere else the two classifiers give identical classification results.

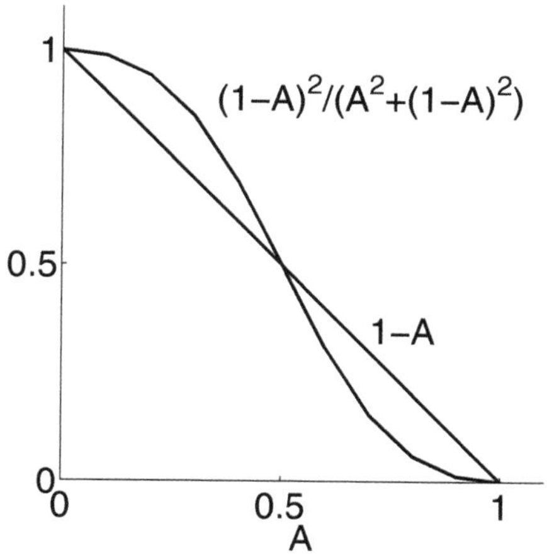

Fig. 13.3 Shows the curves $B = 1 - A$ and $B = (1-A)^2/(A^2 + (1-A)^2)$ as a function of A for the plug-in MAP classifier and the NBC. The only region where the classifiers give different results are regions where one curve is above the other. Everywhere else the two classifiers give identical results.

13.3.3 Likelihood

Several different methods are available for estimating the (one-dimensional) likelihoods $p(y^{(l)}|C=c_k,I)$ from a training set D of labelled objects $O_i, i \in \{1,2,\ldots,N\}$. Some commonly used methods are listed in Table 13.2.

Table 13.2 Principal Methods Used for Estimating the Likelihood $p(y|C=c_k,I)$.

Method	Description
Standard Parametric Function	Fit $p(y^{(l)}|C=c_k,I)$ to a one-dimensional parametric form using a maximum likelihood, or maximum *a posteriori*, algorithm.
Gaussian Mixture Model	Fit $p(y^{(l)}|C=c_k,I)$ to a mixture of one-dimensional Gaussian functions using the EM algorithm: $\hat{p}(y^{(l)}|C=c_k,I) \approx \sum_{m=1}^{M} \alpha_{km}^{(l)} \mathscr{N}(\mu_{km}^{(l)}, \Sigma_{km}^{(l)})$, where $\alpha_{km}^{(l)} > 0$ and $\sum_{m=1}^{M} \alpha_{km}^{(l)} = 1$. See Sect. 9.4.4.
Histogram Binning	Convert the feature values $y^{(l)}$ into a finite set of discrete values $\alpha_m, m \in \{1,2,\ldots,M_l\}$. See Sect. 5.3.1.
Kernel Density Estimation	Calculate $p(y^{(l)}|C=c_k,I)$ using a one-dimensional kernel density estimation procedure. See Sect. 5.3.2.

The most common method for estimating the (one-dimensional) likelihoods $p(y^{(l)}|C=c_k,I)$ is to model them using a standard parametric distribution. Traditionally, we use a one-dimensional Gaussian pdf, $\mathscr{N}\left(\mu_k^{(l)}, \Sigma_k^{(l)}\right)$, for this purpose, where the parameters, $\mu_k^{(l)}$ and $\Sigma_k^{(l)}$, are estimated using the training set D (which contains N labeled objects $O_i, i \in \{1,2,\ldots,N\}$, with instantiated values $y_i^{(l)}$). For example, the unbiased maximum likelihood estimates of $\mu_k^{(l)}$ and $\Sigma_k^{(l)}$ are given by

$$\mu_k^{(l)} = \sum_{i=1}^{N} I_i y_i^{(l)} / n_k \,, \tag{13.10}$$

$$\Sigma_k^{(l)} = \sum_{i=1}^{N} \left(y_i^{(l)} - \mu_k^{(l)}\right)^2 / (n_k - 1) \,, \tag{13.11}$$

where n_k is the number of objects O_i which belong to the class $C = c_k$. *Note*: Although the Gaussian pdf is the most popular distribution, we sometimes find it preferable to use a non-Gaussian pdf [2].

If the likelihood cannot be modeled using a standard parametric distribution we use a non-parametric method e. g. using a histogram or a kernel density estimator. The following example describes modeling the likelihood with a histogram.

Example 13.5. Naive Bayes in Information Retrieval [1]. The naive Bayes classifier is widely used in information retrieval applications: Given the document class and length, we assume the probability of occurrence of a word does not depend on its position or other words in the document. In spite of being completely unrealistic, this assumption greatly simplifies classifier training. In particular, estimation of the likelihoods reduces to a simple normalization of word counts. Let $\{y^{(1)}, y^{(2)}, \ldots, y^{(L)}\}$ denote a set of L specific words, then

$$p(y^{(l)}|C = c_k, I) = \frac{N_{k,l}}{\sum_{h=1}^{L} N_{k,h}} ,$$

where $N_{k,l}$ is the number of occurrences of the word $y^{(l)}$ in training data from class $C = c_k$. To avoid null estimates we often add a "pseudo-count" δ, $\delta > 0$, to every $N_{k,l}$:

$$p(y^{(l)}|C = c_k, I) = \frac{N_{k,l} + \delta}{\sum_{h=1}^{L} (N_{k,h} + \delta)} ,$$

The default value for δ is $\delta = 1$ and is referred to as the Laplace correction.

Alternatively, instead of using a pseudo-count δ we may avoid null estimates by "discounting": We discount a small constant b, $0 < b < 1$, from every occurence of a word. The probability mass gained by discounting is then distributed among all the words. In this case, the likelihood is:

$$p(y^{(l)}|C = c_k, I) = \max\left(0, \frac{N_{k,l}(1-b)}{\sum_{h=1}^{L} N_{k,h}(1-b)}\right) + \beta ,$$

and

$$\beta = \frac{1}{L} \sum_{l=1}^{L} \max\left(0, \frac{N_{k,l}(1-b)}{\sum_{h=1}^{L} N_{k,h}(1-b)}\right) .$$

The preferred method for non-parametric likelihood estimate is kernel density which is explained in the following example.

Example 13.6. Kernel Density Estimation for NBC [28]. Ref. [28] compared different non-parametric density estimates for use in a NBC. Their experiments showed that the highest classification performance was obtained using a kernel density estimate. Surprisingly, however, the most accurate kernel density estimators did *not* give the best classification performance. The kernel density bandwidth which was found to give the best overall classification performance was

$$H = \frac{R}{2(1 + \log_2(N))} ,$$

where R is the range of the values of the training data and N is the number of samples in the training data.

In estimating the likelihoods we have assumed that we have used all of the training samples. Recently, [56, 57] suggested using only a subset of the training data in a "lazy" learning approach: Given an object O we construct a NBC using a set of *weighted* training samples, where the weights fall as we move away from O in feature space. The idea behind the lazy NBC is that by locally learning the likelihoods $\hat{p}(y^{(l)}|C = c_k, I)$ we may reduce the effect of any feature-feature correlations.

The following example illustrates a simple, but very effective, lazy NBC.

Example 13.7. A Selective Neighbourhood Naive Bayes' Classifier [52]. The selective neighbourhood NBC is a lazy learning algorithm which works as follows. Let D denote training set consisting of N objects $O_i, i \in \{1, 2, \ldots, N\}$, which belong to the class $C = c_k$. Each object O_i has a label z_i, where $z_i = k$ if O_i belongs to the class $C = c_k$. For a given test object O, we define a subset $S_l^{(k)}$ of the l objects O_i in D which are most similar to O. Then for each value of l we learn a naive Bayes' classifier, H_l, using the subsets $S_l^{(k)}$. We select the NBC with the highest accuracy (as estimated on the training data) to classify the test object O.

13.4 Variants

The NBC has a good performance over a wide variety of domains, including many domains where there are clear dependencies between the attributes. The NBC is also robust to noise and irrelevant attributes. As a result, the NBC is widely used in many practical applications. It has also attracted much attention from many researchers who have introduced various modifications in an effort to improve its performance. Some of these variants are listed in Table 13.3 lists some of the variants.

Table 13.3 Model Modifications in the Naive Bayes Classifier

Method	Description
Tree Augmented Naive Bayes (TAN)	Relax the NBC assumption of feature independence [8, 21].
Adjusted Probability Model (APM)	Give different weights to each feature [16].
Homologous Naive Bayes (HNB)	Classify multiple objects which belong to the same (unknown) class [17].

13.4.1 Feature Selection

In the original naive Bayes' classifier all of the input features are used. However, in practice we often find it beneficial to reduce the number of features. Table 13.4 lists

Table 13.4 Feature Space Modifications

Method	Description
Feature Selection	Select a small number of the input features. Two methods for doing this are: (i) *Filters* [3, 10, 15, 25, 35] and (ii) *Wrappers* [20, 22, 31, 32].
Feature Extraction	Create a small number of new independent features. Methods for doing this include the subspace techniques discussed in Sect. 4.5 [15, 27, 29, 36, 43].

the two principal ways this is performed. The following example illustrates feature selection in a DNA experiment (see also Ex. 13.2)

Example 13.8. Gene Selection: An Application of Feature Selection in DNA Microarray Experiments [18]. DNA microarray experiments, generating thousands of gene expression measurements, are used to collect information from tissue and cell samples regarding gene expression differences that can be useful for diagnosis disease, distinction of the specific tumor type etc. One important application of gene expression microarray data is the classification of samples into known categories.

From a classification point of view, the biological samples can be seen as objects and the genes as features used to describe each object. In a typical microarray dataset the number of samples is small (usually less than 100) but the number of genes is several thousand, with many of the genes being correlated or irrelevant. For diagnostic purposes it is important to find small subsets of genes that are sufficiently informative to distinguish between cells of different types. Studies show that most genes measured in a DNA microarray experiment are not relevant for an accurate distinction among the different classes of the problem and simple classifiers with few genes (less than 15-20) achieve better accuracies. To avoid the "curse of dimensionality" feature selection is required to find the gene subset with the best classification accuracy for a given classifier.

In feature selection, our goal is to find an optimal subset of the input features that optimizes some criteria $\mathscr{J}$. The simplest feature selection approach is to evaluate each feature individually and select the L' features with the highest scores. Unfortunately this approach ignores feature correlations and it will rarely find an optimal subset. A brute force approach would be to evaluate *all* possible subsets of L' features and select the global optimum, but the number of combinations $C(L,L') = \binom{L}{L'}$ becomes impractical even for moderate values of L' and L. A compromise between these two approaches is to selectively examine a limited number of feature subsets. Some ways of doing this are listed in Table 13.5.

Table 13.5 Feature Selection Methods

Method	Description
Exhaustive Search	Evaluate all $C(L,L')$ possible subsets. Guaranteed to find the optimal subset. Not feasible for even moderately large values of L' and L.
Best Individual Features	Evaluate all the m features individually. Select the best L' individual features. Computationally simple but not likely to lead to an optimal subset.
Sequential Forward Selection	Select the best single feature and then add one feature at a time which in combination with the selected features maximizes the criterion function. Although computationally attractive its main drawback is that once a feature is retained it cannot be discarded.
Sequential Backward Selection	Start with all the L features and successively delete one feature at a time. Computationally attractive although it requires more computation than the sequential forward selection. Its main drawback is that once a feature is deleted it cannot be brought back into the optimal subset.
Sequential Forward Floating Search	First enlarge the feature subset by m features using forward selection and then delete n features using backward selection. This technique attempts to eliminate the drawbacks of the sequential forward and backward selection methods. The technique provides close to optimal solution at an affordable computational cost.

Given a subset of features two different strategies are available to evaluate the degree to which the subset is optimal. In the first strategy we evaluate the degree to which a subset of features is optimal indirectly: it is based on a filter in which we evaluate the optimality of a feature subset on the basis of its "information content" Φ. The second strategy follows a more direct approach and evaluates the optimality of a feature subset on the basis of their estimated classification accuracy, or error probability, P_ε [22].

The following examples illustrate the two different feature selection strategies.

Example 13.9. mRMR: A Filter Feature Selection Algorithm [35]. The mRMR is a powerful filter feature selection algorithm in which our aim is to select a subset of features with *maximum relevancy* (i. e. high discriminatory power) and *minimum redundancy* (i. e. low inter-feature correlations). The mRMR filter relies on the concept of mutual information (Sect. 5.3) to define Φ:

$$\Phi = \frac{1}{L}\sum_{l=1}^{L} MI(C, y^{(l)}) - \frac{2}{L(L-1)}\sum_{l=1}^{L-1}\sum_{h=l+1}^{L} MI(y^{(l)}, y^{(h)}) \ .$$

This equation consists of two terms after the equals sign. The first term, $\frac{1}{L}\sum MI(C,y^{(l)})$, represents the average amount by which the uncertainty in the class variable C is reduced after having observed a feature $y^{(l)}$. The second term, $\frac{2}{L(L-1)}\sum_l\sum_h MI(y^{(l)},y^{(h)})$, represents the average amount by which the uncertainty in a given feature $y^{(l)}$ is reduced after having observed another feature $y^{(h)}, h \neq l$.

Example 13.10. Wrapper Feature Selection [22]. In wrapper-based feature selection, we select feature subsets on the basis of their estimated classification accuracy, or probability of error, P_{ε}. Sophisticated methods for accurately estimating P_{ε} are listed in Table 13.8.

Although theoretically optimal, the procedure is liable to *overfit*. This happens when a feature subset has a low P_{ε} value (as measured on the training data) but which generalizes poorly on the test data [4, 22, 31]. Some effective ways of preventing overfitting are given in [32].

13.4.2 Feature Extraction

In feature selection, we select a small number of the input features for use in the NBC. In *feature extraction*, we follow a different approach and create a small number of new features from the input features. Mathematically, we define feature extraction as creating new features, $y^{(l)}, l \in \{1,2,\ldots,L'\}$, from the original set of features $x^{(l)}, l \in \{1,2,\ldots,L\}$. Our aim is to extract a small number of new features which contain the maximal information concerning the class variable C.

As in the case of feature selection, we may also perform feature extraction by directly optimizing the classifier performance, i. e. by minimizing the estimated probability of classification error P_{ε}.

Example 13.11. Feature Extraction Based on Mutual Information [25]. We perform feature extraction by searching for a set of linear combinations of the original features, whose mutual information with the output class is maximized. The mutual information between the extracted features and the output class is calculated by using a kernel probability density estimate (Sect. 5.3.2).

13.4.3 Tree-Augmented Naive Bayes' Classifer

In the tree-augmented naive Bayes' classifier (TAN) [8, 21]. we relax the independence assumption as used in the NBC. In the Tree Augmented Naive (TAN) Bayes' classifier we allow each feature in the NBC to depend on C, and at most, one additional feature. Mathematically, the TAN classifier therefore has: $pa(C) = 0$ and $|pa(y^{(l)})| \leq 2$. Graphically, the TAN is a restricted Bayesian network which uses a tree-structure imposed on the NBC structure (Fig. 13.2(b)). The corresponding TAN classifier rule is

$$c_{\text{TAN}} = \arg\max_k \big(\pi(C = c_k|I) \prod_{l=1}^{L} p_l(y^{(l)}|pa(y^{(l)}), I)\big) .$$

Many different methods are available for building the TAN classifier. One method is the super-parent algorithm [21]: The super-parent algorithm consists of three steps:

1. Search for a "super"-parent that has the best estimated classification performance[2]. A super-parent is a node with arcs pointing to all other nodes without a parent except for the class label.
2. Determine one favourite child for the super-parent chosen in the first step, based on its estimated classification accuracy.
3. Add the appropriate arc to the tree structure.

The entire process is repeated until no improvement is achieved, or $L-1$ arcs are added into the tree.

The following two examples illustrate two further NBC variants.

Example 13.12. Adjusted Probability Model (APM) [16]. In the Adjusted Probability Model (APM) we allow the features to have different weights, or importances, $w_l, l \in \{1,2,\ldots,L\}$. The corresponding APM classification rule is

$$c_{\mathrm{APM}} = \arg\max_k \left[\pi(C=c_k|I)\prod_{l=1}^{L}\left(\frac{p(C=c_k|y^{(l)},I)}{\pi(C=c_k|I)}\right)^{w_l}\right] .$$

Different authors have restricted the weights in various ways. For example, in an unrestricted APM [16] the authors place no resrictions on the w_l, and allow both postive and negative values.

Example 13.13. Homologous Naive Bayes' Classifier [17]. In the Homologous Naive Bayes' (HNB) classifier we consider the problem of classifying multiple objects which belong to the same (unknown) class. The homologous classification rule is

$$c_{\mathrm{HNB}} = \arg\max_k \left(\pi(C=c_k|I)\prod_{i=1}^{N}\prod_{l=1}^{L} p_l(y_i^{(l)}|C=c_k,I)\right) ,$$

where $y_i^{(l)}$ is the instantiated value of the lth feature in the ith object.

13.5 Multiple Naive Bayes' Classifiers

The fourth set of modifications involve the use of multiple NBC's instead of a single classifier. Three such algorithms are listed in Table 13.6. The following example illustrates the Naive Bayes' Classifier Committee algorithm.

---

2 Classification performance is conventionally measured using a cross-validation procedure (Sect. 13.6).

Table 13.6 Multiple Naive Bayes' Classifiers

Method	Description
Naive Bayes' Classifier Committee	Use an ensemble of K NBC's, in which each classifier is trained on a different subset of features. Final classification is obtained by allowing the NBC's to vote, each classifier having an equal weight [55].
Boosted Naive Bayes	Use ensemble of K NBC's, in which each classifier is trained on a boosted training set (see Sect. 14.10). Final classification is obtained by allowing the NBC's to vote, each classifier having a weight according to the quality of its predictions [6, 39].
Bayesian Model Averaging NBC	Similar to NBCC except classifier has a weight according to the Bayesian evidence that it is the true, or correct, classifier.

Example 13.14. Naive Bayes' Classifier Committee (NBCC) [55]. In the Naive Bayes' Classifier Committee (NBCC) we use an ensemble of L NBC's, where L is the number of independent features. Let H_0 denote the first NBC using all L features, and let E_0 be its estimated error rate. Then we repeatedly create candidate NBC's by randomly selecting $L/2$ features. Assuming the candidate NBC is not already a member of the committee, then it becomes a member of the committee if its estimated error is less than E_0.

To classify a test object O, each NBC in the committee is invoked to produce the probability that O belongs to a class $C = c_k$. Then for each class c_k, the probabilities provided by all committee members are summed up. The optimum classification of O is equal to the class with the largest summed probability.

13.6 Error Estimation

The classification error P_ε is the ultimate measure of the performance of a classifier. Apart from using P_ε to measure the performance of a given classifier we may use it as a general optimization criterion. In Sect. 13.4 we used it in selecting an optimum subset of features ("wrapper") and more generally we may use it to select an optimum classifier.

Mathematically, the classification error of a Bayesian classifier is defined as

$$P_\varepsilon = 1 - \sum_{k=1}^{K} \int_{A_k} \pi(C = c_k|I) p(y|C = c_k, I) dy \;, \tag{13.12}$$

where A_k is the region where the class $C = c_k$ has the highest *a posteriori* probability. In general, it is often very difficult to perform the integration in (13.12) and obtain a closed-form expression for P_ε. However, for the case of two classes $C = c_1$ and $C = c_2$ with equal *a priori* probabilities and Gaussian likelihoods, $\mathcal{N}(\boldsymbol{\mu}_1, \boldsymbol{\Sigma}_1)$ and $\mathcal{N}(\boldsymbol{\mu}_2, \boldsymbol{\Sigma}_2)$, closed-form expression for the bounds on P_ε are available (Table 13.7).

Table 13.7 Bounds on the Two-Class Probability of Error P_ε

Bound	Description
Mahalanobis	$P_\varepsilon \leq 2/(4+\Delta)$; where Δ is the Mahalanobis distance $(\boldsymbol{\mu}_1-\boldsymbol{\mu}_2)^T\boldsymbol{\Sigma}^{-1}(\boldsymbol{\mu}_1-\boldsymbol{\mu}_2)$ [9].
Bhattacharyya	$\frac{1}{2}(1-\sqrt{1-e^{-2B}}) \leq P_\varepsilon \leq \frac{1}{2}e^{-B}$; where B is the Bhattacharyya distance $\frac{1}{8}(\boldsymbol{\mu}_1-\boldsymbol{\mu}_2)^T(\frac{1}{2}(\boldsymbol{\Sigma}_1+\boldsymbol{\Sigma}_2))^{-1}(\boldsymbol{\mu}_1-\boldsymbol{\mu}_2)+(\ln\lvert(\boldsymbol{\Sigma}_1+\boldsymbol{\Sigma}_2)/2\rvert)/(2\sqrt{\lvert\boldsymbol{\Sigma}_1\rvert\lvert\boldsymbol{\Sigma}_2\rvert})$ [9].
Chernoff	$P_\varepsilon \leq \int \mathcal{N}(\mathbf{y}\lvert\boldsymbol{\mu}_1,\boldsymbol{\Sigma}_1)^s\mathcal{N}(\mathbf{y}\lvert\boldsymbol{\mu}_2,\boldsymbol{\Sigma}_2)^{1-s}d\mathbf{y}$, where $s\in[0,1]$ is a parameter [9].
Lee and Choi	$P_\varepsilon \approx 0.040219-0.70019B+0.63578B^2-0.32766B^3+0.087172B^4-0.0091875B^5$ [26].
Hsieh *et. al.*	$P_\varepsilon \approx \frac{1}{2}e^{-D}$; where D is the modified Bhattacharyya distance $D=\left[\frac{2}{3}(\boldsymbol{\mu}_1-\boldsymbol{\mu}_2)^T(\frac{1}{2}(\boldsymbol{\Sigma}_1+\boldsymbol{\Sigma}_2))^{-1}(\boldsymbol{\mu}_1-\boldsymbol{\mu}_2)\right]^{2/3}+\left[(\ln\lvert(\boldsymbol{\Sigma}_1+\boldsymbol{\Sigma}_2)/2\rvert)/\sqrt{\lvert\boldsymbol{\Sigma}_1\rvert\lvert\boldsymbol{\Sigma}_2\rvert}\right]^{2/3}$ [14].

Table 13.8 Error Estimation Methods

Method	Description
Resubstitution	Training and test databases are the same. Thus we use all the available data both for training and for testing. The P_ε is an optimistic biased estimate especially when the ratio of sample size to sample dimensionality is small.
Holdout	Half the data is used for training and the other half is used for testing. Pessimistic biased estimate. Different partitions will give different estimates.
Leave-One-Out	A classifier is designed using $(M-1)$ samples and evaluated on the one remaining sample. This is repeated M times with different training sets of size $(M-1)$. Estimate is unbiased but it has a large variance. When used for model selection it is asymptotically equivalent to the Akaike Information Criterion (AIC) when the number of observations is large.
K-Fold Cross Validation	A compromise between the holdout and leave-one-out methods. Divide the available samples into K disjoint subsets, $1\leq K\leq M$. Use $(K-1)$ subsets for training and the remaining subset for testing. Estimate has lower bias than the holdout method. Recommended value for K is in the range $[M/5, M/2]$. When used for model selection it is asymptotically equivalent to Bayesian Information Criterion (BIC) for an appropriate K when the number of observations is large.
Bootstrap	Generate many bootstrap sample sets of size M by sampling all the available data with *replacement*. Bootstrap estimates can have lower variance than the leave-one-out method. Useful in small sample situations.

In practice, we cannot be sure that the above conditions hold and then the only option is to estimate P_ε from the measurements $\mathbf{y_i}, \mathbf{i}\in\{\mathbf{1},\mathbf{2},\ldots,\mathbf{N}\}$, as follows. We split the $\mathbf{y_i}$ into a set of training data D and a set of test data T. We design the classifier on D and measure its generalization ability, or $1-P_\varepsilon$, on T. If the number of measurements in D is small the classifier will not be robust and its generalization ability will be low. On the other hand, if the number of measurements in T is small, then the confidence in P_ε will be low. In Table 13.8 we list several methods for dividing the $\mathbf{y}_i$ into D and T and for estimating P_ε.

13.7 Pairwise Naive Bayes' Classifier

Until now we have placed no restrictions on the number of possible classes K. However, in practice, finding features which are effective for all of the classes $c_k, k \in \{1,2,\ldots,K\}$, becomes increasingly difficult as K increases. For this reason, we often decompose a K-class pattern recognition problem into $K(K-1)/2$ two-class pattern recognition problems. In each two-class, or pairwise problem, we may use different features and a different number of features. The final classification is then obtained by combining the $K(K-1)/2$ pairwise results. The following example illustrates the the combination of $K(K-1)/2$ pairwise NBC's using a simple voting scheme (see also Sect. 14.8).

Example 13.15. Multi-Class Classification Using Multiple Pairwise Naive Bayes' Classifiers [30, 42]. We consider the classification of an object O into one of K classes $c_k, k \in \{1,2,\ldots,K\}$. For each pair of classes, $c_k, c_l, l \neq k$, we obtain an optimal pairwise classification, or vote V_{kl}, using a pairwise NBC, where

$$V_{kl} = \begin{cases} 1 \text{ if } p(C = c_k|y,I) > p(C = c_l|y,I) \, , \\ 0 \text{ otherwise} \, . \end{cases}$$

Then the optimal multi-class classification is $C = c_{\mathrm{OPT}}$, where

$$c_{\mathrm{OPT}} = \arg\max_k \left(\sum_{\substack{l=1 \\ l \neq k}}^{K} V_{kl} \right) ,$$

and $V_{lk} = 1 - V_{kl}$.

13.8 Software

BNT. A matlab toolbox for performing Bayesian inference. Author: Kevin Patrick Murphy.

GPML. A collection of m-files. Authors: Carl Edward Rasmussen, Chris Williams.

KDE. A matlab toolbox for performing kernel density estimation.

NETLAB. A matlab toobox for performing neural network pattern recognition. Author: Ian Nabney.

STPRTOOL. A matlab toolbox for performing statistical pattern recognition. Authors: Vojtech Franc, Vaclav Hlavac.

13.9 Further Reading

Bayesian decision theory and classification are discussed in detail in [9, 12, 24, 50]. Specific references on Bayesian classification with Gaussian processes are given in [37]. The naive Bayes' classifier is discussed in many publications, including: [13, 40]. Extensions to the NBC include the references listed in Sect. 13.4 and [51, 54]. An entire book devoted to feature extraction is [11]. The NBC has been used in many applications, including text classification, fraud diagnosis [48] and other applications [44]. Finally, the issue of performing statistical analysis using a small number of samples is discussed in detail in [33, 34].

Problems

13.1. What is the curse of dimensionality?

13.2. Define the naive Bayes classifier. Explain why it is called naive.

13.3. In the naive Bayes the likelihood function is written as a product of individual features. Justify this approximation and suggest alternative approximation which could bee used.

13.4. In the context of the naive Bayes classifier compare and contrast feature selection, feature extraction and feature joining.

13.5. Explain the mRMR feature selection algorithm.

13.6. Define wrapper feature selection. What are its advantages and disadvantages.

13.7. Compare and contrast tree augmented naive Bayes (TAN), adjusted probability model (APM) and homologous naive Bayes classifiers.

References

1. Andres-Ferrer, J., Juan, A.: Constrained domain maximum likelihood estimation for naive Bayes text classificaion. Patt. Anal. Appl. 13, 189–196 (2010)
2. Bennett, P.N.: Using asymmetric distributions to improve text classifier probability estimates. In: SIGIR 2003, Toronto, Canada, July 28–August 1 (2003)
3. Chow, T.W.S., Huang, D.: Estimating optimal feature subsets using efficient estimation of high-dimensional mutual information. IEEE Trans. Neural Networks 16, 213–224 (2005)
4. Cunningham, P.: Overfitting and diversity in classification ensembles based on feature selection. Tech. Rept. TCD-CS-2005-17, Dept. Computer Science, Trinity College, Dublin, Ireland (2005)

5. Domingos, P., Pazzani, M.: On the optimality of the simple Bayesian classifier under zero-one loss. Machine Learning 29, 103–130 (1997)
6. Elkan, C.: Boosting and naive Bayes' learning. Tech Rept CS97-557. University of California, San Diego (September 1997)
7. Friedman, J.H.: On bias, variance 0/1-loss and the curse of dimensionality. Data Mining Knowledge Discovery 1, 55–77 (1997)
8. Friedman, N., Geiger, D., Goldszmidt, M.: Bayesian network classifiers. Mach. Learn. 29, 131–163 (1997)
9. Fukunaga, K.: Introduction to Statistical Pattern Recognition, 2nd edn. Academic Press (1990)
10. Guyon, I., Elisseeff, A.: An introduction to to variable and feature selection. Mach. Learn. Res. 3, 1157–1182 (2003)
11. Guyon, I., Gunn, S., Nikravesh, M., Zadeh, L.: Feature Extraction, Foundations and Applications. Springer, Heidelberg (2006)
12. Hastie, T., Tibshirani, R., Friedman, J.: The elements of statistical learning. Springer, Heidelberg (2001)
13. Hoare, Z.: Landscapes of naive Bayes' classifiers. Patt. Anal. App. 11, 59–72 (2008)
14. Hsieh, P.-F., Wang, D.-S., Hsu, C.-W.: A linear feature extraction for multi-class classification problems based on class mean and covariance discriminant information. IEEE Trans. Patt. Anal. Mach. Intell. 28, 223–235 (2006)
15. Holder, L.B., Rusell, I., Markov, Z., Pipe, A.G., Carse, B.: Current and future trends in feature selection and extraction for classification problems. Int. J. Patt. Recogn. Art Intell. 19, 133–142 (2005)
16. Hong, S.J., Hosking, J., Natarajan, R.: Multiplicative adjustment of class probability: educating naive Bayes. Research Report RC 22393 (W0204-041), IBM Research Division, T. J. Watson Research Center, Yorktown Heights, NY 10598 (2002)
17. Huang, H.-J., Hsu, C.-N.: Bayesian classification for data from the same unknown class. IEEE Trans. Sys. Man Cybern. -Part B: Cybern. 32, 137–145 (2003)
18. Inza, I., Larranaga, P., Blanco, R., Cerrolaza, A.J.: Filter versus wrapper gene selection approaches in DNA microarray domains. Art Intell. Med. 31, 91–103 (2004)
19. Jain, A.K., Duin, R.P.W., Mao, J.: Statistical Pattern Recognition: A Review. IEEE Trans. Patt. Anal. Mach. Intell. 22, 4–37 (2000)
20. Jain, A., Zongker, D.: Feature Selection: Evaluation, application and small sample performance. IEEE Trans. Patt. Anal. Mach. Intell. 19, 153–158 (1997)
21. Keogh, E.J., Pazzani, M.J.: Learning the structure of augmented Bayesian classifiers. Int. Art Intell. Tools 11, 587–601 (2002)
22. Kohavi, R., John, G.: Wrappers for feature subset selection. Art. Intell. 97, 273–324 (1997)
23. Kittler, J., Hatef, M., Duin, R.P.W., Matas, J.: On combining classifiers. IEEE Trans. Patt. Anal. Mach. Intell. 20, 226–239 (1998)
24. Kuncheva, L.I.: Combining Pattern Classifiers. John Wiley and Sons (2004)
25. Kwak, N., Choi, C.-H.: Input feature selection using mutual information based on Parzen window. IEEE Patt. Anal. Mach. Intell. 24, 1667–1671 (2002)
26. Lee, C., Choi, E.: Bayes error evaluation of the Gaussian ML classifier. IEEE Trans. Geosci. Rem. Sense 38, 1471–1475 (2000)
27. Liang, Y., Gong, W., Pan, Y., Li, W.: Generalized relevance weighted LDA. Patt. Recogn. 38, 2217–2219 (2005)
28. Liu, B., Yang, Y., Webb, G.I., Boughton, J.: A Comparative Study of Bandwidth Choice in Kernel Density Estimation for Naive Bayesian Classification. In: Theeramunkong, T., Kijsirikul, B., Cercone, N., Ho, T.-B. (eds.) PAKDD 2009. LNCS, vol. 5476, pp. 302–313. Springer, Heidelberg (2009)
29. Loog, M., Duin, R.P.W.: Linear dimensionality reduction via a heteroscedastic extension of LDA: The Chernoff criterion. IEEE Trans. Patt. Anal. Mach. Intell. 26, 732–739 (2004)
30. Lorena, A.C., de Carvalho, A.C.P.L., Gama, J.M.P.: A review on the combination of binary classifiers in multiclass problems. Art Intell. Rev. 30, 19–37 (2008)

31. Loughrey, J., Cunningham, P.: Overfitting in wrapper-based feature subset selection: the harder you try the worse it gets. In: 24th SGAI Int. Conf. Innovative Tech. App. Art Intell., pp. 33–43 (2004); See also Technical Report TCD-CS-2005-17, Department of Computer Science, Trinity College, Dublin, Ireland
32. Loughrey, J., Cunningham, P.: Using early-stopping to avoid overfitting in wrapper-based feature selection employing stochastic search. Technical Report TCD-CS-2005-37, Department of Computer Science, Trinity College, Dublin, Ireland (2005)
33. Martin, J.K., Hirschberg, D.S.: Small sample statistics for classification error rates I: error rate measurements. Technical Report No. 96-21, July 2, Department of Information and Computer Science, University of California, Irvine, CA 9297-3425, USA (1996a)
34. Martin, J.K., Hirschberg, D.S.: Small sample statistics for classification error rates II: confidence intervals and significance tests. Technical Report No. 96-22, July 7, Department of Information and Computer Science, University of California, Irvine, CA 9297-3425, USA (1996b)
35. Peng, H., Long, F., Ding, C.: Feature selection based on mutual information: criteria of max-dependency, max-relevance and min-redundancy. IEEE Trans. Patt. Anal. Mach. Intell. 27, 1226–1238 (2005)
36. Qin, A.K., Suganthan, P.N., Loog, M.: Uncorrelated heteoscedastic LDA based on the weighted pairwise Chernoff criterion. Patt. Recogn. 38, 613–616 (2005)
37. Rasmussen, C.E., Williams, C.K.I.: Gaussian Processes for machine learning. MIT Press, MA (2006)
38. Raudys, S.J., Jain, A.K.: Small sample size effects in statistical pattern recognition: recommendations for practitioners. IEEE Trans. Patt. Anal. Mach. Intell. 13, 252–264 (1991)
39. Ridgeway, G., Madigan, D., Richardson, T., O'Kane, J.W.: Interpretable boosted naive Bayes classification. In: Proc. 4th Int. Conf. Knowledge Discovery Data Mining, New York, pp. 101–104 (1998)
40. Rish, I., Hellerstein, J., Jayram, T.S.: An analysis of data characteristics that affect naive Bayes performance. IBM Technical Report RC2, IBM Research Division, Thomas J. Watson Research Center, NY 10598 (2001)
41. Silverman, B.: Density estimation for statistical data analysis. Chapman-Hall (1986)
42. Sulzmann, J.-N., Fernkranz, J., Hellermeier, E.: On Pairwise Naive Bayes Classifiers. In: Euro. Conf. Mach. Learn., pp. 371–381 (2007)
43. Tang, E.K., Suganthan, P.N., Yao, X., Qin, A.K.: Linear dimensionality reduction using relevance weighted LDA. Pattern Recognition 38, 485–493 (2005)
44. Thomas, C.S.: Classifying acute abdominal pain by assuming independence: a study using two models constructed directly from data. Technical Report CSM-153, Department of Computing Science and Mathematics, University of Stirling, Stirling, Scotland (1999)
45. Thomaz, C.E., Gillies, D.F.: "Small sample size": A methodological problem in Bayes plug-in classifier for image recognition. Technical Report 6/2001, Department of Computing, Imperial College of Science, Technology and Medicine, London (2001)
46. Thomaz, C.E., Gillies, D.F., Feitosa, R.Q.: A new covariance estimate for Bayesian classifiers in biometric recognition. IEEE Trans. Circuits Sys. Video Tech. 14, 214–223 (2004)
47. Trunk, G.V.: A problem of dimensionality: a simple example. IEEE Trans. Patt. Anal. Mach. Intell. 1, 306–307 (1979)
48. Viaene, S., Derrig, R., Dedene, G.: A case study of applying boosting naive Bayes to claim fraud diagnosis. IEEE Trans. Know Data Enging. 16, 612–619 (2004)
49. Vilalta, R., Rish, I.: A decomposition of Classes Via Clustering to Explain and Improve Naive Bayes. In: Lavrač, N., Gamberger, D., Todorovski, L., Blockeel, H. (eds.) ECML 2003. LNCS (LNAI), vol. 2837, pp. 444–455. Springer, Heidelberg (2003)
50. Webb, A.: Statistical pattern Recognition. Arnold, England (1999)
51. Webb, G.I., Boughton, J., Wang, Z.: Not so naive Bayes: aggregating one-dependence estimators. Mach. Learn. 58, 5–24 (2005)
52. Xie, Z., Hsu, W., Liu, Z., Li Lee, M.: SNNB: A Selective Neighborhood Based Naïve Bayes for Lazy Learning. In: Chen, M.-S., Yu, P.S., Liu, B. (eds.) PAKDD 2002. LNCS (LNAI), vol. 2336, pp. 104–114. Springer, Heidelberg (2002)

53. Xie, Z., Zhang, Q.: A study of selective neighbourhood-based naive Bayes for efficient lazy learning. In: Proc. 16th IEEE Int. Conf. Tools Art Intell. ICTAI, Boca Raton, Florida (2004)
54. Zhang, H.: Exploring conditions for the optimality of naive Bayes. Int. J. Patt. Recog. Art Intell. (2005)
55. Zheng, Z.: Naive Bayesian Classifier Committee. In: Nédellec, C., Rouveirol, C. (eds.) ECML 1998. LNCS, vol. 1398, pp. 196–207. Springer, Heidelberg (1998)
56. Zheng, Z., Webb, G.I.: Lazy learning of Bayesian rules. Mach. Learn. 41, 53–87 (2000)
57. Zheng, Z., Webb, G.I., Ting, K.M.: Lazy Bayesian rules: a lazy semi-naive Bayesian learning technique competitive to boosting decision trees. In: Proc. ICML 1999. Morgan Kaufmann (1999)

Chapter 14
Ensemble Learning

14.1 Introduction

The subject of this chapter is ensemble learning in which our system is characterized by an ensemble of M models. The models may share the same common representational format or each model may have its own distinct common representational format. To make our discussion more concrete we shall concentrate on the (supervised) classification of an object O using a *multiple classifier system* (MCS) . Given an unknown object O, our goal is to optimally assign it to one of K classes, $c_k, k \in \{1,2,\ldots,K\}$, using an ensemble of M (supervised) classifiers, $S_m, m \in \{1,2,\ldots,M\}$. The theory of multiple classifier systems suggests that if the *pattern of errors* made by one classifier, S_m, is different from the pattern of errors made by another classifier, S_n, then we may exploit this difference to give a more accurate and more reliable classification of O. If the error rates of the classifiers are less than $\frac{1}{2}$, then the MCS error rate, E_{MCS}, should decrease with the number of classifiers, M, and with the mean diversity, $\bar{\sigma}$, between the classifiers[1]. Mathematically, if $\bar{E}$ is the mean error rate of the S_m, then

$$E_{\mathrm{MCS}} \sim \alpha \bar{E} \ , \tag{14.1}$$

where $\alpha \sim 1/(\bar{\sigma} M)$ and $0 \leq \alpha \leq 1$.

The theory of ensemble learning deals with exploiting these differences and this will be the main emphasis in this chapter. In Table 14.1 we list some examples of these applications together with their input/output classification.

14.2 Bayesian Framework

In this section we present a general Bayesian framework [14, 33] for combining an ensemble of classifiers $S_m, m \in \{1,2,\ldots,M\}$. Given an unknown object O, we

---

[1] The concept of diversity is discussed in Sect. 14.5. For the present we may regard $\bar{\sigma}$ as being equal to the inverse of the mean correlation between the S_m.

Table 14.1 Applications in which Ensemble Learning is the Primary Fusion Algorithm

Class		Application
DhI-DeO	Ex. 11.15	Target tracking initialization using a Hough transform.
	Ex. 14.7	Weighted majority vote.
	Ex. 14.8	Image thresholding using an unsupervised weighted majority vote rule.
	Ex. 14.9	BKS look-up table.
	Ex. 15.7	An adaptive multi-modal biometric management algorithm.
DsI-DeO	Ex. 3.5	Tire pressure monitoring device.
	Ex. 3.6	Multi-modal biometric identification scheme.
	Ex. 3.12	Audio-visual speech recognition system.
	Ex. 5.17	Multi-modal cardiac imaging.
	Ex. 10.6	Non-intrusive speech quality estimation using GMM's.
	Ex. 13.14	Naive Bayes' classifier committee.
	Ex. 14.1	Bayesian model averaging of the naive Bayes classifier.
	Ex. 14.11	Image orientation using Borda count.
	Ex. 14.17	Boosting a classifier.

The designations DsI-DeO and DhI-DeO refer, respectively, to the (modified) Dasarathy input/output classifications: "Soft Decision Input-Decision Output", and "Hard Decision Input-Decision Output" (see Sect. 1.3.3).

suppose each classifier S_m makes a measurement $\mathbf{y}^{(m)}$ on O and returns an estimate of the *a posteriori* class probability $p(C = c_k|\mathbf{y}^{(m)},I)$:

$$\hat{p}(C = c_k|\mathbf{y}^{(m)},I) = p(C = c_k|\mathbf{y}^{(m)},I) + \varepsilon^{(m)} , \tag{14.2}$$

where $\varepsilon^{(m)}$ is the error made by S_m in estimating $p(C = c_k|\mathbf{y}^{(m)},I)$. We may improve the classification accuracy of O we estimating the joint *a posteriori* probability, $p(C = c_k|\mathbf{y}^{(1)},\mathbf{y}^{(2)},\ldots,\mathbf{y}^{(M)},I)$, which we do by combining the approximate $\hat{p}(C = c_k|\mathbf{y}^{(m)},I)$ values.

We distinguish two extreme cases [33].

Shared Representation. In this case, the S_m share the same common representational format. It follows that the classifiers use the same input data $\mathbf{y}^{(1)} = \mathbf{y}^{(2)} = \ldots = \mathbf{y}^{(M)}$ and as a result,

$$p(C = c_k|\mathbf{y}^{(1)},\mathbf{y}^{(2)},\ldots,\mathbf{y}^{(M)},I) = p(C = c_k|\mathbf{y}^{(m)},I) \qquad m \in \{1,2,\ldots,M\} , \tag{14.3}$$

where $p(C = c_k|\mathbf{y}^{(m)},I)$ is estimated by $\hat{p}(C = c_k|\mathbf{y}^{(m)},I)$. If we assume zero-mean errors for $\varepsilon^{(m)}$, we may average the $\hat{p}(C = c_k|\mathbf{y}^{(m)},I)$ to obtain a less error-sensitive estimate. This leads to the mean, or average, rule

$$p(C = c_k|\mathbf{y}^{(1)},\mathbf{y}^{(2)},\ldots,\mathbf{y}^{(M)},I) \approx \frac{1}{M}\sum_{m=1}^{M} \hat{p}(C = c_k|\mathbf{y}^{(m)},I) . \tag{14.4}$$

We may give each classifier S_m a different weight w_m (see Sect. 13.12), where $0 \leq w_m \leq 1$ and $\sum w_m = 1$. In this case, the weighted mean, or average, rule is

$$p(C = c_k|\mathbf{y}^{(1)}, \mathbf{y}^{(2)}, \ldots, \mathbf{y}^{(M)}, I) \approx \sum_{m=1}^{M} w_m \hat{p}(C = c_k|\mathbf{y}^{(m)}, I) . \tag{14.5}$$

Different Representations. In this case the S_m use different common representational formats which we assume are independent (see Sect. 13.3):

$$p(\mathbf{y}^{(1)}, \mathbf{y}^{(2)}, \ldots, \mathbf{y}^{(M)}|C = c_k, I) = \prod_{m=1}^{M} \hat{p}(\mathbf{y}^{(m)}|C = c_k, I) . \tag{14.6}$$

Assuming equal *a priori* probabilities, $\pi(C = c_k|I)$, and small errors, $\varepsilon^{(m)}$, this reduces [14, 16] to the product rule:

$$p(C = c_k|\mathbf{y}^{(1)}, \mathbf{y}^{(2)}, \ldots, \mathbf{y}^{(M)}, I) \approx \frac{\prod_{m=1}^{M} \hat{p}(C = c_k|\mathbf{y}^{(m)}, I)}{\sum_{l=1}^{K} \prod_{m=1}^{M} \hat{p}(C = c_l|\mathbf{y}^{(m)}, I)} . \tag{14.7}$$

In many applications we do not use (14.7) but use one of the following approximations:

$$p(C = c_k|\mathbf{y}^{(1)}, \mathbf{y}^{(2)}, \ldots, \mathbf{y}^{(M)}, I) \sim \begin{cases} \min_l \big(\hat{p}(C = c_k|y^{(l)}, I)\big) , \\ \frac{1}{L} \sum \hat{p}(C = c_k|y^{(l)}, I) , \\ \mathrm{median}_l \big(\hat{p}(C = c_k|y^{(l)}, I)\big) , \\ \max_l \big(\hat{p}(C = c_k|y^{(l)}, I)\big) . \end{cases} \tag{14.8}$$

Although these approximations are not very accurate they are often preferred because they are less sensitive to noise and outliers [14, 22, 33]. We may also give each classifier, S_m, a different weight w_m, where $0 \leq w_m \leq 1$. In this case the weighted product rule is

$$p(C = c_k|\mathbf{y}^{(1)}, \mathbf{y}^{(2)}, \ldots, \mathbf{y}^{(M)}, I) \approx \frac{\prod_{m=1}^{M} \hat{p}(C = c_k|\mathbf{y}^{(m)}, I)^{w_m}}{\sum_{l=1}^{K} \prod_{m=1}^{M} \hat{p}(C = c_k|\mathbf{y}^{(m)}, I)^{w_m}} . \tag{14.9}$$

In the following example, we illustrate the weighted mean rule (14.5), where the weights w_m are calculated using Bayesian model averaging (see Sect. 9.6.2).

Example 14.1. Bayesian Model Averaging of the Naive Bayes' Classifier [2, 10]. We consider Bayesian model averaging of the Naive Bayes' classifier (NBC) (see Sect. 13.3). We assume an unknown object O with an L-dimensional measurement, or feature, vector $\mathbf{y} = (y^{(1)}, y^{(2)}, \ldots, y^{(L)})^T$. We may generate several different naive Bayes' classifiers $S_m, m \in \{1, 2, \ldots, M\}$, by using different subsets of features $y^{(l)}$. Let

$$a_{ml} = \begin{cases} 1 \text{ if } S_m \text{ includes the } l\text{th feature } y^{(l)} , \\ 0 \text{ otherwise} , \end{cases}$$

then, by definition, the *a posteriori* classification probability, as estimated by S_m, is

$$\hat{p}_m(C = c_k|\mathbf{y},I) = \frac{\pi(C = c_k|I)\prod_{l=1}^{L}(\hat{p}(y^{(l)}|C = c_k,I))^{a_{ml}}}{\hat{p}(\mathbf{y}|I)} .$$

For each S_m we assume an *a priori* probability, $\pi(S_m|I)$, which describes our prior belief in S_m. Given a training data set D, we use Bayes' rule to calculate the *a posteriori* probability $\hat{p}(S_m|D,I) = \pi(S_m|I)\hat{p}(D|S_m,I)/\hat{p}(D|I)$. In Bayesian model averaging (BMA) we use $\hat{p}(S_m|D,I)$ to weight the predictions $\hat{p}_m(C = c_k|\mathbf{y},I)$. The predicted BMA *a posteriori* probability that the object O belongs to the class $C = c_k$ is

$$\hat{p}_{\text{BMA}}(C = c_k|\mathbf{y},I) \propto \sum_{m=1}^{M} p(S_m|D,I)\hat{p}_m(C = c_k|\mathbf{y},D,I) .$$

The object O is assigned to the class $C = c_{\text{OPT}}$, where

$$c_{\text{OPT}} = \arg\max_k \big(\hat{p}_{\text{BMA}}(C = c_k|\mathbf{y},I)\big) .$$

Although useful, the Bayesian framework is restricted to classifiers whose output is an *a posteriori* probability $\hat{p}_m(C = c_k|\mathbf{y},I)$ [2]. Unfortunately, most classifiers do not directly measure the *a posteriori* probability $p_m(C = c_k|\mathbf{y},I)$. Instead they produce a belief, $\mu_m^{(k)}$, that $C = c_k$ is the true assignment. Although theoretically it may be possible to convert $\mu_m^{(k)}$ to an *a posteriori* probability, it is rarely performed in practice (see Sect. 8.9). As a consequence, recent research on the MCS has relied less on a formal Bayesian framework, and more on an empirically-based framework.

14.3 Empirical Framework

The empirical framework is based on a wealth of experimental studies which all suggest that it is possible to reduce the classification error in a MCS by a factor which increases with the mean diversity $\bar{\sigma}$. In order to realize this reduction in classification error, the empirical framework introduces the following four innovations in the Bayesian framework:

---

[2] The above Bayesian framework is subject to additional restrictions [10]. The most important restriction is that the S_m are required to be mutually exclusive and exhaustive, i.e. the S_m must cover all possibilities concerning how the input data was generated.

Diversity Techniques. Empirical techniques are used to create a diverse ensemble of classifiers.

Diversity Measures. Probabilistic and non-probabilistic formulas are used to measure the ensemble diversity.

Classifier Types. Classifiers are not restricted to those which generate a probabilistic output.

Combination Strategies. Probabilistic and non-probabilistic techniques are used to combine the outputs of the individual classifiers.

We shall now consider each of these innovations in turn.

14.4 Diversity Techniques

At the heart of a MCS is an ensemble of diverse classifiers, each classifier having a "reasonable" performance. One way of creating such an ensemble is to devise a set of M classifiers $S_m, m \in \{1,2,\ldots,M\}$, which rely on different physical principles. The following example illustrates an important application of this technique.

Example 14.2. Thresholding techniques for the measurement of bread-crumb features [11]. In bread baking, the principal factors which determine the quality of the bread are: water, enzyme activity, starch, protein properties and heat. Recently bread quality has been measured by analyzing the corresponding bread crumb images. For this purpose, the bread crumb images were thresholded using a global thresholding algorithm (see Sect. 8.5). In [11], seven physically different thresholding algorithms were considered.

The more common way of creating such an ensemble is to train a parametric classifier S on multiple training sets $D_m, m \in \{1,2,\ldots,M\}$. If the training sets are diverse, we shall in turn obtain a diverse set of *fixed* classifiers S_m which do not have any adjustable parameters. Some common methods for constructing the training sets $D_m, m \in \{1,2,\ldots,M\}$, are

Disjoint Partitions. Sample a original training set D without replacement. If D contains N labeled objects, then on average, each D_m contains N/M labeled objects. Also the D_m share a common representational format. In general, the S_m which are produced will be diverse but will have high error rates (due to the small size of the D_m).

Different D_m. Construct the $D_m, m \in \{1,2,\ldots,M\}$, from different input sources. In this case, the D_m may, or may not, share a common representational format. In general, the S_m will be diverse.

Sampling D. Sample D with replacement. Two important sampling techniques are "boostrap" or "bagging" and "boosting"[3]. The D_m share a common representational format and, in general, the S_m will be moderately diverse.

Subspace Techniques. Sample D after feature extraction or after applying a subspace transformation. In many cases, the D_m will have different common representational formats [4]. In general, the S_m will be diverse.

Switching Class Labels. The training sets D_m are generated using the original training set D with randomized class labels [19]. The class labels of each sample is switched according to a probability that depends on an overall switching rate (defined as the proportion of training samples that are switched on average) and on the proportions of the different class labels in D.

14.4.1 Bagging

Given the original training set D we create training sets $D_m, m \in \{1,2,\ldots,M\}$, by randomly sampling D with replacement. On average, each training set D_m only contains two-thirds of the original samples. The matlab code in Ex. 4.22 may be used to generate a set of M bagged training sets D_m.

Bagging is widely used in supervised and unsupervised learning algorithms. The following example illustrates the use of bagging in unsupervised learning situation (see also Ex. 7.12).

Example 14.3. Bagging in a DNA microarray cluster algorithm [5]. An important topic is tumor classification is the identification of new tumor classes using gene expression profiles which are obtained from DNA microarray experiments. An important part of the identification process is the accurate partition of the tumor samples into clusters each cluster corresponding to a different class of tumor. Given the number of clusters K it is important to estimate for each measurement its cluster label and a measure of confidence of the cluster assignment.

In [5] the DNA microarray measurements are bootstrapped. The cluster algorithm is then applied to each set of bootstrap measurements. For each microarray measurement the final partition is obtained by taking a majority cluster label. Mathematically, the steps in the procedure are:

(1) Form a "reference" decision map, $D^* = \{d_1^*, d_2^*, \ldots, d_N^*\}$, by clustering the input measurements $X = \{x_1, x_2, \ldots, x_N\}$.

(2) Form B bootstrap sets of measurements: $X^{(b)} = \{x_1^{(b)}, x_2^{(b)}, \ldots, x_N^{(b)}\}, b \in \{1,2,\ldots,B\}$.

[3] See [3] and Table 4.10 for additional sampling techniques.

[4] This is not however guaranteed and depends on the sampling technique which is used. For further details see Sect. 4.6.

(3) Separately apply the clustering procedure to $X^{(b)}, b \in \{1,2,\ldots,B\}$. Let $D^{(b)} = \{d_1^{(b)}, d_2^{(b)}, \ldots, d_N^{(b)}\}$, denote the corresponding decision, or cluster, map.

(4) Perform semantic alignment on the bootstrap decision maps $D^{(b)}$. In [5] authors performed semantic alignment by permuting the cluster labels assigned to $X^{(b)}$ so there is maximum overlap with the reference decision map D^* (obtained in step (1)).

(5) Use a majority vote algorithm to assign an optimal cluster label to each measurement $x_n, n \in \{1,2,\ldots,N\}$.

14.4.2 Boosting

Boosting is a complete classification system and is discussed in detail Sect. 14.10. However, for the moment, we may regard boosting as simply a method of creating an ensemble of training sets D_m by sampling D [27]. In boosting, we create the $D_m, m \in \{1,2,\ldots,M\}$, one at a time by sampling D as follows: Let S_m be a classifier which was learnt on D_m. Then, we use the classification results obtained with S_m to create D_{m+1} by sampling D such that samples which are misclassified by S_m have a higher chance of being chosen than samples which were correctly classified by S_m.

An interesting variant of boosting is multi-boosting [38] which is illustrated in the next example.

Example 14.4. Multi-Boosting a Common Training Set D [38]. We consider multi-boosting a common training set, D.We first boost D generating M training sets D_m as explained above. We then perform bagging or wagging (see Sect. 4.6) on each of the D_m.

14.4.3 Subspace Techniques

There are a wide variety of different sub-space techniques. The following example illustrates a novel subspace technique due Skuichina and Duin [30]

Example 14.5. Skurichina-Duin Subspace Technique [30]. We perform principal component analysis on the common training set D. Let $\mathbf{U} = (\mathbf{u}_1, \mathbf{u}_2, \ldots, \mathbf{u}_L)^T$ denote the corresponding principal component axes. Then we create the training set $D_m, m \in \{1,2,\ldots,M\}$, by applying $(\mathbf{u}_{(m-1)K+1}, \mathbf{u}_{(m-1)K+2}, \ldots, \mathbf{u}_{mK-1})$ to D, where $K = L/M$. Thus D_1 is formed by applying $(\mathbf{u}_1, \mathbf{u}_2, \ldots, \mathbf{u}_K)$ to D, D_2 is formed by applying $(\mathbf{u}_{K+1}, \mathbf{u}_{K+2}, \ldots, \mathbf{u}_{2K})$ to D, and so on. Fig. 14.1 illustrates the Skurichina-Duin technique.

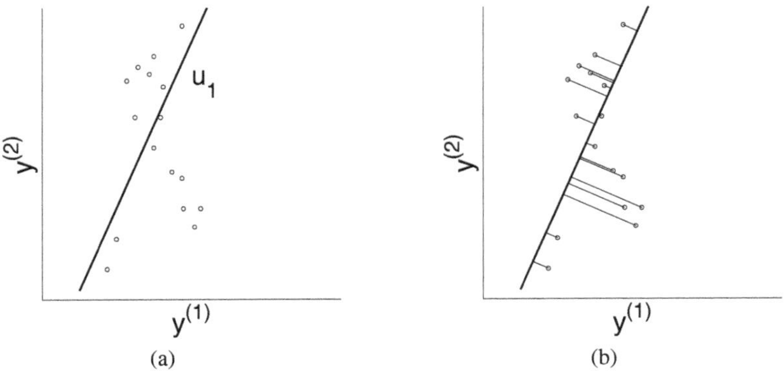

Fig. 14.1 Shows the Skurichina-Duin subspace technique. **(a)** Shows the common training set D which consists of N two-dimensional points $\mathbf{y}_i = (y_i^{(1)}, y_i^{(2)}), i \in \{1,2,\ldots,N\}$. Overlaying the $\mathbf{y}_i$ is the main principal axis $\mathbf{u}_1$. **(b)** Shows the training set D_1 which is formed by projecting the $\mathbf{y}_i$ onto $\mathbf{u}_1$.

14.4.4 Switching Class Labels

In the method of switching class labels [19] we generate the training sets D_m using the original training set D with randomized class labels. The class labels of each sample is switched according to the transition matrix

$$P_{i\to j} = \omega P_j \ ,$$
$$P_{i\to i} = 1 - \omega(1 - P_i) \ ,$$

where $P_{i\to j}$ is the probability that an example which belongs to the class $C = c_i$ is placed in the class $C = c_j$, P_i is the proportion of elements of class $C = c_i$ in the training set and ω is proportional to the switching rate (average fraction of switched examples), p

$$\omega = \frac{p}{1 - \sum_j P_j^2} \ .$$

This form of the transition matrix is ensures that the class proportions remain approximately constant. In order for the method to work, the value of the switching rate p should be small enough to ensure that the training error tends to zero as the size of the ensemble grows. In binary classification, this requires $p < P_{\min}$, where $P_{\min}$ is the proportion of examples that belong to the minority class.

14.5 Diversity Measures

In principle, the techniques discussed in Sect 14.4 may generate an unlimited number of classifier ensembles. In order to select the most appropriate ensemble we need to compare the diversity of one ensemble with the diversity of another ensemble. In Table 14.2 we list several quantitative diversity measures [32] which are commonly used for this purpose [5]. The simplest diversity measures are pairwise measures (P), which are defined between a pair of classifiers. If the ensemble contains $M > 2$ classifiers, we may estimate its diversity by averaging all of the $M(M-1)/2$ pairwise diversity measures. Apart from the pairwise measures, there are also non-pairwise measures (N) which work on an entire ensemble.

Table 14.2 Diversity Measures

Name	Type	Formula
Yule statistic (Q)	P	$Q = (ad - bc)/(ad + bc)$, where a, b, c, d are, respectively, the number of objects $O_i, i \in \{1,2,\ldots,N\}$, that are correctly classified by S_1 and S_2; are correctly classified by S_1 and incorrectly classified by S_2; are incorrectly classified by S_1 and correctly classified by S_2; are incorrectly classified by S_1 and S_2.
Correlation Coefficient (ρ)	P	$\rho = (ad - bc)/\sqrt{(a+b)(c+d)(a+c)(b+d)}$.
Disagreement Measure (D)	P	$D = (b+c)/N$.
Double Fault Measure (DF)	P	$DF = d/N$.
Entropy E	N	$E = \sum_{i=1}^{N} \min(\xi_i, (M - \xi_i))/(N(M - \lceil M/2 \rceil))$, where $\lceil x \rceil$ is the ceiling operator and ξ_i is the number of classifiers that misclassify $O_i, i \in \{1,2,\ldots,N\}$.
Kohavi-Wolpart Variance (KW)	N	$KW = \sum_{i=1}^{N} \xi_i (M - \xi_i)/(NM^2)$.
Measure of Difficulty (θ)	N	$\theta = \sum_{i=1}^{N} (\xi_i - \bar{\xi})^2/(NM^2)$, where $\bar{\xi} = \sum_{i=1}^{N} \xi_i / N$.

14.5.1 *Ensemble Selection*

Given an accurate diversity measure we may select the optimal ensemble by exhaustive enumeration, that is, by comparing the diversity measure of all ensembles and choosing the ensemble with the largest diversity measure. However, this technique is only feasible if the number of ensembles is relatively small. When the number of ensembles is large, we may search for the optimal ensemble using any of the search algorithms used in feature selection and listed in Table 13.5.

Alternatively, and in practice this is the preferred method, we may select the optimal ensemble by directly assessing the classification accuracy of each

[5] In Sect. 14.1 we defined the diversity of an ensemble as the degree to which the pattern of errors made by one classifier, S_m, differs from the pattern of errors made by another classifier, $S_n, n \neq m$. This definition is not, however, very useful in comparing the diversity of one ensemble with another ensemble. For this purpose, we use a quantitative diversity measure.

ensemble (as measured on a given test set using any of the methods listed in Table 13.8).

14.6 Classifier Types

Let O denote an unknown object. In the empirical framework we assume that each classifier $S_m, m \in \{1,2,\ldots,M\}$, generates a *belief* vector, $\boldsymbol{\mu}_m = (\mu_m^{(1)}, \mu_m^{(2)}, \ldots, \mu_m^{(K)})^T$, where $\mu_m^{(k)}$ is the belief that $C = c_k$ is the true assignment. The concept of a belief vector is more general than a probability vector. In particular, belief vectors need not sum to one. Instead we only require that $\mu_m^{(h)} > \mu_m^{(k)}$ if $\hat{p}(C = c_h|\mathbf{y}, I) > \hat{p}(C = c_k|\mathbf{y}, I)$, where $\hat{p}(C = c_k|\mathbf{y}, I)$ is the *a posteriori* probability that $C = c_k$ is the true assignment given the input measurement $\mathbf{y}$ and the background information.

We may identify several types of "belief" vectors [37]:

Probability Vector. The belief vector is equal to the *a posteriori* pdf vector $\mathbf{p}_m = (\hat{p}(C = c_1|\mathbf{y}, I), \hat{p}(C = c_2|\mathbf{y}, I), \ldots, \hat{p}(C = c_K|\mathbf{y}, I)^T$.

Measurement Vector. The belief vector $\boldsymbol{\mu}_m$ is a measurement vector $\mathbf{s}_m$. A measurement vector $\mathbf{s}_m$ is often generated as follows. Off-line we create a set of *prototype* measurements $\mathbf{Y}^{(1)}, \mathbf{Y}^{(2)}, \ldots, \mathbf{Y}^{(K)}$ which correspond to the classes $C = c_1, C = c_2, \ldots, C = c_K$. Then, $\mu_m^{(k)}$ is obtained by measuring the similarity between $\mathbf{y}$ and $\mathbf{Y}^{(k)}$ using the mth (proximity) degree-of-similarity operator S_m:

$$s_m^{(k)} = S_m\big(\mathbf{y}, \mathbf{Y}^{(k)}\big) .$$

*Note.*In principle, the measurement vector may be converted into a probability vector (Sect. 8.9). However, in practice, this is rarely done.

Fuzzy Membership Vector. The belief vector is a vector of fuzzy membership function values.

Rank Vector. The belief vector is a vector of preferences, or ranks, $\mathbf{R}_m = (r_m^{(1)}, r_m^{(2)}, \ldots, r_m^{(K)})^T$, where

$$r_m^{(k)} = K - l \qquad \text{if } \mu_m^{(k)} \text{ is } l\text{th smallest value in } b_m^{(l)}, l \in \{1,2,\ldots,K\} \ . \tag{14.10}$$

Identity Vector. The belief vector $\boldsymbol{\mu}_m$ is an identity vector $\mathbf{I}_m = (I_m^{(1)}, I_m^{(2)}, \ldots, I_m^{(K)})^T$, where

$$I_m^{(l)} = \begin{cases} 1 \text{ if } \mu_m^{(l)} = \max(\mu_m^{(1)}, \mu_m^{(2)}, \ldots, \mu_m^{(K)}) , \\ 0 \ \text{ otherwise} . \end{cases} \tag{14.11}$$

The matlab code in Ex. 8.21 may be used to convert the belief vector $\boldsymbol{\mu}$ into a rank vector $\mathbf{R}$. The following matlab code converts $\boldsymbol{\mu}$ into an identity vector $\mathbf{I}$.

Example 14.6. Matlab Code for Identity Vector.
Input
$MU = (mu(1), mu(2), \ldots, mu(N))$ is the input belief vector of length N.
Output
$IVEC$ is the corresponding identity vector.
Code.

```
IVEC = zeros(1,N);
[junk,kmax] = max(MU);
IVEC(kmax) = 1;
```

14.7 Combination Strategies

Let O denote an unknown object. Suppose each classifier $S_m, m \in \{1,2,\ldots,M\}$, generates a *belief* vector regarding the classification of O. Then the MCS generates a new belief vector, $\widetilde{\boldsymbol{\mu}}$, by combining the $\boldsymbol{\mu}_m$:

$$\widetilde{\boldsymbol{\mu}} = f(\boldsymbol{\mu}_1, \boldsymbol{\mu}_2, \ldots, \boldsymbol{\mu}_M)^T , \tag{14.12}$$

where f denotes a MCS combination function.

Broadly speaking we may divide the combination functions, f, into two groups:

Simple Combiners (S). The simple combiners are simple aggregation funcions such as product, mean, maximum and minimum. The functions, f, may also include weights and parameters which are learnt on the training data D. In general, the simple combiners are best suited for problems where the individual classifiers perform the same task and have comparable performances.

Meta-Learners (M). The meta-learners include the methods of stacked generalization and mixture of experts, where the combination function is a meta-classifier, S_0, which acts on the outputs of the individual $S_m, m \in \{1,2,\ldots,M\}$, or on the input measurement vector $\mathbf{y}$. Meta-learners are more general than the simple combiners but they are more vulnerable to problems of overfitting.

In Table 14.3 we list the most common simple combiners and meta-learners.

Table 14.3 Combination Strategies

Name	Type	Description
Mean Vote	S	Used when each classifier S_m produces a belief vector μ_m. In this case, the mean vote for the kth class is $\frac{1}{M}\sum_{m=1}^{M}\mu_m^{(k)}$. The class with the highest average belief value in the ensemble wins.
Maximum, Minimum, Median Vote	S	Used when each classifier S_m produces a belief vector μ_m. In this case, the maximum, minimum or median vote for the kth class is $\max_m(\mu_m^{(k)})$, $\min_m(\mu_m^{(k)})$ or $\mathrm{median}_m(\mu_m^{(k)})$. The class with the highest maximum, median or minimum belief value in the ensemble wins.
Borda count	S	Used when each classifier S_m produces a rank vector $\mathbf{R}_m$. In this case, the Borda count for the kth class is $B(k)=\sum_{m=1}^{M}B_m(k)$, where $B_m(k)$ is the number of classes ranked below class $C=c_k$ by the mth classifier. The class with the highest Borda count wins.
Majority Vote	S	Used when each classifier S_m produces an identity vector $\mathbf{I}_m$. In this case, the majority vote for the kth class is $\sum_{m=1}^{M}I_m^{(k)}$. The class with the majority vote in the ensemble wins.
Weighted Simple Combiners	S	In weighted versions of the above simple combiners, the output of each classifier S_m is given a weight w_m, where $0\leq w_m\leq 1$ and $\sum_m w_m=1$. The weight w_m is often a measure of the performance of S_m performance as measured on a separate validation test set.
Stacked Generalization	M	The output of the ensemble of $S_m, m\in\{1,2,\ldots,M\}$, serves as a feature vector to a meta-classifier.
Mixture of Experts (ME)	M	The feature space is partitioned into different regions, with one expert (S_m) in the ensemble being responsible for generating the correct output within that region. The experts in the ensemble and the partition algorithm are trained simultaneously which can be performed using the EM algorithm.
Hierarchical ME	M	Mixture of experts is extended to a multi-hierarchical structures where each component is itself a mixture of experts.

14.8 Simple Combiners

The choice of combination function f depends very much upon the nature of the belief vectors $\mu_m, m\in\{1,2,\ldots,M\}$, upon which it acts. For example, the maximum combination function cannot be used if the μ_m are identity vectors. We shall therefore divide the simple combiners into three sub-classes: identity, rank and belief combiners.

14.8.1 Identity Combiners

The identity combiner only uses the class information contained in the identity vectors $\mathbf{I}_m, m\in\{1,2,\ldots,M\}$. The *majority voter* is probably the most popular identity combiner. It finds the optimal class c_{OPT} by selecting the class which is chosen most often by different classifiers. Mathematically, the optimal class is

$$c_{\mathrm{OPT}} = \arg\max_k \left(\sum_m I_m^{(k)} \right) , \qquad (14.13)$$

where $I_m^{(k)} = 1$ if the mth classifier S_m assigns O to the class $C = c_k$; otherwise $I_m^{(k)} = 0$.

A variant of the majority voter is the *weighted majority voter*. This is a majority voter in which each vector $\mathbf{I}_m$ is given a weight, w_m, which is often learnt on a training data set D.

Example 14.7. Weighted Majority Vote [23]. Each classifier $S_m, m \in \{1,2,\ldots, M\}$ is given a weight w_m in proportion to its estimated performance. Then the optimal class is

$$c_{\mathrm{OPT}} = \arg\max_k \left(\sum_{m=1}^{M} w_m I_m^{(k)} \right) .$$

If the S_m are class independent, then the optimal weights are

$$w_m = \log \frac{p_m}{1-p_m} ,$$

where p_m is the probability of correct classification if we were to use S_m by itself.

In some applications the weights are not learnt on a training data set D. The following example illustrates one such application.

Example 14.8. Image Thresholding Using an Unsupervised Weighted Majority Vote Rule [21]. The effectiveness of a thresholding algorithm is strongly dependent on the input image characteristics. Experimental results confirm that for one input image, a given thresholding algorithm may appear the best, while it may fail completely for another image. This makes it difficult to choose the most appropriate algorithm to binarize a given image. One way to solve this problem is to binarize the input image using M different thresholding algorithms and then fuse the thresholded images together using an *unsupervised* weighted majority vote rule as follows. Suppose t_m is the optimal threshold found by the mth thresholding algorithm. If $g^{(k)}$ denotes the gray-level of the kth pixel, then $I_m^{(k)} = 1$ if the $g^{(k)} \geq t_m$, otherwise $I_m^{(k)} = 0$. For each pixel, we form the weighted sum

$$S^{(k)} = \frac{\sum_{m=1}^{M} I_m^{(k)} w_m^{(k)}}{\sum_{m=1}^{M} w^{(k)}} ,$$

where $w_m^{(k)}$ denotes the weight associated with the mth thresholding algorithm when applied to the kth pixel. The final binary image $B^{(k)}$ is

$$B^{(k)} = \begin{cases} 1 \text{ if } S^{(k)} \geq \frac{1}{2} , \\ 0 \text{ otherwise} . \end{cases}$$

The formula for the weights $w_m^{(k)} = 1 - \exp(-\gamma |g^{(k)} - t_m|)$ is based on the idea that the larger the difference $|g^{(k)} - t_m|$, the higher the degree of confidence in the decision of the mth algorithm, where γ is a real positive constant which controls the steepness of the weight function.

The behaviour-knowledge space (BKS) [12] is a supervised identity-level combiner which takes into account inter-classifier correlations. In the BKS combiner we record how often the classifiers $S_m, m \in \{1,2,\ldots,M\}$, produce a given combination of class labels. If there are K classes, then altogether there are K^M different combinations of class labels. The true class for which a particular combination of class labels is observed most often, is chosen every time that a given combination of class labels occurs during testing. Let $n(k,l)$ denote the number of objects $O_i, i \in \{1,2,\ldots,N\}$, in D which belong to class $C = c_k$ and which have the lth combination of class labels ω_l. Suppose the test object is found to belong to a given combination ω_{l^*}. Then we classify it as belong to the class

$$c_{\text{OPT}} = \arg\max_k \Big(n(k, l^*) \Big) .$$

Although, theoretically, BKS has a high performance, it is limited by the very large amounts of training data which are required [6].

Example 14.9. BKS Look-up Table. We consider the construction of a BKS look up table using a training set D containing N labeled objects $O_i, i \in \{1,2,\ldots,N\}$. We assume $M = 3$ classifiers $S_m, m \in \{1,2,3\}$, and $K = 2$ classes $C \in \{c_1, c_2\}$. Altogether there are $L = K^M = 2^3 = 8$ different label combinations, $\omega_l, l \in \{1,2,\ldots,8\}$, where

$$\begin{aligned} \omega &= (\omega_1, \omega_2, \ldots, \omega_L)^T , \\ &= ((1,1,1),(1,1,2),(1,2,1),(1,2,2),(2,1,1),(2,1,2),(2,2,1),(2,2,2))^T , \end{aligned}$$

where $\omega_l = (\alpha, \beta, \gamma)$ denotes that in the lth combination of class labels, ω_l, the classifiers S_1, S_2 and S_3 returned, respectively, the class labels $C = c_\alpha, C = c_\beta$

[6] Given K classes and M classifiers we have K^M label combinations $\omega_l, l \in \{1,2,\ldots,K^M\}$. If each combination requires a minimum number $n_{\min}$ of training samples, then the BKS algorithm requires at least $K^M \times n_{\min}$ training samples.

and $C = c_\gamma$. A possible distribution of the objects $O_i, i \in \{1,2,\ldots,N\}$, among the $\omega_l, l \in \{1,2,\ldots,L\}$, is given in table 14.4. Suppose a test object, O, has a combination of class labels $\Omega = \omega_l = (1,2,1)$, then we assign O to the class $C = c_{OPT}$:

$$c_{OPT} = \max_k \left(p(C = c_k|\Omega = (1,2,1)\right) = c_1 ,$$

where

$$p(C = c_1|\Omega = (1,2,1)) = \frac{76}{76+17} = 0.82 ,$$
$$p(C = c_2|\Omega = (1,2,1)) = \frac{17}{76+17} = 0.18 .$$

Table 14.4 Distribution of training objects O_i among label combinations $\{\omega_l\}$

$C = c_k$	$n(k,l)$							
c_1	100	50	76	89	54	78	87	5
c_2	8	88	17	95	20	90	95	100

$n(k,l)$ is the number of objects $O_i, i \in \{1,2,\ldots,N\}$, in D which belong to class $C = c_k$ and which have the lth combination of class labels ω_l.

A modification of the BKS look-up table which requires much less training data is the pairwise fusion matrix [15]. In the pairwise fusion matrix we divide the M classifiers $S_m, m \in \{1,2,\ldots,M\}$, into an ensemble $\mathbf{P} = (P_1, P_2, \ldots, P_N)^T$ of all possible pairs of classifiers, where $P_1 = (S_1,S_2)$, $P_2 = (S_1,S_3), \ldots, P_N = (S_{M-1},S_M)$ and $N = M(M-1)/2$. Then for each pair of classifiers we perform a BKS-type classification where each P_i has only K^2 label combinations $\omega_l, l \in \{1,2,\ldots,K^2\}$, associated with it. Finally the set of N classifications are fused together using an appropriate single combiner. The following example illustrates the principle of the pairwise fusion matrix.

Example 14.10. Pairwise Fusion Matrix [15]. We reconsider Ex. 14.9. The training set D contains N labeled objects $O_i, i \in \{1,2,\ldots,N\}$. We assume $M = 3$ classifiers $S_m, m \in \{1,2,3\}$, and $K = 2$ classes $C \in \{c_1,c_2\}$. We arrange the classifiers into $N = 3$ pairs: $P_1 = (S_1,S_2)$, $P_2 = (S_1,S_3)$ and $P_3 = (S_2,S_3)$. For each pair of classifiers there are $L = K^2 = 4$ different label combinations, $\omega_l, l \in \{1,2,3,4\}$, where $\boldsymbol{\omega} = (\omega_1,\omega_2,\omega_3,\omega_4)^T = ((1,1),(1,2),(2,1),(2,2))^T$ and $\omega_l = (\alpha,\beta)$ denotes that in the lth combination of class labels the pair of classifiers returned respectively, the class labels

$C = c_\alpha$ and $C = c_\beta$. In Table 14.8.1 we give, for each pair of classifiers, the distribution of the objects $O_i, i \in \{1,2,\ldots,N\}$, among the labels ω_l. Suppose a test object O has the following combination of class labels

$$\Omega = \omega_3 = (2,1) \text{for classifier pair } P_1 = (S_1,S_2) ,$$
$$\Omega = \omega_4 = (2,2) \text{for classifier pair } P_2 = (S_1,S_3) ,$$
$$\Omega = \omega_2 = (1,2) \text{for classifier pair } P_3 = (S_2,S_3) .$$

Then the corresponding class assignments are:

$$C = c_2 \text{ from classifier pair } P_1 = (S_1,S_2) ,$$
$$C = c_2 \text{ from classifier pair } P_2 = (S_1,S_3) ,$$
$$C = c_1 \text{ from classifier pair } P_3 = (S_2,S_3) .$$

The final global class assignment is found by fusing these class assignments. For example, if we use a majority vote combiner, then the final class assignment $C = c_{\text{OPT}}$ is c_2.

Table 14.5 Distribution of training objects O_i among label combinations $\{\omega_l\}$ for each pair of classifiers

Pair of Classifiers	$C = c_k$	$n(k,l)$			
(S_1,S_2)	c_1	90	6	3	1
	c_2	0	9	21	70
(S_1,S_3)	c_1	70	10	15	5
	c_2	1	6	3	90
(S_2,S_3)	c_1	84	13	3	0
	c_2	10	5	14	74

$n(k,l)$ is the number of objects $O_i, i \in \{1,2,\ldots,N\}$, in D which belong to class $C = c_k$ and which have the lth combination of class labels ω_l.

14.8.2 Rank Combiners

Rank combiners use the information contained in the rank vectors $\mathbf{R}_m, m \in \{1,2,\ldots,M\}$. These combiners are used when we do not believe the *a posteriori* probability distributions generated by a classifier but we do believe the relative ordering of the distributions. The Borda count is probably the most popular rank-level combiner. It works as follows: For any class c_k, let $B_m(k)$ be the number of classes ranked below the class c_k by the mth classifier S_m. Then the Borda count[7] for the class $C = c_k$ is

[7] If there are no ties, then by definition, $B_m(k) = R_m^{(k)}$. In the case of ties we may let $B_m(k)$ be equal to the average number of classes below c_k if we randomly break the ties.

$$B(k) = \sum_{m=1}^{M} B_m(k) \ . \tag{14.14}$$

The optimum class, c_{OPT} is defined as the class with the largest Borda count:

$$c_{\mathrm{OPT}} = \arg\max_k \left(\sum_{m=1}^{M} B_m(k) \right) \ . \tag{14.15}$$

Example 14.11. Image Orientation Using Borda Count [18]. We describe an image orientation algorithm based on the Borda count. The orientation of an input image is measured in $M = 20$ different ways. Each classifier $S_m, m \in \{1,2,\ldots,M\}$, measures the orientation, θ, of an input image and returns a rank vector $\mathbf{R}_m = (R_m^{(1)}, R_m^{(2)}, \ldots, R_m^{(K)})^T$, where $R_m^{(k)}$ is the degree-of-support (expressed as a rank) that θ falls in the interval $[2\pi(k-1)/K, 2\pi k/K]$. In [18] the authors used $K = 4$.

A variant of the Borda count is the *weighted Borda count*. This a Borda count combiner in which each vector $\mathbf{R}_m$ is given a weight w_m which is usually learnt on a training data set D.

14.8.3 Belief Combiners

The belief combiners use all of the information contained in the belief vectors $\mu_m, m \in \{1,2,\ldots,M\}$. The *mean*, or average, vote is a popular belief combiner. It finds the optimal class c_{OPT} by selecting the class with the highest average value. Mathematically, the optimal class is

$$c_{\mathrm{OPT}} = \arg\max_k \left(\frac{1}{M} \sum_m \mu_m^{(k)} \right) , \tag{14.16}$$

where $\mu_m^{(k)}$ represents the belief as generated by the mth classifier for the kth class.

A variant of the mean vote is the *weighted mean (average) vote* in which each belief vector is given a weight w_m which is usually learnt.

Example 14.12. Weighted Mean Vote. The weighted mean approach is similar to the mean vote. However, the outputs of the various classifiers are multiplied with a weighting factor or

$$c_{\mathrm{OPT}} = \arg\max_k \left(\frac{1}{M} \sum_m^M w_m \mu_m^{(k)} \right) \tag{14.17}$$

The weights $w_m, m \in \{1,2,\ldots,M\}$, can be derived by minimizing the error of the different classifiers on the training set.

In principle the belief vector $\boldsymbol{\mu}_m$ can be transformed into an *a posteriori* pdf. In this case we can interpret the weights w_m as the *a posteriori* probabilities $p(S_m|\mathbf{y},I)$ and use the Bayesian model averaging technique (see Ex. 14.1). In the following example we compare the action of several different simple combiner functions.

Example 14.13. Combination Function: A Numerical Experiment [23]. Given an object O we assume the following decision profile:

$$\mathrm{DP} = \begin{pmatrix} 0.85 & 0.01 & 0.14 \\ 0.30 & 0.50 & 0.20 \\ 0.20 & 0.60 & 0.20 \\ 0.10 & 0.70 & 0.20 \\ 0.10 & 0.10 & 0.80 \end{pmatrix},$$

where the (m,k)th element of DP is defined to be the belief $\mu_m^{(k)}$ as calculated for the class, $C = c_k$, by the mth classifier, S_m. The corresponding results obtained when we use the following combination functions: product, maximum, mean, median, minimum, Borda count and majority vote, are listed in the table 14.6 together with their optimal class assignment $C = c_{\mathrm{OPT}}$.

Example 14.14. Weighted Combination Functions: A Numerical Experiment [23] We repeat Ex. 14.13 assuming the classifiers S_m are assigned the following weights: $w_1 = 0.30, w_2 = 0.25, w_3 = 0.20, w_4 = 0.10, w_5 = 0.15$. In Table 14.7 we list the results obtained using the weighted majority vote and the weighted mean combination functions.

Table 14.6 Combination Strategies for Ex. 14.13

Combiner f		$\widetilde{\mu}^T$		$C = c_{\mathrm{OPT}}$
Majority Vote	1	3	1	c_2
Borda Count	4	6.5	4.5	c_2
Product	0.0179	0.0002	0.009	c_1
Maximum	0.85	0.70	0.80	c_1
Mean	0.310	0.382	0.308	c_2
Median	0.20	0.50	0.20	c_2
Minimum	0.10	0.01	0.14	c_3

Table 14.7 Combination Strategies for Ex. 14.14

Combiner f	$\widetilde{\mu}^T$			$C = c_{OPT}$
Weighted Majority Vote	0.30	0.55	0.15	c_2
Weighted Mean	0.395	0.0333	0.272	c_1

14.9 Meta-learners

In the meta-learners, the combination function is a meta-classifier which we denote by $\widetilde{S}$. In the stacked generalization model [16, 34] the classifiers $S_m, m \in \{1,2,\ldots,M\}$, are trained on the training sets D_m. The outputs of these classifiers and the corresponding true classes are then used as input/output training pairs for the meta-classifier $\widetilde{S}$ (see Fig. 14.2). However, by training the meta-classifier on the same training data used to train the pairwise classifiers, $\widetilde{S}$ may suffer from overfitting. One way of reducing the problem of overfitting is to train the meta-classifier $\widetilde{S}$ via cross-validation. This is explained in Ex. 14.15. *Note.* In place of cross-validation bootstrap may be used.

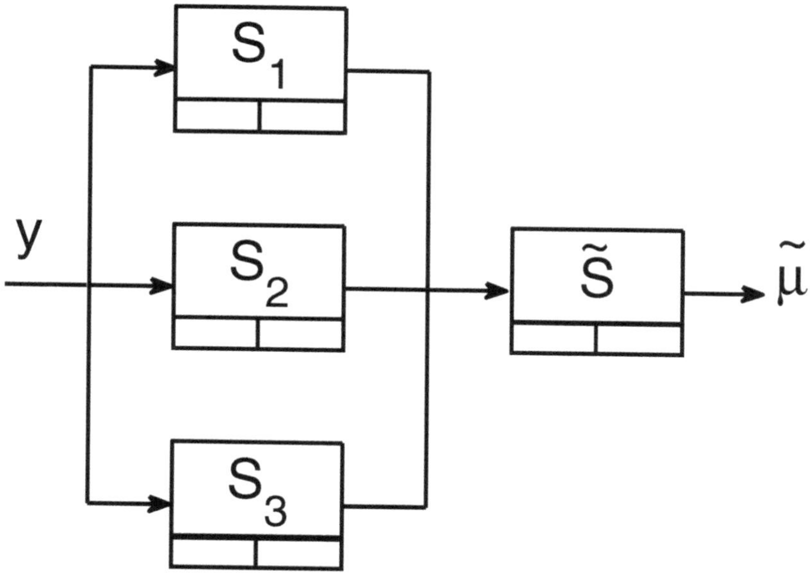

Fig. 14.2 Shows the stacked generalization. This consists of an ensemble of $M = 3$ classifiers $S_m, m \in \{1,2,3\}$, and a meta-classifier $\widetilde{S}$. The input to $\widetilde{S}$ is the belief vector $\mu_m, m \in \{1,2,3\}$, generated by the S_m. The final output, as generated by $\widetilde{S}$, is a new belief vector $\widetilde{\mu}$, or in some cases, a new identity vector $\widetilde{\mathbf{I}}$.

Example 14.15. Pairwise Classification Using Stacking [29]. We consider the multi-class classification of an object O into one of K classes using $K(K-1)/2$ pairwise classifiers. The outputs of the pairwise classifiers and the corresponding true classes are used as input/output training pairs for the stacked generalization model.

We partition the training data D into L separate training sets $D_1, D_2, \ldots, D_L$. Then we define $\widetilde{D}_l$ as the training set obtained when we remove D_l from D. We train each binary pairwise classifier on $\widetilde{D}_l, l \in \{1,2,\ldots,L\}$. We thus obtain $M = LK(K-1)/2$ pairwise classifiers $S_m, m \in \{1,2,\ldots,M\}$, which are then used as input to the stacked meta-classifier $\widetilde{S}$.

Another meta-learner is the mixture of expert model. This is illustrated in the next example.

Example 14.16. Mixture of Experts Model [13, 16]. The mixture of expert model (see Fig. 14.3) is conceptually similar, but different, to the stacked generalization model. In the mixture of experts model, the input to $\widetilde{S}$ is the measurement vector $\mathbf{y}$ and the output is a weight vector $\mathbf{w} = (w_1, w_2, \ldots, w_M)^T$. The final output of the mixture of experts model is a belief vector, $\widetilde{\boldsymbol{\mu}}$, which is formed by combining the belief vectors $\boldsymbol{\mu}_m, m \in \{1,2,\ldots,M\}$, with $\mathbf{w}$ using a simple combiner F. The mixture of experts model is often interpreted as a classifier selection algorithm in which individual classifiers are regarded as experts in some portion of the feature space. In this case, for each measurement vector $\mathbf{y}$, $\widetilde{S}$ selects the most appropriate classifier, S_n, by selecting a weight vector $\mathbf{w} = (w_1, w_2, \ldots, w_M)^T$, where $w_n = 1$ and $w_m = 0, m \neq n$.

14.10 Boosting

A recent development in ensemble learning is the idea of "boosting" [27]. Boosting has proved itself to be an effective ensemble learning algorithm. At present, several different versions of the boosting algorithm exist, some of which are listed in Table 14.8.

We first introduced the concept of boosting in Sect. 4.6, where we described it as a technique for generating an ensemble of training sets $D_m, m \in \{1,2,\ldots,M\}$. Boosting is, however, much more than this. Boosting is a *complete* technique for ensemble learning which includes both an iterative method for generating multiple training sets D_m and an algorithm for combining the corresponding classifiers S_m [8].

[8] From this point of view, bagging and wagging are also complete ensemble learning techniques, since they both contain a method for creating multiple training sets $D_m, m \in \{1,2,\ldots,M\}$, on which the classifiers S_m are learnt, as well as an algorithm (majority vote) for combining the S_m.

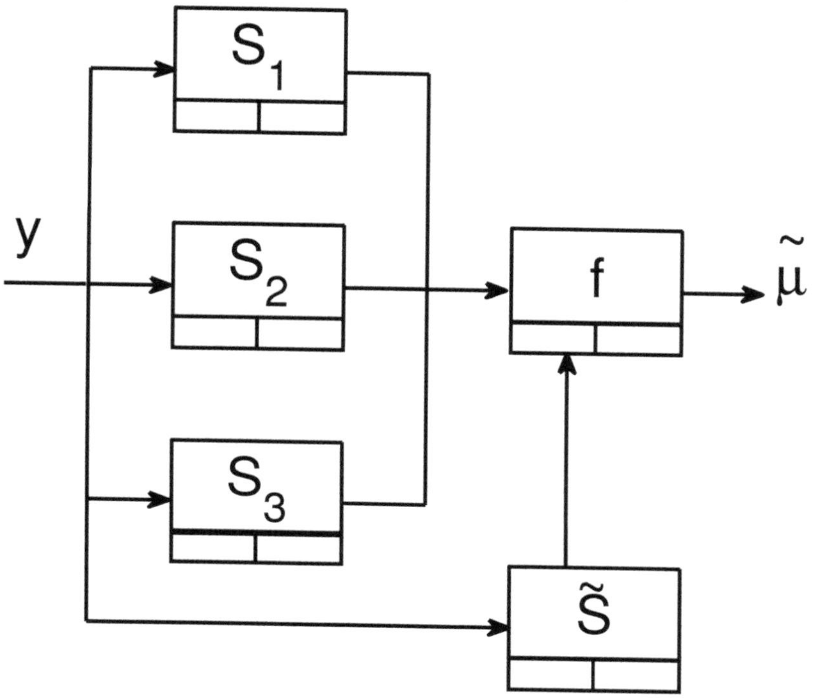

Fig. 14.3 Shows the mixture of experts model. This consists of an ensemble of $M = 3$ classifiers $S_m, m \in \{1,2,3\}$, and a meta-classifier $\tilde{S}$. The meta-classifier outputs a weight vector $\mathbf{w} = (w_1, w_2, w_3)^T$. The final output is formed by combining the belief vectors $\mu_m, m \in \{1,2,3\}$, with the weight vector $\mathbf{w}$, using a simple combiner f.

Table 14.8 Boosting Algorithms

Name	Description
AdaBoost	The most popular boosting algorithm [27].
Real AdaBoost	A generalization of AdaBoost. Real Adaboost is now regarded as the basic boosting algorithm [27].
Modest AdaBoost	A regularized version of AdaBoost [35].
Gentle AdaBoost	A more robust and stable version of Real AdaBoost [7].
FloatBoost	An improved algorithm for learning a boosted classifier [17].
SmoteBoost	A boosting algorithm which improves the performance of classes which are under-represented in the training set D [4].
MadaBoost, EtaBoost, LogitBoost	Various robust versions of AdaBoost.
SpatialBoost	An algorithm which adds spatial reasoning to AdaBoost [1].
MutualBoost	See [28].

The most common boosting algorithm is “Adaboost” which is based on incrementally build a set of classifiers that are combined to yield a more powerful decision rule. At each boosting step a new classifier is generated, and the training samples are re-weighted according to the classification rules. The weights are used to generate the classifier for the next step. In mathematical terms it works on a two class problem as follows. Let D denote a common training set containing N samples $\mathbf{y}_i, i \in \{1,2,\ldots,N\}$, and let S denote a parametric classifier. In the mth iteration, we create a training set D_m, by randomly sampling the common training set D with replacement, using a non-uniform probability distribution p_i, where $p_i = w_i / \sum w_i$. We then use D_m to learn the parameters of S. The result is a fixed classifier S_m, i. e. a classifier which does not have any free parameters. In boosting we adjust the weights $w_i, i \in \{1,2,\ldots,N\}$, in accordance with the performance of the previous classifier, S_{m-1}, which was trained on the previous training set, D_{m-1}. By increasing the weights of misclassified samples we cause the learning algorithm to focus on different samples and thus lead to different classifiers S_m. The sequence of classifiers $S_m, m \in \{1,2,\ldots,M\}$, are then fused together which should have a superior performance.

The following example describes the pseudo-code for creating a set of fixed classifiers $S_m, \in \{1,2,\ldots,M\}$, by boosting a parametric classifier S.

Example 14.17. Boosting a Parametric Classifier S [6, 25]. In boosting a parametric classifier, S, we learn a series of fixed classifiers $S_m, m \in \{1,2,\ldots,M\}$, on the training sets D_m. The steps in the boosting algorithm are:

1. Set $w_i^{(1)} = 1, i \in \{1,2,\ldots,N\}$.
2. For $m = 1$ to M perform the following steps.
   (a) Create a training set D_m by randomly sampling the common training set D with replacement and according to the non-uniform probability distribution $p_i^{(m)} = w_i^{(m)} / \sum_{i=1}^{N} w_i^{(m)}$.
   (b) Learn the fixed classifier S_m on D_m.
   (c) Calculate the normalized error ε_m for the mth classifier S_m. If δ_i is a suitable error function, then
   $$\varepsilon_m = \sum_{i=1}^{N} p_i^{(m)} \delta_i .$$
   (d) Update the weights according to
   $$w_i^{(m+1)} = w_i^{(m)} \beta_m^{1-\delta_i} ,$$
   where $\beta_m = \varepsilon_m / (1 - \varepsilon_m)$.
3. Output the final boosted *a posteriori* probability that O belongs to class $C = c$:

$$p(\mathbf{y}|C=c,I) = \frac{1}{1+\prod_{m=1}^{M}\beta_m^{2r-1}},$$

where $p_m(C=c|\mathbf{y},I)$ denotes the *a posteriori* probability that O belong to $C=c$ as estimated by S_m; and

$$r = \sum_{m=1}^{M} p_m(C=c|\mathbf{y},I)\ln(1/\beta_m) / \sum_{m=1}^{M}\ln(1/\beta_m).$$

The basic idea behind adaboosting is to build a strong classifier from a set of weak classifiers. At each round, we select the weak classifier with the lowest weighted classification error and adjusts the weights based on this error. Weight adjustment aims to alter the distribution of training samples such that the weak classifier selected at current round m is "uncorrelated" with the class label in the next round $m+1$. However the classifier selected at round $m+1$ could be similar to the one selected in a previous round $m', 0 < m' < t$. As a result, many classifiers selected by the adaboost algorithm might be similar, and are redundant. A better solution is to adjust the weights in such a way that all of the m selected weak classifiers are "uncorrelated" with the class label. However, such a M-dimensional optimization problem might be practically intractable, and the solution might not even exist.

Instead the mutualboost [28] algorithm uses mutual information to eliminate non-effective weak classifiers. Before a weak classifier is selected, the mutual information between the new classifier and each of the previously selected ones is examined to make sure that the information carried by the new classifier has not been captured before.

Although originally developed for applications involving classification, boosting has also been used in non-classification applications as we illustrate in the following example.

Example 14.18. Boosting Kernel Density Estimation [20]. Given a random sample of observations $y_i, i \in \{1,2\ldots,N\}$ from an unknown density $f(y)$, the kernel density estimate of f at the point y is

$$\hat{f}(y) = \frac{1}{NH}\sum_{i=1}^{N} K\left(\frac{y-y_i}{H}\right),$$

where H is the bandwidth and $\int K(y) = 1$. For an accurate estimate of $f(y)$ we must carefully choose the bandwidth H (see Table 5.3). Alternatively, we may boost a simple rule-of-thumb bandwidth estimator by comparing $\hat{f}(y)$ with the cross-validation "leave-one-out" density estimate:

$$\hat{f}_i(y_i) = \frac{1}{N-1} \sum_{j=1 j \neq i}^{N} \frac{1}{H} K\left(\frac{y_i - y_j}{H}\right).$$

1. Set $w_i^{(i)} = 1/N, i \in \{1,2,\ldots,N\}$.
2. For $m = 1$ to M perform the following steps.
   (a) Obtain a weighted kernel estimate

$$\hat{f}^{(m)}(y) = \sum_{i=1}^{N} \frac{w_i^{(m)}}{H} K\left(\frac{y - y_i}{H}\right),$$

   where $H = 1.06\sigma N^{-1/5}$ is the simple "rule-of-thumb" bandwidth and σ is the sample standard deviation.
   (b) Update the weights $w_i^{(m)}$ according to

$$w_i^{(m+1)} = w_i^{(m)} + \log\left(\frac{\hat{f}^{(m)}(y_i)}{\hat{f}_i(y_i)}\right).$$

3. Output the final boosted estimate of f:

$$f_B(y) = \prod_{m=1}^{M} \hat{f}^{(m)}(y) / \int (\prod_{m=1}^{M} \hat{f}^{(m)}(y)) dy$$

   where the denominator is a normalization constant which ensures that $f_B(y)$ integrates to one.

14.11 Recommendations

The various ensemble generation functions have been compared for a wide variety of different classification problems [23]. The consensus is that boosting usually achieves better generalization than bagging and wagging but it is somewhat more sensitive to noise and outliers. Regarding the combination functions: In general, the preferred functions are majority vote, Borda count and averaging.

14.12 Software

ENTOOL. A matlab toolbox for performing ensemble modeling. Authors: C. Merkwirth, J. Wichard.

GML. A matlab toolbox for performing variants of the AdaBoost algorithms: Real, Modest and Gentle-AdaBoost algorithms. Authors: A. Vezhnevets, V. Vezhnevets.

HME. A matlab toobox for implementing hierarchical mixture of experts algorithm. Author: David R. Martin.

MIXLAB. A matlab toolbox for implementing mixture of experts algorithm. Author: Perry Moerland.

STPRTOOL. A matlab toolbox for performing statistical pattern recognition. Authors: Vojtech Franc, Vaclav Hlavac. The toolbox contains m-files for boosting and cross-validation.

Problems

14.1. Describe the basic idea of ensemble learning.

14.2. Describe the Bayesian framework for ensemble learning.

14.3. Compare and contrast the empirical and Bayesian frameworks for ensemble learning.

14.4. Compare and contrast two methods for constructing an ensemble of diverse training sets.

14.5. Compare and contrast the following diversity measures: Yule statistic, correlation coefficient, disagreement measure, double fault measure, entropy, Kohavi-Wolpart variance and measure of difficulty.

14.6. Compare and contrast the following combination strategies: mean vote, Borda count, majority vote, mixture of experts.

14.7. Describe the BKS (behaviour-Knowledge space) combiner.

14.8. Describe the method of boosting.

14.13 Further Reading

Modern up-to-date reviews of ensemble learning are [23, 24, 26]. In addition, Kuncheva [16] has written a full length book on the subject. For a good overview of boosting see the review article by Schapire [27]. For additional modern references see the references listed in Table 14.8. Recently boosting has been successfully combined with the naive Bayes' classifier [6, 25, 36]. Recently encouraging results have been obtained in multi-class classification by using meta-classifiers to combine pairwise classifiers [29]. See also [8, 9]

References

1. Avidan, S.: SpatialBoost: Adding spatial reasoning to Adaboost. In: Proc. 9th Euro. Conf. Comp. Vis., pp. 780–785 (2006)
2. Boulle, M.: Regularization and averaging of the selective naive Bayes classifier. In: Proc. 2006 Int. Joint Conf. Neural Networks, pp. 2989–2997 (2006)

3. Breiman, L.: Random forests. Mach. Learn. 45, 5–32 (2001)
4. Chawla, N.V., Lazarevic, A., Hall, L.O., Bowyer, K.W.: SmoteBoost: Improving Prediction of the Minority Class in Boosting. In: Lavrač, N., Gamberger, D., Todorovski, L., Blockeel, H. (eds.) PKDD 2003. LNCS (LNAI), vol. 2838, pp. 107–119. Springer, Heidelberg (2003)
5. Dudoit, S., Fridlyand, J.: Bagging to improve the accurcay of a clustering procedure. Bioinformatics 19, 1090–1099 (2003)
6. Elkan, C.: Boosting and naive Bayes learning. Tech Rept CS97-557. University of California, San Diego (September 1997)
7. Friedman, J., Hastie, T., Tibshirani, R.: Additive logistic regression: a statistical view of boosting. Ann. Stat. 38, 337–374 (2000)
8. Galar, M., Fernandez, A., Barrenechea, E., Bustince, H., Herrero, F.: Patt. Recogn. 44, 1761–1776 (2011)
9. Garcia-Pedrajas, N., Ortiz-Boyer, D.: An empirical study of binary classifier fusion methods for multi-class classification. Inf. Fusion. 12, 111–130 (2011)
10. Ghahramani, Z., Kim, H.-C.: Bayesian classifier combination. Gatsby Tech Rept, University College, University of London, UK (2003)
11. Gonzales-Barron, U., Butler, F.: J. Food Engng. 74, 268–278 (2006)
12. Huang, Y.S., Suen, C.Y.: A method of combining multiple experts for the recognition of unconstrained handwritten numerals. IEEE Trans. Patt. Anal. Mach. Intell. 17, 90–94 (1995)
13. Jordan, M.I., Jacobs, R.A.: Hierarchical mixture of experts and the EM algorithm. Neural Comp. 6, 181–214
14. Kittler, J., Hatef, M., Duin, R.P.W., Matas, J.: On combining classifiers. IEEE Trans. Patt. Anal. Mach. Intell. 20, 226–239 (1998)
15. Ko, A.H.R., Sabourin, R., de Souza Britto Jr., A., Oliveria, L.: Pairwise fusion matrix for combining classifiers. Patt. Recogn. 40, 2198–2210 (2007)
16. Kuncheva, L.I.: Combining Pattern Classifiers. John Wiley and Sons (2004)
17. Li, S.Z., Zhang, Z.-Q.: Floatboost learning and statistical face detection. IEEE Trans. Patt. Anal. Mach. Intell. 26, 1112–1123 (2004)
18. Lumini, A., Nanni, L.: Detector of image orientation based on Borda count. Patt. Recogn. 27, 180–186 (2006)
19. Martinez-Munoz, G., Suarez, A.: Switching Class Labels to Generate Classification Ensembles. Patt. Recogn. 38, 1483–1494 (2005)
20. Marzio, M., Taylor, C.C.: On boosting kernel density methods for multivariate data: density estimation and classification. Stat. Meth. Appl. 14, 163–178 (2005)
21. Melgani, F.: Robust image binarization with ensembles of thresholding algorithms. Elec. Imag. 15, 023010 (2006)
22. Minka, T.P.: The "summation trick" as an outlier model. Unpublished article. Available from Minka's homepage (2003)
23. Polikar, R.: Ensemble based systems in decision making. IEEE Circuit Syst. Mag. 6, 21–45 (2006)
24. Ranawana, R.: Multiclassifier systems - review and a roadmap for developers. Int. J. Hybrid Intell. Syst. 3, 35–61 (2006)
25. Ridgeway, G., Madigan, D., Richardson, T., O'Kane, J.W.: Interpretable boosted naive Bayes classification. In: Proc. 4th Int. Conf. Know. Discovery Data Mining, pp. 101–104 (1998)
26. Rokarch, L.: Taxonomy for characterizing ensemble methods in classification tasks: A review and annotated bibliography. Comp. Stat. Data Anal. 53, 4046–4072 (2009)
27. Schapire, R.E.: The boosting approach to machine learning: An overview. In: Proc. MSRI Workshop Nonlinear Estimation and Classification (2002)
28. Shen, L., Bai, L.: MutualBoost learning for selecting Gabor features for face recognition. Patt. Recogn. Lett. 27, 1758–1767 (2006)
29. Shiraishi, Y., Fukumizu, K.: Statistical approaches to combining binary classifiers for multi-class classification. Neurocomp. 74, 680–686 (2011)
30. Skurichina, M., Duin, R.P.W.: Combining Feature Subsets in Feature Selection. In: Oza, N.C., Polikar, R., Kittler, J., Roli, F. (eds.) MCS 2005. LNCS, vol. 3541, pp. 165–175. Springer, Heidelberg (2005)

31. Tang, E.K., Suganthan, P.N., Yao, X., Qin, A.K.: Linear dimensionality reduction using relevance weighted LDA. Patt. Recogn. 38, 485–493 (2005)
32. Tang, E.K., Suganthan, P.N., Yao, X.: An analysis of diversity measures. Mach. Learn. 65, 247–271 (2006)
33. Tax, D.M.J.: One class classification. PhD thesis, Delft University, The Netherlands (2001)
34. Ting, K.M., Witten, I.H.: Stacked generalization: when does it work? In: Proc. 15th Int. Joint Conf. Art. Intell. (1997)
35. Vezhnevets, A., Vezhnevets, V.: Modest AdaBoost-teaching AdaBoost to generalize better. In: 15th Int. Conf. Comp. Graph. Appl. (2005)
36. Viaene, S., Derrig, R., Dedene, G.: A case study of applying boosting naive Bayes to claim fraud diagnosis. IEEE Trans. Know. Data Engng. 16, 612–619 (2004)
37. Xu, L., Krzyzak, A., Suen, C.Y.: Several methods for combining multiple classifiers and their applications in handwritten character recognition. IEEE Trans. Syst. Man Cybern. 22, 418–435 (1992)
38. Webb, G.I.: Multiboosting: A technique combining boosting and wagging. Mach. Learn. 40, 159–197 (2000)

Chapter 15
Sensor Management

15.1 Introduction

In this chapter we complete our study of multi-sensor data fusion by analyzing the control application/resource management block shown in Fig. 1.1. To make our discussion more concrete, we shall consider the case when decisions made by the control application are fed back to the sensors. In this case, the control block is more commonly known as a *sensor management* (SM) block. Formally we define sensor management as "a process that seeks to manage, or coordinate, the use of a set of sensors in a dynamic, uncertain environment, to improve the performance of the system" [12].

Example 15.1. Active Vision System [12]. Fig. 15.1 shows a control loop in an active vision system. It consists of three blocks that interact in a cyclical manner. The sensor block is responsible for acquiring sensory measurements of the external world or environment. The sensor observations are then passed to the data fusion block which fuses the last observation with previously acquired measurements in order to make an assessment about the environment taking into account the various sources of uncertainty. Based on the task requirements and on the current output of the data fusion block, the control application loop controls the data acquisition process by acting on the sensors.

In Fig. 15.1 we show our multi-sensor data fusion system with the sensor, data fusion and sensor management blocks. The sensor observations $O_i, i \in \{1,2,\ldots,N\}$, are sent from the sensors $S_m, m \in \{1,2,\ldots,M\}$, to the data fusion block for processing. Information from the data fusion block is then passed, via the human-machine interface (HMI), to the human operator who monitors the entire scene. The inputs to the Sensor Manager may come from the data fusion block and also from operator via the HMI.

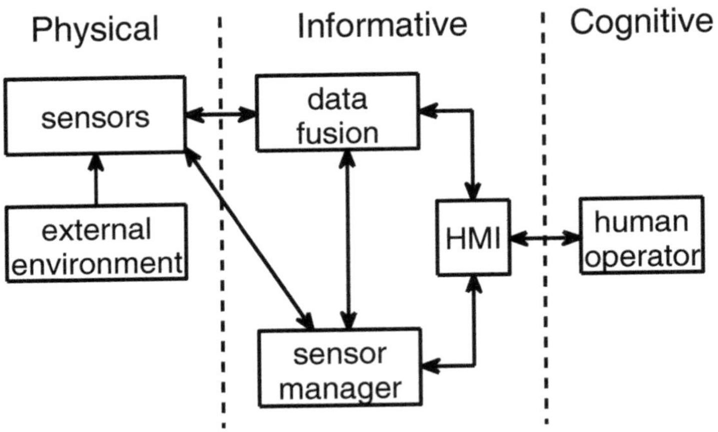

Fig. 15.1 Shows an active fusion system.

In general, the sensor observations, $\mathbf{y}$, are transmitted with minimal delay from the sensor block to the data fusion block via a real-time service (RS) interface. On the other hand, the information which is passed between the sensor manager and the sensor and data fusion blocks may be subjected to a significant delay. In this case, the information is transmitted via the slower diagnostic and management (DM) and configuration and planning (CP) interfaces. See Sect. 2.4.1.

15.2 Hierarchical Classification

Sensor management is clearly a very broad concept which involves many different issues which may help enhance the performance of the multi-sensor data fusion system. Broadly speaking, we may divide the field of sensor management into a hierarchy consisting of three levels, or categories: sensor control, sensor scheduling and resource planning [6]. We shall now consider each hierarchical level in turn.

15.2.1 Sensor Control

Sensor control is the lowest hierarchical level in sensor management. In it which we focus on the individual control of each sensor, including the specification of its parameters. At the sensor control level the SM acts as a classic feedback, or control, mechanism whose goal is to optimize the performance of the sensor and data fusion blocks given the present sensor measurements $\mathbf{y}$.

The following example illustrates this aspect of the SM.

Example 15.2. Building Security [7, 13]. The security of buildings is an increasing concern nowadays. One aspect of this concerns the level of access required by employees, clients and customers. For example, access to security, financial and sensitive information is usually restricted to specific employees who work in one, or more, exclusive regions of the building. On the other hand, other regions of the building must be open to employees, clients and cuustomers without any restrictions.

One approach to this problem is to employ a network of biometric sensors interfaced to door-locking mechanisms. The security level corresponding to the job description and biometric features constitute the information gathered when an employee is enrolled into the system. The biometric features are collected and matched against the enrollment features when the individual requests access. Control of the door-locking mechanisms is optimized for a given criteria, e. g. to minimize the number of mistakes, i. e. to minimize the sum of the number of genuine employees who are refused access and the number of imposters who are allowed access.

15.2.2 Sensor Scheduling

Sensor scheduling is the middle hierarchical in sensor management. In it we focus on the actual tasks performed by the sensors and on the different sensor modes. At the sensor scheduling level the SM prioritizes the different tasks which need to be performed and determines when, and how, a sensor should be activated. An important task at this level is the sensor schedule.

The following example illustrates this aspect of the SM.

Example 15.3. Meteorological Command and Control [14]. Distributed Collobarative Adaptive Sensing (DCAS) is a new paradigm for detecting and predicting hazardous weather. DCAS uses a dense network of low-powered radars which periodically sense a search volume V which occupies the lowest few kilometers of the earth's atmosphere. At the heart of the DCAS system is a meteorological Command and Control (MCC) unit which performs the systems' main control loop. An important function performed by the MCC is the allocation/optimization processes that determines the strategy for taking radar measurements during the next radar scan. Within each voxel in V each meteorological object which has been tracked until now in V has an utility that represents the value of scanning that voxel/object during the next scan. The utility value weights considerations such as the time since the voxel/object was last scanned, the object type (e. g. scanning an area with a tornado vortex will have higher utility than sensing clear air) and user-based considerations, such as the distance from a population center (e. g. among two objects with identical features, the one closer to a population center will have higher utility).

15.2.3 Resource Planning

Resource planning is the highest hierarchical level in sensor management. In it we focus on tasks such as the placement of the sensors, or the optimal mixture of the sensors required for a given task.

The following example illustrates this aspect of the SM.

Example 15.4. Mobile Sensors Observing an Area of Interest [12]. In this application we consider the coordination of several autonomous sensors $S_m, m \in \{1,2,\ldots,M\}$. The sensors S_m, are required to co-operatively monitor an area of interest (Fig. 15.2). Each sensor S_m has its own dynamics (specified by a velocity vector $\mathbf{V}_m$) and can only perceive a limited local area A_m. The local areas can be shared by the sensors. Thus, a local picture from one sensor can be used to direct the attention of other sensors. The sensor manager is responsible for coordinating the movements and sensing actions of the sensors so that we develop an optimal picture of the entire surveillance area with minimal consumption of time and resources.

15.3 Sensor Management Techniques

Over the years a variety of SM techniques have been used for solving the problem of SM. We conclude the book by briefly reviewing two of these techniques.

15.3.1 Information-Theoretic Criteria

From the information-theoretic point of view, multi-sensor data fusion is concerned with increasing the information, i. e. reducing the uncertainty, about the state of the external world, or environment. In this case, the task of sensor management is to optimize the multi-sensor data fusion process such that the greatest possible amount of information is obtained whenever a measurement is made.

In the following example we consider the optimal selection of a sensor, using an information-theoretic criteria.

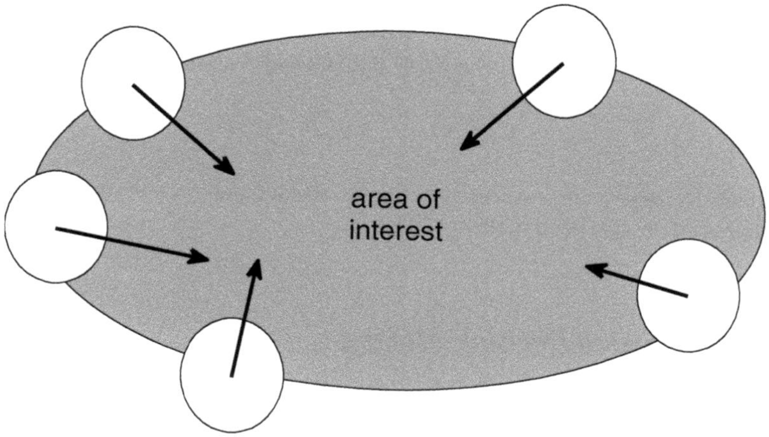

Fig. 15.2 Shows five mobile sensors $S_m, m \in \{1,2,\ldots,5\}$ cooperatively monitoring an area of interest (shown shaded). Each sensor S_m has a velocity $\mathbf{V}_m$ (denoted by an arrow) and can only monitor a local area A_m (denoted by a white circle).

Example 15.5. Sensor Selection Using Information-Theoretic Criteria [2, 11]. We assume a target is known to be present in a given surveillance area. Let $\pi(x|I)$ denote the *a priori* location of the target. Suppose $S_m, m \in \{1,2,\ldots,M\}$, denotes a set of M sensors, whose observation likelihoods are $p(\mathbf{y}_m|x,I)$. Our aim is to select the sensor S_m whose observation will maximize the mutual information $MI(x,\mathbf{y}_m)$, where x denotes the unknown location of the target and $\mathbf{y}_m$ denotes the observation $\mathbf{y}_m$ predicted for the sensor S_m. According to (5.5), the mutual information, $MI(\mathbf{x},\mathbf{y}_m)$, is given by

$$MI(\mathbf{x},\mathbf{y}_m) = \int p(\mathbf{x},\mathbf{y}_m)\log\frac{p(\mathbf{x},\mathbf{y}_m|I)}{p(\mathbf{x}|I)p(\mathbf{y}_m|I)}d\mathbf{x}d\mathbf{y}_m\,,$$

where $p(\mathbf{x},\mathbf{y}_m|I) = p(\mathbf{y}_m|\mathbf{x})p(\mathbf{x}|I)$ and $p(\mathbf{y}_m|I) = \int p(\mathbf{x},\mathbf{y}_m)d\mathbf{x}$. In this case, we choose the observation, i. e. the sensor, which maximizes the mutual information $MI(\mathbf{x},\mathbf{y}_m)$:

$$m_{\mathrm{OPT}} = \arg\max MI(\mathbf{x},\mathbf{y}_m)\,. \tag{15.1}$$

The above expression for the mutual information receives a simple form if we assume Gaussian distributions for the state of the target.

Example 15.6. Sensor Selection Using Mutual Information. Assuming a Gaussian distribution for the state of a given target O, then

$$MI(\mathbf{x},\mathbf{y}) = \frac{1}{2}\log(|P_{\mathbf{x}}|/|P_{\mathbf{y}}|) , \qquad (15.2)$$

where $P_{\mathbf{x}}$ and $P_{\mathbf{y}}$ are, respectively the covariance matrices before and after a measurement has been made.

15.3.2 Bayesian Decision-Making

From the viewpoint of decision theory (see Chapt. 13), we may regard sensor management as a decision-making task in which our aim is to minimize a given loss function. Given M sensors $S_i, i \in \{1,2,\ldots,M\}$, we assume each sensor gives a local decision u_i regarding the environment, where

$$u_i = \begin{cases} H_0 \text{ if } S_i \text{ favours hypothesis } H_0 , \\ H_1 \text{ if } S_i \text{ favours hypothesis } H_1 . \end{cases} \qquad (15.3)$$

The u_i are transmitted to the multi-sensor data fusion center which acts as a sensor manager and makes the optimal global decision, i. e. the global decision which minimizes the given loss function. We now consider the case of sensor management in which the local decisions are correlated. In general the solutions lead to complicated iterative algorithms where detailed knowledge of the statistical correlation among the u_i are required. A simple model for the statistical dependence among the u_i is to use two correlation parameters r_0 and r_1 as follows.

Let $\mathbf{u} = (u_1,u_2,\ldots,u_M)^T$ be the vector formed by the set of local decisions ($u_i = 0$ when the ith sensor S_i is in favour of H_0 and $u_i = 1$ when S_i is in favour of H_1), corresponding to M local detectors. We consider that m out of M local detectors are in favour of H_0 (i. e. there are m 0's in vector u). Then [1] the likelihood ratio is given by

$$\begin{aligned}\Lambda(u) &= \frac{p(u|H_1)}{p(u|H_0)} , \\ &= \sum_{i=0}^{m}(-1)^i C_i^m p_m^D \prod_{k=0}^{L-m+i-2} \frac{r_1(k+1-p_m^D)+p_m^D}{1+kr_1} \Big/ \\ &\qquad \sum_{i=0}^{m}(-1)^i C_i^m p_m^F \prod_{k=0}^{L-m+i-2} \frac{r_0(k+1-p_m^F)+p_m^F}{1+kr_0} , \end{aligned} \qquad (15.4)$$

and the correlation coefficients $0 \le r_k \le 1, k \in \{0,1\}$ are defined as

$$r_k = \frac{E(u_i u_j|H_k) - E(u_i|H_k)E(u_j|H_k)}{\sqrt{E((u_i - E(u_i)^2|H_k)E((u_j - E(u_j))^2|H_k)}} , \qquad (15.5)$$

The optimum rule i. e. maximizing the global PD, GPD, for a given global probability of false alarm, GPFA) is obtained by means of the likelihood ratio test [1, 9]

$$H_{OPT} = \begin{cases} H_1 \text{ if } \Lambda(m) \geq \lambda \ , \\ H_0 \text{ if } \Lambda(m) < \lambda \ , \end{cases} \tag{15.6}$$

where $\lambda > 0$ fit the global PFA.

An equivalent counting rule is defined as

$$H_{OPT} = \begin{cases} H_1 \text{ if } m \leq m_0 \ , \\ H_0 \text{ if } m > m_0 \ , \end{cases} \tag{15.7}$$

As an example we shall consider the sensor control of the biometric sensors in Ex. 15.2.

Example 15.7. An Adaptive Multimodal Biometric Management Algorithm [10]. We consider M independent biometric sensors $S_m, m \in \{1,2,\ldots,M\}$. The task of identifying an unknown person O as a hypothesis testing problem, with the following two hypotheses:

$H = h_1$. The unknown person O is an imposter.
$H = h_2$. The unknown person O is genuine.

We suppose each sensor S_m receives a measurement vector $\mathbf{y}_m$ from O and outputs the decision variable $U_m \in \{u_1, u_2\}$, where

$$U_m = \begin{cases} u_1 \text{ if } p(U_m = u_1|H = h_1) \geq \lambda_m p(U_m = u_2|H = h_2) \ , \\ u_2 \text{ otherwise} \ . \end{cases}$$

where λ_m is an appropriate threshold.

Assuming each of the biometric sensors are idependent, then the optimal fusion rule [8] can be implemented by forming a weighted sum of the incoming local decisions $U_m, m \in \{1,2,\ldots,M\}$, and then comparing it with a threshold t. The weights and the threshold are determined by the reliability of the decisions, i. e. by the probabilities of false alarm and missed detection of the sensors S_m. In mathematical terms, the output decision variable is $\widetilde{U} = u_{OPT}$, where

$$u_{OPT} = \begin{cases} u_1 \text{ if } \left[\sum_{m=1}^{M} \left(z_m \log \frac{1-p_m^M}{p_m^F} + (1-z_m) \log \frac{p_m^M}{1-p_m^F}\right)\right] \geq t \ , \\ u_2 \text{ otherwise} \end{cases}$$

where

$$z_m = \begin{cases} 1 \text{ if } U_m = u_1 \ , \\ 0 \text{ otherwise} \ , \end{cases}$$

and p_m^F, and p_m^M are, respectively, the probabilities of false alarm and missed detection for the sensor S_m,

$$\begin{aligned} p_m^F &= p_m(U = u_2 | H = h_1) , \\ p_m^M &= p_m(U = u_1 | H = h_2) . \end{aligned}$$

We optimally choose the threshold t in order to minimize the cost of a output decision $\widetilde{U}$. This cost depends on the *a priori* probabilities $\pi(H = h_1|I)$ and $\pi(H = h_2|I)$ and on the loss function which is used [10].

15.4 Further Reading

Recent comprehensive reviews of sensor management are [3, 6, 12]. Ref. [5] describes a novel approach to sensor management based on e-commerce. A detailed description of sensor management and in particular sensor scheduling, in a multi-function radar system is given in [4].

15.5 Postscript

In writing this book our aim was to provide the reader with an introduction to the concepts and ideas in multi-sensor data fusion. The theories and techniques may be formulated in many different frameworks, but for reasons explained in the preface, we have, by and large, restricted ourselves to the Bayesian framework. The reader should, however, keep in mind that multi-sensor data fusion is a pragmatic activity which is driven by practicalities. When circumstances dictate the engineer may abandon the Bayesian framework and adopt an alternative framework instead.

During the time this book was being written we have witnessed a growing interest in multi-sensor data fusion: Many commercial applications now employ multi-sensor data fusion in one form or another. We have tried to convey this growing interest by giving examples drawn from a wide range of real-world applications and the reader is now encouraged to review these examples.

Problems

15.1. Sensor management is often described as a hierarchy of three levels: sensor control, sensor scheduling and resource planning. Compare and contrast these three hierarchical levels of sensor management.

15.2. Compare and contrast the information-theoretic and Bayesian decision-making methods for sensor management.

References

1. Drakopoulos, E., Lee, C.C.: Distributed decision fusion using empirical estimation. IEEE Trans. Aero. Elec. Syst. 27, 593–605 (1991)
2. Kreucher, C., Kastella, K., Hero III, A.O.: Sensor management using an ative sensing approach. Sig. Proc. 85, 607–624 (2005)
3. McIntyre, G.A., Hintz, K.J.: A comprehensive approach to sensor management, Part I: A survey of modern sensor management systems. Avaliable on the Internet (1999)
4. Miranda, S., Baker, C., Woodbridge, K., Griffiths, H.: Knowledge-based resource management for multi-function radar. IEEE Sig. Proc. Mag. 67, 66–77 (2006)
5. Mullen, T., Avasarala, V., Hall, D.L.: Customer-driven sensor management. IEEE Intell. Syst., 41–49 (March-April 2006)
6. Ng, G.W., Ng, K.H.: Sensor management - what, why and how. Inform. Fusion 1, 67–75 (2000)
7. Osadciw, L., Veeramachaneni, K.: A controllable sensor management algorithm capable of learning. In: Proc. SPIE Defense and Security Symposium (2005)
8. Varshney, P.K.: Distributed detection and data fusion. Springer, Heidelberg (1996)
9. Vergara, L.: On the equivalence between likelihood ratio tests and counting rules in distributed detection with correlated sensors. Sig. Proc. 87, 1808–1815 (2007)
10. Veeramachaneni, K., Osadciw, L.A., Varshney, P.K.: An adaptive multimodal biometric management algorithm. IEEE Trans. Sys. Man Cybern. - Part C 35, 344–356 (2005)
11. Wang, H., Yao, K., Pottie, G., Estrin, D.: Entropy-based sensor selection heuristic for target localization. In: Proc. IPSN 2004 (2004)
12. Xiong, N., Svensson, P.: Multisensor management for information fusion: issues and approaches. Inf. Fusion 3, 163–186 (2002)
13. Yilmazer, N., Osadciw, L.A.: Sensor management and Bayesian networks. In: Proc. SPIE, vol. 5434 (2004)
14. Zink, M., Westbrook, D., Abdallah, S., Horling, B., Lakamraju, V., Lyons, E., Manfredi, V., Kurose, J., Hondl, K.: Meteorological command and control. In: Proc. EESR 2005: Workshop on End-to-End, Sense-and-Respond Systems, Applications and Services (2005)

Appendix A
Background Material

Table A.1 contains a summary of basic formulas used in Bayesian probability theory.

Table A.1 Basic Formulas in Bayesian Probability Theory

Name	Formula
Probability density function (pdf)	$p(y\|I)dy = P(y \leq Y \leq y + dy\|I)$
Normalization	$\int p(y\|I)dy = 1$
Expectation of $f(Y)$	$E(f(Y)) = \int f(y)p(y\|I)dy$
Expected value	$E(Y) = \int yp(y\|I)dy$
Expected value of order r	$M_r(Y) = \int y^r p(y\|I)dy$
Variance	$\sigma^2 = \int (y - E(Y))^2 p(y\|I)dy$
Product rule	$p(x,y\|I) = p(x\|y,I)p(y\|I)$
Independence	$p(x,y\|I) = p(x\|I)p(y\|I)$
Marginalization	$p(x\|I) = \int p(x,y\|I)dy$
Decomposition	$p(x\|I) = \int p(x\|y,I)p(y\|I)dy$
Bayes' rule	$p(x\|y,I) = p(y\|x,I)p(x\|I)/p(y\|I)$
	$p(x\|y,z,I) = p(x\|I)p(y\|x,I)p(z\|x,y,I)/(p(y\|I)p(z\|I))$
Likelihood	$L = p(y\|x,I)$

Table A.2 contains a summary of elementary results in linear algebra.

Table A.2 Elementary Results in Linear Algebra

Name	Formula
Definitions	x is a 1×1 scalar; $\mathbf{x} = (x_1, x_2, \ldots, x_m)^T$ is a $m \times 1$ column vector; and $\mathbf{A} = \begin{pmatrix} A_{11} & A_{12} & \ldots & A_{1n} \\ A_{21} & A_{22} & \ldots & A_{2n} \\ \vdots & \vdots & \ddots & \vdots \\ A_{m1} & A_{m2} & \ldots & A_{mn} \end{pmatrix}$ is a $m \times n$ matrix; $\mathbf{A} = (\mathbf{a}_1, \mathbf{a}_2, \ldots, \mathbf{a}_n)$, where $\mathbf{a}_i = (A_{1i}, A_{2i}, \ldots, A_{mi})^T$
Transpose	$(\mathbf{A}^T)_{ij} = A_{ji}$
Multiplication	$(\mathbf{A}\mathbf{A}^T)_{ij} = \sum_k A_{ik}(\mathbf{A}^T)_{kj} = \sum_k A_{ik}A_{jk}$
Dimensions	$\mathbf{x}^T\mathbf{x}$ is a 1×1 scalar; $\mathbf{x}\mathbf{x}^T$ is a $m \times m$ matrix; $\mathbf{A}\mathbf{x}$ is a $m \times 1$ vector; $\mathbf{A}\mathbf{A}^T$ is a $m \times m$ matrix; $\mathbf{A}\mathbf{A}^T$ is a $n \times n$ matrix; $\mathbf{x}^T\mathbf{A}\mathbf{x}$ is a 1×1 scalar
Combinations	$(\mathbf{AB})\mathbf{C} = \mathbf{A}(\mathbf{BC})$; $\mathbf{A}(\mathbf{B}+\mathbf{C}) = \mathbf{AB}+\mathbf{AC}$;$(\mathbf{A}+\mathbf{B})^T = \mathbf{A}^T + \mathbf{B}^T$; $(\mathbf{AB})^T = \mathbf{B}^T\mathbf{A}^T$

Table A.3 contains a summary of elementary results on square matrices.

Table A.3 Basic Formulas Used in Square Matrix Theory

Name	Formula
Definitions	$\mathbf{B} = \begin{pmatrix} B_{11} & B_{12} & \dots & B_{1n} \\ B_{21} & B_{22} & \dots & B_{2n} \\ \vdots & \vdots & \ddots & \vdots \\ B_{n1} & B_{n2} & \dots & B_{nn} \end{pmatrix}$ is a $n \times n$ square matrix; $\mathbf{D} = \begin{pmatrix} D_{11} & 0 & \dots & 0 \\ 0 & D_{22} & \dots & 0 \\ \vdots & \vdots & \ddots & \vdots \\ 0 & 0 & \dots & D_{nn} \end{pmatrix} = diag(D_{11}, D_{22}, \dots, D_{nn})$ is a $n \times n$ diagonal matrix; and $\mathbf{I} = \begin{pmatrix} 1 & 0 & \dots & 0 \\ 0 & 1 & \dots & 1 \\ \vdots & \vdots & \ddots & \vdots \\ 0 & 0 & \dots & 1 \end{pmatrix} = diag(1, 1, \dots, 1)$ is a $n \times n$ identity matrix
Identity Results	$\mathbf{Ix} = \mathbf{x}$; $\mathbf{IB} = \mathbf{B} = \mathbf{BI}$; $\mathbf{X}^T\mathbf{I} = \mathbf{x}^T$
Inverse	$\mathbf{B}^{-1}\mathbf{B} = \mathbf{BB}^{-1} = \mathbf{I}$; $(\mathbf{B}^{-1})^{-1} = \mathbf{B}$; $(\mathbf{BC})^{-1} = \mathbf{C}^{-1}\mathbf{B}^{-1}$; $(\mathbf{B}^{-1})^T = (\mathbf{B}^T)^{-1}$
Determinant	$\lvert\mathbf{B}\rvert = \Pi_{i=1}^n \lambda^{(i)}$; $\lvert\mathbf{BC}\rvert = \lvert\mathbf{B}\rvert\lvert\mathbf{C}\rvert$; $\lvert x\rvert = x$; $\lvert x\mathbf{B}\rvert = x^n\lvert\mathbf{B}\rvert$; $\lvert\mathbf{B}^{-1}\rvert = 1/\lvert\mathbf{B}\rvert$
Trace	$trace(\mathbf{B}) = \Sigma_{i=1}^n B_{ii} = \Sigma_{i=1}^n \lambda^{(i)}$; $trace(\mathbf{BCD}) = trace(\mathbf{DBC}) = trace(\mathbf{CDB})$; $\mathbf{x}^T\mathbf{Bx} = trace(\mathbf{x}^T\mathbf{Bx}) = trace(\mathbf{xx}^T\mathbf{B})$
Eigenvector equation	$\mathbf{Bu}^{(i)} = \lambda^{(i)}\mathbf{u}^{(i)}$, where $\mathbf{u}^{(i)}$ is the ith $n \times 1$ (eigen)vector

Index

Lightning Source UK Ltd.
Milton Keynes UK
UKHW010812170919
349923UK00013B/126/P

9 783642 437304